To Tom,

Our common interest is remarkable across the void stretching between disciplines, where our worlds had to be crossed by personal relationships. I hope this book helps bridge the divide.

With aloha,

The

MARVELOUS
LEARNING
ANIMAL

The
MARVELOUS
LEARNING
ANIMAL

What Makes Human Nature Unique

ARTHUR W. STAATS

 **Prometheus Books**

59 John Glenn Drive
Amherst, New York 14228-2119

Cover image © 2012 Media Bakery
Cover design by Nicole Sommer-Lecht

Inquiries should be addressed to
Prometheus Books
59 John Glenn Drive
Amherst, New York 14228–2119
VOICE: 716–691–0133
FAX: 716–691–0137
WWW.PROMETHEUSBOOKS.COM

16 15 14 13 12   5 4 3 2 1

Library of Congress Cataloging-in-Publication Data

Staats, Arthur W.
   The marvelous learning animal : what makes human nature unique / by Arthur W. Staats.
     p. cm.
   Includes bibliographical references and index.
   ISBN 978–1–61614–597–2 (cloth : alk. paper)
   ISBN 978–1–61614–598–9 (ebook)
   1. Learning, Psychology of. 2. Learning ability. 3. Human evolution. I. Title.

LB1060.S73 2012
370.15'23—dc23

2012000381

Printed in the United States of America on acid-free paper

To those who believe in a destiny of human progress

*To those who want a scientific explanation of human behavior and
Human nature, unobstructed by traditional belief*

*To those who want good lives for everyone in the world,
not just our own people*

To those who draw from the past but look for what can be better

CONTENTS

PART 1: THE GREAT SCIENTIFIC ERROR

Chapter 1: Of Paradigms and Fallacies,   11
    of Sciences and Things

PART 2: THE HUMAN ANIMAL

Chapter 2: Such a Body! The Versatile Animal   47

Chapter 3: Such a Brain!   67
    100,000,000,000 Neurons Buy a Lot of Learning

PART 3: CHILD DEVELOPMENT AND THE MISSING LINK

Chapter 4: Learning Child Development   97

Chapter 5: Marvelous Learning: The Missing Link   147

PART 4: LEARNING HUMAN NATURE

Chapter 6: Marvelous Learning of Personality   171

Chapter 7: Marvelous Learning of Abnormal Personality   209

PART 5: HUMAN EVOLUTION AND MARVELOUS LEARNING

Chapter 8: On the Origin of the Human Species   255

Chapter 9: Who We Are   293

PART 6: A HUMAN PARADIGM

Chapter 10: The Time Has Come,   327
    the Walrus Said, to Talk of Many Things

ACKNOWLEDGMENTS   355

REFERENCES   359

INDEX   371

Part 1

THE GREAT SCIENTIFIC ERROR

Chapter 1

OF PARADIGMS AND FALLACIES, OF SCIENCES AND THINGS

Once obscure, the word *paradigm* was made a catchy concept a few decades ago by Thomas Kuhn (1962). There was even a *New Yorker* cartoon with a beggar asking a potential client, "Can you paradigm?" in place of the Depression-era original "Can you spare a dime?" Kuhn's definition bedazzled the history of science and philosophy of science worlds. Science wasn't just the objective search for truth via fact and theory. Scientific paradigms consisted of much more—methodology, beliefs, convictions, technology, and problems to be solved. Paradigms garnered adamant loyalty. A new paradigm—like Einstein's initially was—couldn't expect an open-armed reception. It had to scratch and fight for acceptance through obstacles of disbelief. For scientists, contrary to popular belief, exhibit human motivations that interfere with that legendary objectivity of their stereotype (Toulmin 1972).

Scientific paradigms—despite their glamorous fruits of problem solution, application, and research creativity—may also contain errors large and costly, to society as well as science. Although improving on the previous paradigm, a new one generally has its own problems and limitations. When those problems are fundamental or extensive enough, and this becomes recognized, another paradigm may be put forth that better deals with the field—abandoning unproductive endeavors, indicating important problems, and projecting new avenues of application and research. Its arrival produces a period of doubt and struggle with the old paradigm, a scientific revolution that may culminate with the acceptance of the new paradigm. This then becomes the primary framework that generates the normal science activities that elaborate its expectations and produce its advances.

A paradigm exists with respect to human behavior and human nature. It is not clean-cut, systematically constructed like advanced theories, but rather is composed of many poorly articulated bits and pieces. The paradigm consists of a mix. If a history of the paradigm existed, it would show that it began long ago, with animistic belief that everything from atolls to zebras contained spirits that determined their nature and behaviors. Volcanoes erupt and oceans storm when their spirits become angry. A human behaves badly when his or her spirit is evil or when an evil spirit invades him or her. As animistic beliefs advanced in many steps to yield modern religions, the spirit concept for humans became the soul. When the search for knowledge brought forth various science endeavors and then science fields, including an interest in human actions, the concept of the soul morphed into the concept of the mind, as well as the personality, the brain, and other internal causal processes or structures. The name *psychology* literally means the study of the mind, the Greek term *psyche* meaning either "soul" or "mind."

A theory of human behavior and human nature exists in the common language that we all learn. We can see elements of the common conception of human behavior and human nature in religious and literary writing, as in the Bible and Shakespeare and Dostoyevsky, in descriptions of types of human behavior and their causes. We can see that people behave peculiarly, destructively, because of a mental disease. We can see numberless explanations of behavior—behaviors occur because of inner states such as *intelligence, anxiety, honesty, ambition, laziness, greed*, and *cruelty*. Those trait concepts and many others are the conceptual tools by which people explain their own behaviors and the behaviors of others. A child does well in school because she is intelligent, a boy picks on others because he is a cruel bully, and another works all the time because he has ambition. Why does an adult protect children? Human nature. Why have people gone to war throughout history? Human nature. What gave humans the nature that determines their behavior—God, Mother Nature, or genetic evolution? All of these concepts are used to explain human actions.

There are differences in people's complex sets of beliefs, so the paradigm varies for different people. But we all grew up absorbing the explanations of human behavior our language provides for us. That language is part of each of us, rock-hard belief, so natural it is not open to consideration, giving all of us a common core theory. We accept things that fit with these beliefs, and are poised to reject things that do not. That informal theory of human behavior and human nature functions covertly, not realized for what it is and what it does. It affects the beliefs, interpretations, and plans of scientists as well as of lay people, setting gross guidelines for what to study and how to do so. I call this paradigm "The Great Scientific Error" because although it serves many of our purposes—much better than the conceptions of bygone days—it contains profound error and misdirects us in many, costly ways. Billions-of-dollars ways.

This paradigm is implicit, unrealized, and should be brought to light systematically, described, made evident. With such an account it would be possible to consider and work on the paradigm, improve it, address its various problems, as well as understand the widespread effects it has on everyone. No books have been written to describe the Great Scientific Error paradigm. I have come to recognize the paradigm because my fifty-plus-year program of work has on so many fronts taken a contrary view. My aim in this book centers on presenting a new paradigm, so I cannot devote this book to the full consideration of the old paradigm that is needed. But I can introduce the Great Scientific Error and I can additionally flesh out that introduction here and throughout the book and thereby call for a more complete work.

DARWIN'S FALLACY

Explanation of origin is central in a conception of human behavior and human nature, humanness. Before Charles Darwin, the dominant explanation of origin lay in the biblical account of God's creation of humans. Of course, that explanation of origin had many predecessors in the beliefs of many different peoples. A belief of an Australian Aborigine people had a beautiful simplicity: a pool exists of little fish, some of which are selected by spirits to become people. When a person dies he again becomes a little fish in that pool, waiting another turn.

Darwin's theory of evolution raised a fury by calling into question the truth of theological creationism. Various science-oriented thinkers, including Erasmus Darwin, previously had suggested an evolutionary process. The church successfully opposed such views, and they were believed by relatively few. Charles Darwin, however, shook that balance with his systematic collection of observations that confirmed evolution, and with his theory of natural selection. That provided the principles by which evolution took place, and his theory had the weight of demonstrable evidence. Darwin's evolution was translated into many languages and read by a wide audience, with a profound impact. Evolution—with its findings, scientific methods, theory, and productivity for many fields—constituted a new paradigm that challenged the old religious paradigm of divine creation. This development gives a good example of the paradigm clash; the new one is ensconced in science fields but the old one is still continuing in some religious circles.

The impact of Darwin's theory of evolution and Mendel's genetics provided the foundation for various sciences from biology to paleoanthropology, from psychology and the social sciences to the philosophy of science. Of special interest to the present book, this synthesis contributes heavily to traditional beliefs about humans, about origin, about what human nature is, and about why humans behave as they do. These developments enormously affected the human-behavior paradigm.

Given the proven value of the theory of evolution, wherein lies the Darwin fallacy? Interestingly, as tends to be the case with scientific theories, the power of a theory when it is true matches the power of the error it projects when it is in error. Darwin's theory did contain error, egregious error, and that has had great impact too. The following words of Darwin are simple and seemingly innocent, but, as with the iceberg, we must look for the much larger, lurking danger.

> As man is a social animal, it is almost certain that he would inherit a tendency to be faithful to his comrades, and obedient to the leader of his tribe, for these qualities are common to most social animals. He would consequently possess some capacity for self-command. He would from an inherited tendency be willing to defend, in concert with others, his fellow-men; and be willing to aid them in any way. . . . (See Watson 2005, 689)

Such words are common in the belief of the society, that the causes of human behavior come from within, in biologically inherited traits of various kinds. What must be recognized is that being faithful, self-commanding, and obedient, along with defending and aiding fellow humans, describe *behavioral traits*. They are not physical traits like a bird's beak, a horse's hooves, or a human's hand. That lack of distinction is understandable; his beliefs regarding human behavior and human nature were like everyone else's, a pastiche of belief learned like everyone else in the acquisition of language. Of course he did not distinguish physical and behavioral traits; no one else did. In the religious paradigm, God created man physically *and* behaviorally. Darwin's lack of distinction between physical and behavioral traits of humans threw their explanation into the same mold. That was an enormous mistake, providing the foundation for expecting behavioral traits to be explained by the biological sciences, from genetics to neurology to paleoanthropology. The power and breadth of this conceptualization can't be overestimated.

No question, Darwin made a monumental study of the physical features of various species. He systematically worked out how physical traits possessed by some members of the species were selected because the traits aided survival and reproduction. Called *natural selection*, his work on this principle was scientifically impeccable—the hallmark of his theory was massive evidence. That same careful science analysis, however, was not extended to the evolution of human behavioral traits. He had no evidence of that, no evidence that "humans would from inherited tendency be willing to defend, in concert with others, his fellowmen; and be willing to aid them in any way" (Charles Darwin, quoted in Watson 2005, 689). Darwin said such things on simple assumption, following the beliefs of his culture, without any evidence, because he didn't distinguish between physical and behavioral traits.

Because of the power of his approach, that error continued with those who followed Darwin's evolution theory; after all, they also had the same language theory he did. His cousin, Sir Francis Galton, for example, interpreted intelligence as a trait that had evolved because it aided survival and reproduction. He attempted to prove intelligence was inherited by tracing the number of exceptionally able relatives in the families of geniuses. Ignoring the possible effects of the environment, wide individual and group differences in intelligence were interpreted as evidence that some groups of people had evolved more, or better, than others. One derivation of this view was that human intelligence could be maximized by eugenics, by selecting only intelligent people for reproduction. Such extensions helped fix the very strong belief in the biological explanation of intelligence.

That was not all. Various traits of behavior were analyzed in evolutionary terms, and a field of social Darwinism formed. It enjoyed a period of popularity, but social Darwinism did not lead anywhere, since it was comprised of "after-the-fact interpretation," assuming things that could not be proven or tested or used profitably. In science, developments that lead nowhere may be pursued for a time but will be abandoned eventually because support by evidence constitutes science's bottom line.

So social Darwinism faded away. But the underlying belief did not die; like a cat with nine lives, Darwin's view that human behavior traits are genetically evolved later gave rise to the field of *sociobiology*. For example, Edward O. Wilson—selected a while back as one of the small group of top scientists of the twentieth century—studied the behavior of ants as insects whose social behavior enables them to survive and procreate. His observations showed that soldier ants fight intruders into the nest, perhaps losing life in the process. He called the behavior "altruistic," a seemingly simple act of naming. Oops, error! "Altruism" is a human act. Calling the behavior of the ants altruistic equates it with the human behavior. That is sleight of hand that brings on a host of other beliefs. Is the behavior of an ant fighting and dying under attack by foreign ants to be considered the same as the altruism of a male passenger giving a woman his lifeboat seat on the *Titanic*? Are the two to be explained by the same genetically evolved trait? On consideration we might question that, and more generally the value of Wilson's equating of animal behavior to human behavior. Is it really productive to assume that human behavior traits have evolved in the way that animal behavior traits have?

Perhaps the answer to that question is there to see, for sociobiology really came to the same end as social Darwinism. After a run of a quarter century, it began fading away; sociobiology too hadn't been able to explain anything; it merely *inferred*, after the fact, evolutionary processes that could not be tested and proven. In 1997, the professional journal of the field dropped the title of *Sociobiology* and adopted that of *Evolution of Human Behavior*. This title, of course, shows that Darwin's position is still retained. It would have been constructive if it had been recognized that the field of sociobiology had failed along with its progenitor social Darwinism because both were built on Darwin's fallacy. Oh well.

The Selfish Gene

With that history of two burials—of social Darwinism and sociobiology—why would the underlying belief be resurrected again? The answer: because the underlying belief is strong in the culture everyone learns, so strong its failure under one rubric leaves room for its resurrection under a different name, inspiring renewed efforts. That present name is *evolutionary psychology*, and it proceeds with the usual verve and fanfare (see Workman and Reader 2004), adding conceptual elements and making analyses of new phenomena. For example, with more conceptual "sleight of hand" Richard Dawkins added the concept of the "selfish gene" to the theory of how behavior is passed on via evolution (1976). According to his theory, genes are selfishly motivated to reproduce themselves. Somehow, with a biological mechanism he does not indicate, the gene motivation he assumes determines the organism's behavior. That was why Wilson's soldier ant risked its life, because saving the nest and the queen means saving its own genes for reproduction. All kinds of behavior are supposedly "explained" by this motivation in evolutionary psychology (see Workman and Reader 2004). Popular as it is to many biological scientists—as indicated by various studies and articles that use the conception—it has no more explanatory power than the old "instinct of self-preservation," also popular as an "explanation" of behavior a century ago. Has anyone ever found a selfish gene in an ant, let alone in a human? Has anyone ever manipulated a selfish gene and changed any behavior thereby? By what *mechanism* would a gene activate behavior such that the gene's reproductive selfishness is fulfilled? Does the brain have neurons sensitive to gene motivations, whatever they are? Can those neurons, once activated, select the behaviors needed? To my knowledge, such mechanisms have never been found. Let's face it, without such evidence, the concept of the selfish gene is pure assumption and titillating speculation. When we ask "Where's the beef?" the *evidence* turns out to be like that of Wilson's "altruistic" ant soldiers, namely absent. Without such evidence, the conception is empty, leading nowhere in our understanding not only of ants but also of humans.

The field of evolutionary psychology is full of studies and conceptual analyses such as these (see Workman and Reader 2004). Our society is transfixed with the deeply ingrained biological/evolutionary explanation of human nature and human behavior. As a consequence, any plausible-sounding biological/evolutionary explanation of some human behavior is swallowed without question, overwhelming even normal scientific skepticism. With a good ring, ring-a-ding-ding sound and simple enough to be understood by anyone, eminent journals are loaded with spurious evolutionary explanations of behaviors ranging from anorexia nervosa (Guisinger 2008) to leadership-followership behavior (Van Vugt, Hogan, and Kaiser 2008). And the conception is not only in science and intellectual pursuit, it is in our "blood" for all of us and it is expressed widely in the popular media. For example, an article in the *New York Times* quotes Olivia Judson, the biologist and author, saying a show by rap artist Baba Brinkman is "one of the most astonishing and brilliant lectures on evolution I have ever seen" (2011, D3). In the show, Brinkman says "Don't sleep with mean people [whether] pretty or handsome," in essence because your grandchildren will inherit those genes for meanness.

The cost in science of that error is great. Spread the same over the popular media and popular thought and the cost to lay culture is also great. Billions? Oh yes! The fact is that evolutionary psychologists do not study the evolution of human behavior, actual human behavior. So what they do must be judged on other grounds; those grounds must be valuable to a much greater extent than either social Darwinism or sociobiology.

Good Things Can Be Carried Too Far

Like the proctologist who gave a rectal examination to a patient complaining of a cold, science that is very valuable for one problem may be in error when applied in another. Darwin, despite the genius of his theory of evolution with respect to physical traits and the origins of species, erred profoundly in not distinguishing between physical and behavioral traits. The proctologist has difficulty with respiratory disease because he lacks the necessary knowledge, the same with Darwin. Worse, to becloud the distinction between behavioral and physical traits, some animal *behaviors* are indeed products of evolution. There is a blue crab, for example, where male-female pairs perform a ritualistic dance before mating. No blue crab needs to learn the dance, so that can be ruled out. Similarly, the cheetah performs an elaborate mating ritual where a male chases the female for a long period prior to mating, the chase functional in producing the female's ovulation. The female, in essence, runs until she gets "hot." This behavior, too, appears to have a biological explanation, as do many other behaviors in the animal kingdom, including the aggressiveness of the common fruit fly.

To enhance the mixing of cause, sometimes physical and behavioral traits involve the same organ. As an example, hummingbirds have long beaks, typical of the species. Sticking their beaks in flowers is also typical of the species. It is true that the hummingbird's behavioral skill could be determined by an evolved, genetically wired-in brain characteristic, like the blue crabs' dance. However, there is another possibility. Although the long beak evolved and is carried in the genes of the species, *that might not be the case for the behavioral skill of using the beak.* The birds actually might have to *learn* the behavior.

The confusion of the two types of causation occurs in all animals, including humans. Being ferocious behaviorally is considered a lion's inherited trait. But consider, lions have hugely strong bodies, powerful legs, and paws and claws and teeth and jaws that can wound, maim, and kill. The change of the physical creature in evolution can be traced by paleontology. Clearly the lion inherits its structure through its genes. But are the lion's characteristic traits of behavior, its ferocity—its stalking and brutal killing of prey, its lack of fear as it struts across the savanna—also inherited? A lion raised in a zoo and fed pieces of meat might not be able to survive by hunting and killing if introduced to an African game preserve. I remember reading an article about lions that described how the cubs *learn* to hunt and kill. Maybe, *by nature*, lions do not have the natural killer "instinct" or "trait" of brutality. Perhaps their behavior results from what their physical traits—great strength, speed, size, and deadly weapons, and a digestive system made for processing meat—allow them to learn. But their behavior, generally exhibited in the species, depends upon the experiences provided by a mother and others that train them in using their body for hunting and killing. Perhaps the lion must learn to act in accord with its evolved body, thereby gaining its behavioral trait of ferocity. Giraffes, built as they are, eat leaves from trees; horses, built as they are, run; some primates built with grasping feet swing in trees. Always there are the two possibilities. A behavior might have evolved, wired into the brain by genes, along with the special physical features that "enable" the behavior. But evolution may not explain the behavior trait itself. *A physical feature and a related behavioral feature may be quite independent with respect to evolution; the behavior may be strictly learned.*

Does this apply to humans? Certainly sensory-motor skills, like walking, must be learned, even though humans have the necessary body parts. But how about traits, like aggression and altruism? Generally believed to have evolved—to be carried in the genes and brain—the belief also holds that males have a more physically aggressive *nature* than females. It is true that in life over the ages men have exhibited more aggressive behavior than women have. But is aggressive behavior produced from inner biological causes handed down by evolution? Or is it learned? More pointedly, are the gender physical-trait differences the cause of leaning differences in the sexes? The fact is males, in terms of physical traits, are better suited for aggression, being on average larger, more heavily muscled, stronger, and faster than females. Perhaps those differences in physical characteristics result in different types of learning for the sexes. Perhaps the gender differences in aggression do not result from evolved brain differences.

Perhaps only the physical traits of humans are passed on genetically. Perhaps learning principles determine all human *behavioral traits*. Perhaps human language, as a central example—which is frequently considered a defining human trait—is not an inherited ability from evolution as linguists and other scientists believe. Perhaps, in contrast, the human physical structure has evolved such that humans have the wherewithal to *learn* language. Just as the hummingbird may have an evolved beak whose use must be learned, so too humans may have evolved the necessary brain, diaphragm, vocal chords, tongue, jaw, and lips requisite for language. Humans may be *born* with no language skill at all, no wiring for language, and none that comes through biological maturation. Perhaps the infant has to learn language completely, according to the same principles that apply to learning other behaviors. Perhaps child autism, a symptom of which can be the absence of language, comes not from genetic defect but from an unusual set of learning conditions. Perhaps the behavioral development that occurs throughout childhood does not come about through physical maturation, with stages like the terrible twos, but entirely through learning.

Is the endlessly repeated violent behavior of war inherited, a product of evolution from earlier hominid and ape ancestors? Or are the many different behaviors that go into a war all learned? Are sex and gender behaviors, including sexual identity and preference, as in homosexuality, inherited like physical sex differences? Or are sexual behaviors, including what will "turn on" the individual emotionally, purely learned? Are personality traits, like intelligence, not evolved and carried in the genes, as is generally thought? Is there no human mind that came down through natural selection, beginning with the early hominids and culminating with *H. sapiens*? Could all behavioral traits be learned? Clearly assumption will not answer such questions.

Clearly, also, what I am saying involves weighty matters. Much science and many scientists have developed along lines I am rejecting. I will need a lot of backing to be credible. Presenting that backing—presenting a new conception of humanness—is what this book is about, beginning with examples that show the error of treating human behavior and human nature as genetically determined and evolved.

THE MANY FACES OF THE GREAT SCIENTIFIC ERROR

I do not say that Charles Darwin's confusion of physical traits and behavioral traits in evolution constitutes the whole of the Great Scientific Error. Definitely not. Popular consideration as well as scientific consideration of human behavior and human nature has come from various sources. Let's look at a few examples to characterize the Great Scientific Error and show how they occur in science and in other realms. One of the reasons the Great Scientific Error inspires such deep acceptance is that supporting evidence appears to come from widely different areas. But much of that evidence is tainted, accepted because of strong prior belief that we all get in our culture. We can see this in considering cognition.

Gender Differences in Ability

Not all consideration of human cognition is erroneous by any means, some produces valuable knowledge of human behavior. But many findings depend on interpretation and contribute to the Great Scientific Error. I selected the first example because it involves society as well as a contemporary social problem. The problem consists of the gender disparity in educational success, with girls surging ahead of boys, a startling reversal of tradition. "By almost any benchmark, boys across the nation and in every demographic group are falling behind" (see Tyre 2006, 46) in every educational level from primary school through college. Experts, locked within the beliefs of the Great Scientific Error paradigm, still use the same explanation, despite the fact that it represents a U-turn.

> [M]ost five-year-old girls are more fluent than boys and can sight-read more words. Boys tend to have better hand-eye coordination, but their fine motor skills are less developed, making it a struggle for some to control a pencil or a paintbrush. Boys are more impulsive than girls. . . . Thirty years ago feminists argued that classic "boy" behaviors were a result of socialization, but these days scientists believe they are the expression of male brain chemistry . . . the "boy brain". . . [that produces] the kinetic, disorganized, maddening and sometimes brilliant behaviors that scientists now believe are not learned but hardwired. (Tyre 2006, 48)

Here we have a rich set of views, unfortunately all ludicrous, all assumptions, but all widely believed. Consider this: up until recent times, the tables were entirely reversed—boys were more successful in education. And that male advantage was also attributed, with deep conviction, to wired-in gender differences. Not long ago girls' "genetically determined nature" was not even considered suitable for education, let alone for professional and scientific success. What flies by without comment is that if gender differences in nature were the cause of gender differences in ability, there could be no such reversal. Surely that obvious error should not be overlooked and the error in belief that produced it should be discredited. The problem, of course, is that strong beliefs constitute blinders.

This example also shows how the Great Scientific Error's tilt toward biology squeezes out learning. Not even considered is the possibility that boys *learn*, not inherit, those "boys-will-be-boys" behaviors that interfere with school success. By way of exemplifying how beliefs produce actions, the above view of gender differences calls for educational changes to advantage boys' behaviors, rather than for a change of parenting practices in order to train boys to be docile rather than rambunctious students. Just the other day I saw a military ad in a movie theater exhorting young men to be "warriors," along with aggressive pop music, certainly not calculated to produce docility in boys and young men.

Somehow all the things boys *experience* in our culture that produce "boy's behavior" goes unconsidered, unrecognized.

The Aggressive Human Nature

Take the explanation of the human history of wars. The general assumption is that military actions result from an innate, unchangeable, evolved human nature.

> What triggers most wars is not ideology or honor, says a new theory based on evolutionary biology, but a society bottom-heavy with young, unmarried and violence-prone males . . . bent on seizing territory, goods or other resources they need to marry and have offspring. . . . [W]omen's best strategy for repro-ducing—and advancing their own kind—is to choose mates with enough resources to rear children, [so] women have historically been the ones over whom the wars have been fought, rather than the combatants, say the scientists. (Saltus 1998, A6)

This interpretation, stated so definitively, gained widespread publicity even though no evidence was given or needed; most people already believe that wars are part of human nature. Strange, nevertheless, that scientists give any credibility to such evident mistruth since, clearly, wars are not started by young, unmarried, and violence-prone males. Young soldiers can enlist for economic and other rea-sons, but wars are begun by older, well-off leaders—usually already married and with children—for a variety of complex social, political, religious, patriotic, and economic motivations. Beginning the Iraq War, for example, was a complex social-economic-political event involving many mature people—from CIA agents to public-relations firms—but centrally the movers and shakers of our society. Simplistic "explanation" in terms of evolution is ludicrous. Since writing this I ran across the similar view of the noted historian Howard Zinn.

> If wars were the result of human nature it would not be necessary for govern-ments to work so strenuously to mobilize their populations for war. People would naturally, spontaneously, rush to kill. But that is not the case. Govern-ments have to deceive the population, use enormous amounts of propaganda to persuade people to go to war, entice young people of the working class into the military in hope of bettering their lives. If none of that is sufficient, the gov-ernment must coerce the young, draft them, and threaten them with prison if they don't join. (Schivone 2007, 51)

Revealing in this example is how a piece of spurious science is taken seriously enough to gain international publicity. The evolutionary "explanation," along with other parts of the nature approach, sails along, in the face of obvious disproof. Are such poorly based beliefs like this one important? You bet; they exert causal status in leaders' planning for, and the populace's acceptance of, wars. They also perpetuate the erroneous belief that complex human behavior has been wired in by evolution, also misleading in science.

Psychopathology

That same error occurs in the field of psychopathology where the causes of abnormal behavior are assumed to be neurological, genetic, physiological, and chemical—in the brain. A good example occurs with childhood autism. We don't actually know that autism has a biological cause; the evidence is simply not there to turn that belief into fact. Nevertheless, a recent news article indicates that the "epidemic" of childhood autism has stirred interest in research. "While research into potential environmental factors [chemicals, foods, and toxic conditions] is growing, the search for the autism gene commands the bulk of the research dollars" (Lowy 2004, C5).

Such occurrences completely jump the gun. As the article shows, although the belief in the genetic cause of autism is strong, it does not have the needed evidential justification.

Brain Imaging and Behavioral Traits: Confusion of Assumption, Theory, and Method

Biological inferences and interpretations are taken as fact on the belief that supporting evidence surely will be found later. A great boost has been taken from the recent development of the brain-imaging technology using MRI (magnetic resonance imaging), CT (cat scan), and PET (positron emission tomography) scan methods. Brain activity (believed to be mind activity) can be observed while subjects are engaged in different types of tasks, showing what parts of the brain are active in each case. Also, the brains of people with different natures and disorders can be observed and compared. These methods are being taken as proof that the mind/brain is the cause of human behavior and human nature—a belief held for a century and a half even without brain imaging. The use of such methods expands daily in subject matter studied and in researchers involved. But what do such observations mean?

To raise that question concretely let's take a study that compared the brains of a group of teen children with schizophrenia and a group of normal children, using MRI techniques (see Bower 1997c). The brains of the schizophrenic children had larger fluid-filled spaces indicative of loss of brain tissue. The interpretation was that the difference in the behavior of schizophrenics from normals is caused by such brain differences, with genetic difference as the underlying cause. A moment's consideration, however, reveals that this conclusion involves wide jumps of assumption. Couldn't the fluid-filled spaces have arisen because of medication the schizophrenic children took, *or because of different learning experiences*? After all, the evidence only consists of a correlation between a brain feature and a type of behavior. Correlations, however, do not indicate causation; a third condition could be the cause of both the brain feature and the behavior defined as schizophrenia.

Let me give another example. A study done a few years ago found that a part of the brain in homosexual men was a different size than in heterosexual men. The brain difference was interpreted as the *cause* of homosexuality. The study was widely reported in the media—a titillating "explanation" of homosexuality. Crassly erroneous, however, it turned out there was an obvious but crucial methodological flaw in the study. For the homosexual men all had AIDS. That meant, of course, that the disease or its treatment, not genetics, could have produced the brain condition. The interesting thing is how the study was snapped up for publication and received such widespread publicity despite its disqualifying flaw. How could science and society be so misled? Again, the answer: when belief is deep enough, most anything in agreement can be accepted, and most anything in disagreement will be contended, rejected, or ignored.

Twin Studies and Brain Imaging

In another area, MRI technology was employed to study seven-year-old identical twins, one diagnosed as autistic (Kates et al 2001). The other was not so diagnosed, having lesser symptoms than autism, including language and social problems. The MRIs showed the amygdala (involved in emotion) and the hippocampus (important for learning and memory) regions of the autistic child's brain were about half the size of his brother's. The autistic child's cerebellum and caudate nucleus (thought to be involved in shifting attention from one task to another) were also smaller in comparison. These twin brothers were then compared to five same-aged boys who did not have autism or communication and social problems. The comparisons showed the twins had a smaller frontal cortex (a region involved in planning, problem solving, and organizing) and a smaller superior frontal gyrus (responsible for processing language). The authors conclude by explaining their results as though in the twins' brains there might be differences that caused their differences in language and social problems.

This *interpretation* is actually outrageous in several ways. Importantly, identical twins have been widely used in studies—as in the inheritance of intelligence—precisely because it is assumed they have *the same* genetic structure, thus the *same* brains, and thus more similar intelligence than lesser-related pairs. Isn't it clear, thus, that this study has momentous implications. The many twin studies based on *sameness* form a very fundamental part of the fields of developmental psychology, psychological measurement, educational psychology, and behavioral genetics, major sources of the belief in the biological determination of human behavioral traits. Now here comes this study, actually presenting explosively important counter evidence, stating that identical twins do not have the same brains. That blows the twin-study methodology out of the water, taking away its validity. Either this study of schizophrenic brothers is junk or all twin studies of intelligence and other behavioral traits are. Notably, neither the authors of the study, the editors of the journal publishing the study, nor the viewing audience saw its real meaning. That was overlooked in favor of following their preconceived belief that *brain differences cause behavioral trait differences*. When strong beliefs operate, scientists' ability to understand evidence also disappears behind the blinders of the paradigm in which they believe.

Let's make another interpretation, asking a leading *question*: Why did the brains of the identical twins differ? Since identical twins have the same genetic inheritance, the cause must lie elsewhere. How about different learning experiences? Could children have different experiences even though living in the same home and having the same age and bodies? Could the differences between the identical twins and normal children also have stemmed from differences in life experiences? These possibilities carry the general message: *studies that can be interpreted in different ways cannot produce anything more than suggestive conclusions. That's a point of theory-construction methodology to keep in mind, along with the understanding that no matter how many suggestive studies exist, the interpretations adduced from suggestions can only be suggestive.* That principle is not sufficiently understood. The number of suggestive studies doesn't matter; the many suggestive studies made within the Great Scientific Error do not advance the paradigm's truth value. Society and science have to step back and consider that, since a lot rides on it.

Education

Given that education clearly centers on children's learning, we might expect that its science works would focus on the experimentally established basic principles of learning. Just the reverse. There is almost no use of those principles except by behaviorists. Educational psychology focuses on genetic, brain, and mind explanation. As one example, a group of children with normal reading skills has a different pattern of brain activity than a group of nonreading children when both groups engage in a task of distinguishing the speed of two objects on a screen. Demb and associates also found that the slowest readers displayed the least blood flow (brain activity) in the visual cortex during the motion-perception task (Demb et al. 1998).

The usual *interpretation* was made, namely that the brains of reading and nonreading children are different and cause their differences in reading ability as shown on the expeimental task. A study by the Shaywitzes, Yale researchers, also shows the prominence of the Great Scientific Error and its impact on education and educational research.

> Sally and Bennett Shaywitz found that [a particular] neural region remains inactive as [poor readers] grow up. Preliminary evidence from other researchers indicates that this [neural] structure, located near the back of the brain, fosters immediate recognition of familiar written words and is thus crucial for fluent reading. (Bower 2004, 291)

Again the usual interpretation is the same, without consideration of another possibility. Perhaps as children successfully learn to read their brains are changed, and they become more skilled also in tracking moving objects. This interpretation makes the learning the primal cause, the change in the brain the mechanism that produces the increased perceptual skill. The researchers, however, didn't even consider such a possibility. Rather, while it is mentioned that good teaching changes the brain, the authors focus on the brain as the cause of reading ability. Brain imaging becomes the center of interest. The study thus projects further research on the brain, not on analysis of the learning involved, on how it is learning that produces the changes in the brain. No analysis is made that differences in learning conditions could account for dyslexia, with calls for the study of those learning differences. We can see the same contemporary conception in the following.

> [T]he researchers at Haskins . . . find that the brain of someone with dyslexia functions differently from a typical brain as it processes phonemes—the "c," "a," and "t" that come together to form "cat.". . . The [brain] images show different blood flow patterns in the brains of fluent readers and dyslexics—evidence . . . that this research is beginning to reveal dyslexia's neurobiological basis. . . . Some psychologists even believe that [brain] imaging can one day help people shift the way they use their brains to boost their learning functions. (Bridget 2000, 23)

Despite this fundamental problem with these studies the sheer number of them is impressive, and that helps make what they say impressive. Katzir and Pare-Blagoev (2006) employed a number of them in their attempt to formulate a neurobiological theory that aims to link "mind, brain, and educational psychology" (but not learning) in constructing a theory of dyslexia (53). To begin, they state, "Dyslexia is a specific learning disability that is neurobiological in origin." That statement of course defines the limited nature of where their theory comes from and where it is going. Isn't that taking as fact that which remains in question? And doesn't that approach pretty much squeeze out any study of learning as the cause of dyslexia?

A sharp contest of interpretations shows itself here. At the very least, a theory raised on the foundation of such iffy evidence calls for real questioning. That did not arise. Strange, especially since there is such strong evidence of the central importance of learning as the cause of reading ability. Considering that the institution of education bases actions on its science beliefs, and projects public policies on the results of those actions, not much thought is needed to realize that the errors involved are very costly.

Racism

I chose my next example because it concerns a deep-seated social problem that goes back forever and that continues to work its destructive influence, *racism*. The point I wish to make is that racism derives scientific support from the biological focus of the Great Scientific Error. In Darwin's view, human evolution took place differently in geographically different places, depending on local conditions. Again, the trait of intelligence, too, had evolved in humans, more rapidly in some geographic locations, racial locations, than others because of the different environmental conditions involved. Sir Francis Galton systematically tried to collect evidence that intelligence is inherited. Later twin studies showing twins' intelligence was more similar than for other pairs advanced this view in scientific circles and seemingly provided evidence of racial differences. The general belief, that genes will be found that underlie the various personality traits and behavior disorders, also feeds into the scientific foundation of racism. Nazi Germany's treatment of Jews, Gypsies, and Slavic peoples dramatized the social evil of racism. Today, as a consequence, most everyone condemns racism and most everyone denies being racist. But in a mammoth act of inconsistency, albeit unrecognized, most everyone believes in the evolutionary/genetic explanation of human-behavior differences, and those kinds of differences also are used to explain racism.

Clearing up that inconsistency is one of the reasons I began to critically analyze foundations of the Great Scientific Error in my books in the 1960s. Stephen Jay Gould's *The Mismeasure of Man* (1995) and Allan Chase's *The Legacy of Malthus* (1977) also critically treat "scientific racism." In doing so Gould refers to prominent early figures in psychology's intelligence-measurement field, such as H. H. Goddard and statistician-psychologist Karl Pearson, who concluded that the lower intelligence-test scores of blacks compared to whites result from racial genetic difference. That derogatory interpretation continues to be held, supported by twin studies and other types of research just considered. Various authors have composed theories and written articles and books claiming genetically determined differences in human behavior across "racial" groups. Rushton expresses an old evolutionary belief—built on the original evolutionary theory that Africa, because it was warm and "the living was easy" did not select for intelligence in the evolution of its people (1999). In Rushton's view, thus, Asians are the most evolutionarily advanced race, Caucasians are close behind, and Africans are much farther back. That is generally paired with the belief that the less advanced race more generally has an inferior nature, being less moral, less judicious, more impulsive, more promiscuous, and generally more primitive and less human. Such views persist in powerful form, frequently not evident, as in the case of federal research monies having been spent in the attempt to isolate the genetic (read: racial) "causes" of violence.

[James Watson, the Nobel Prize–winning geneticist stated that] all our social policies are based on the fact that their [those of African descent] intelligence is the same as ours—whereas all the testing says not really. . . . There is no firm reason to anticipate that the intellectual capacities of peoples geographically separated in their evolution should prove to have evolved identically. (Hunt-Grubbe 2007)

Nobel Prize or not, Watson, like Darwin, assumes that behavioral traits, like intelligence, evolved via the same principles as physical traits and some groups of people are more intelligent than others. (Let me add that James Watson vigorously denies being racist, which could be true in terms of holding no discrimination practices.)

The central point here is that the same "scientific" evidence that is a fundamental part of the generally believed, biologically oriented conception of human behavior and human nature also is a fundamental part of the racism conception. Racism rests upon the unproven belief in the biological causation of human behavior traits, and racism will not be eliminated until that belief is diminished to equal the nullity of its proof.

Developmental Psychology

The field of child development focuses on making observations of children's development of behaviors as they advance in age and grow physically. Children, for example, generally babble at six months, say a word or two at one year, and begin to learn words more rapidly at a year and a half or two. The field generally *assumes* biological maturation as the cause of the behavior development, without considering learning as a possible cause.

That biological viewpoint dominates the field of child development. Consider the following. In the research of Elizabeth Spelke, who directs the Harvard University's Laboratory for Developmental Studies, she and her colleagues present different situations to infants and young children and, on the basis of the children's reactions, infer the nature of the young mind and the way it must develop through maturation. They believe that babies are born with knowledge about the world already in their minds. For example, they suggest that babies have an innate knowledge about geometry, such as what paths must be traversed to get from one place to another. Babies also are said to have an innate, wired-in ability for numbers and mathematics because six-month-olds can respond differently to eight and sixteen objects and sixteen and thirty-two objects.

Many works in the field of child development fall within that conceptual framework, attempting to infer the nature and cause of the child's mind from observing the behavior of the child, believing that the child's nature is innate. The Harvard Laboratory is a very large and expensive research setup, and that is only the tip of the iceberg of expenditure for such science. So it is appropriate to ask what kind of knowledge can such research produce since even other cognitive psychologists have raised questions about Spelke underestimating how much babies learn through experience, giving a distorted view that babies are much smarter than they are for their age. Sylvain Sirois makes the same point in the *European Journal of Developmental Psychology*, indicating that "the methods [employed] are correct and replicable. . . . It's the interpretation that is the problem" (see Brunton 2007, 95).

The general point is that when the behavior of children is the only thing studied, there is no evidence produced by which to infer the cause. Perhaps the development is due to biological phenomena, such as biological maturation of the brain or whatever. But the development may also be due to the child's learning experiences. There's no way of telling which without actual evidence of cause.

Gerontology

In its science section of September 23, 2008, the *New York Times* reported a study in which the emotional response in the brain of a group of oldsters in their sixties, when receiving a reward, was compared to that of young people in their twenties. The "study found [using an imagery technique] that as people age their brains respond less strongly to rewards. The main difference is in the response of the brain to dopamine, a naturally occurring chemical messenger that plays a central role in the reward system" (Nagourney 2008, D6). A difference in response is attributed to the effects of age on the brain.

It all sounds so convincing. Certainly the researchers who conducted the study were convinced, as was the editor who accepted the study as the subject for inclusion in the science section of the newspaper. Since the nature and strength of the rewards that function for a person are central causes of the person's behavior, having our knowledge of rewards is quite important. So we wouldn't want to make empty assumptions on this topic. But that is what occurred in this study. The researchers, intent on establishing how changes in the ageing brain alter human behavior, completely neglected consideration of learning. But consider this: a sixty-year-old and a twenty-year-old have had much different quantities and types of experience with rewards. Older people, having had much more time, have received many more rewards than younger people, including bigger and more varied rewards. Consequently, through that learning they will have become more "jaded," less excited about a small reward than a young person is. On the basis of learning, older people shouldn't react to small rewards as intensely as a young person. Moreover, a three-year-old should be more excited with a very small gift—like a gumdrop or a little plastic toy car—than will a twenty-year-old—and not because of nervous-system deterioration. Of course learning experiences, as a function of age, would be expected to have effects on response to rewards, including the brain response to rewards.

One has to ask why the researchers missed this weakness in the study, why the newspaper considered the findings solid enough and valuable enough to give it such audience exposure. Why did no one think of the difference in the learning experiences of the two groups, since it raises questions that invalidate the study? The answer lies in the Great Scientific Error paradigm that beclouds everyone's vision, from medical researchers to newspaper editors. Maybe a new paradigm that didn't make such errors would be a great advantage.

Philosophy of Science

I will take my last example from the pen of a well-known philosopher of science, indicating that even those dedicated to understanding the character of science are not immune from the Great Scientific Error. Stephen Toulmin theorizes that science progresses via evolution's natural-selection principles (1972). Scientists produce many new developments and natural selection determines which ones will survive. Consider for a moment how poor this reasoning is. Natural selection works its effects genetically. But there is no genetic change in scientific progress. When Einstein introduced a new theory in physics, he and those who used his theory did not change genetically. There is no evidence that genetic change explained their extraordinary creativity. Making science consists of learned behaviors, and it would make much more sense to use learning principles in studying science than natural selection's genetic principles. Even used metaphorically, evolution makes no sense in considering science progress, so this is yet another example how the Great Scientific Error penetrates society and science in many ways, with no justification, and without being realized and noted. That occurs because the paradigm is carried in the words of our common language, is learned by everyone, and thus is in everyone's way of thinking. The Great Scientific Error is deeply established and needs no justification, even when it is wrong.

Only a few examples were given here to show the error of considering behavioral traits the same as physical traits, too few to ask the reader to abandon an important part of the common theory of human behavior and human nature. These examples, however, are only the beginning of the tale.

WHY IS LEARNING LEFT OUT?

A major problem of the Great Scientific Error resides in its woeful treatment of learning, largely leaving it out. Isn't there another alternative, one that focuses on how external environmental conditions affect human behavior? After all, Aristotle held that the mind is a blank slate until written upon by experiences. That interest in learning has persisted. An experimental science study of learning did indeed arise in psychology. Around 1900, Ivan Pavlov, a Russian physiologist, and Edward Thorndike, an American psychologist, each serendipitously discovered an experimental way of showing with animals one of the two fundamental principles of learning (or conditioning).

These two discoveries established fecund traditions of research. As early as 1913, John B. Watson projected learning as the basis for a general approach, called *behaviorism*, to serve as the framework for the new science of psychology (see Watson 1930). Behaviorism was advanced as a revolution against the existing psychology that focused on studying the internal mind as the cause of human behavior. In this psychology, subjects reported on the state of their own minds, producing a subjective jungle of contradictory findings. Rebelling against this program, Watson proposed instead that psychology should study behavior and how the environment causes behavior. The environment and behavior were material observables, not intangible and inferred, like the mind. He considered reporting on one's mind a capricious type of research—like speculating about the soul—in contrast to studying how observable environmental stimuli produced observable changes in behavior.

Watson thus brought the study of learning front and center. Clark Hull and B. F. Skinner, and many others, led in creating a generation of behaviorism study devoted to laboratory experiments. The leading behaviorists each formulated an independent learning-behavior theory based on the many animal-learning findings. They also furthered Watson's beginning attempts to create a philosophy of science for psychology, identifying it as a natural science based on empirical study.

Did the basic learning principles, found with animals, apply to humans? A number of experiments showed humans learn according to the same principles. In describing his learning theory, Hull states eloquently "that all behavior, individual and social, moral and immoral, normal and psychopathic, is generated from the same primary [learning] laws. . . . Consequently the present work may be regarded as a general introduction to the theory of all the behavioral (social) sciences" (1943, v). Clearly Hull was envisioning a new paradigm, to be based on learning.

The same vision held for all the behaviorists. They specialized in animal research, believing the basic principles would explain human behavior. The problem was that the major behaviorists were animal psychologists; they did their learning research with animals. They projected the learning principles as explanations of human behavior, but none of them researched real human behaviors or indicated how this was to be done. None of them constructed a conception that explicitly treated real human behaviors and how they were learned and how the learning and the behaviors could studied. Skinner, who wrote more on human behavior than any of the others, did not do research on how actual human behavior is learned and did not analyze specifically the learning conditions that produce specified human behaviors. He believed that the methodology he used for research with animals, involving a simple motor behavior—called the experimental analysis of behavior—was the way to approach the study of human behavior, as the research of his students shows (see Bijou 1957, Lindsley 1956). None of the behaviorists called for a broad science of learning that would study *real* human behaviors.

Watson's (and later Skinner's) rejection of the concept of the mind and all things "mentalistic" was extended to a wide range of phenomena—including personality, intelligence, attitudes, interests, or psychological measurement—actually any inferred internal process. Thus, this traditional behaviorism threw out the baby with the bath water. While traditional psychologists concocted inferred concepts of behavior causation, like personality—concepts that could never be observed—nevertheless their conception led them to study real human behaviors. The common concept of intelligence, for example, is inferred and doesn't explain *why* some individuals do better in school than others. But traditional psychologists did attempt to measure intelligence, and doing that was of value. The fact is intelligence involves important behaviors. And intelligence tests can predict school performance. Those things shouldn't be ignored or rejected. Traditional behaviorism in rejecting much of human behavior, because poor concepts were used to describe that behavior, was, and is, an obstacle to considering how that behavior might be learned.

Applied Behaviorism

Despite that weakness and behaviorism's focus on basic animal research, a handful of us in the 1950s broke out of that tradition and began applying animal learning-behavior principles to real human-behavior problems. Joseph Wolpe invented systematic desensitization for treating phobias and other anxiety problems and generally promoted the application of learning principles to clinical problems (1958). Hans Eysenck compiled a number of studies that employed learning principles for treating problems of behavior (1960). I analyzed how the behavior of psychotic patients could be considered and treated via conditioning principles (see Staats 1957), invented the token reinforcer system in 1958 for wide employment in treating behavior problems of children and adults, and innovated other clinical developments, including laying foundations for a behavioral theory of abnormal behavior and a cognitive-behavior therapy (see Staats 1963, 1972). (A publisher's brochure I read recently listed the ten areas covered in a behavior-therapy book sold online. Two of the ten areas dealt with my "time-out" and my "token-economy" methods.) Leonard Krasner and Leonard Ullmann compiled a group of our papers and called the approach "behavior modification," giving strong dissemination to our efforts (1965). Franks did early research and helped give form to some of these developments, including naming the field "behavior therapy" (1958).

Our small group thus provided the foundations of the fields of behavior therapy and behavior analysis. Many works have broadly extended our framework to various different types of behavior problems, creating new treatments and new analyses of behavior disorders for psychologists all over the world to use. Today this movement has well-developed fields producing many studies and new methods and interests. The different number of behavior problems dealt with continues to expand, and the number of people in the field also grows.

In addition to applied significance, the behavioral fields have produced something more general and significant: *extensive evidence that learning principles generally apply to human behaviors, evidence that learning is an important cause of human behavior.* Various kinds of problem behaviors have been analyzed in terms of learning principles. This evidence does not depend upon interpretation, assumption, inference, or belief. It is not a "pie-in-the-sky-by-

and-by" promise. It is right there, already established. Dyslexic children can be trained to read, for example, which is quite different than interpreting data as showing that brain aberrations are the cause. Children have a variety of problem behaviors that can be treated by explicit training. The methods employed are reliable and replicable, and thousands of studies and applications have been made that produce such evidence.

Not all the biological studies together can match this for producing precise effects on human behavior. Genes can't be flipped, the brain can't be given electric impulses, nor can a pill be administered that can produce reading, or any other particular behavior, in a child. An autistic child can't be given language or social skills via biological means. But such behaviors can be produced by the use of learning principles. *The important point is that these many studies provide irrefutable evidence of the tremendous power of learning for determining human behavior.*

Basic animal-learning principles cannot be left out of the paradigm needed for explaining human behavior and human nature. But the *general* significance of our applied behavioral movement is still not understood even in the movement itself. The focus is on correcting problem behaviors, because the behavioral movement today has become an applied science. We need to understand the unique nature of human learning. We also need to understand the astoundingly huge learning prowess of the human. Unlike the beliefs of Charles Darwin, and John Watson in learning, humans do not just learn like other animals; humans also have a hugely enhanced learning ability.

If we humans learn our behaviors, then shouldn't we study the principles involved and the way important human behaviors are learned? I mean study those things deeply and broadly? Shouldn't we have big science for that study? Behaviorism, despite its great contributions, does not contain a conception or framework for a big learning science, a science that extends from the evolution of learning, through the biology of learning and the basic principles of learning, to all kinds of complex human behavior to reach an understanding of humanness in a profound way that yields new developments in science and in applications, a unified framework for vigorous and broad activity. The Great Scientific Error constitutes an obstacle to establishing the importance of learning to human behavior and human nature.

OF PARADIGMS OLD AND NEW

Kuhn (1962) describes the case where an old, accepted paradigm contains errors and leaves central concerns and questions unanswered. In such a case those weaknesses can become more and more exposed, arouse concern, and finally inspire a new paradigm. Many findings of the old paradigm are retained. But the new paradigm gives a new view, supplies new evidence, and opens new questions. Although convincing, the new view and evidence are only the beginning; they outline a framework that must be filled in by a great many additional developments. The new paradigm thus presents fruitful avenues for scientific advancement. The new paradigm only does that after a "paradigm clash," for the old paradigm by its fruitfulness will have gained loyal followers.

I suggest that a process like that has been building with respect to the phenomena of human behavior and human nature. There is an old paradigm that is multifaceted and unsystematic but productive in many areas of study and broadly used in many different ways. But the old paradigm contains many errors and much waste, and it doesn't provide ways to answer many important questions. It does not even consider vast and vastly important aspects of human behavior and human nature, dealing very little and poorly with learning.

The weaknesses of the Great Scientific Error paradigm are there to see and are being seen ever more fully. Various leading thinkers are realizing that biology does not explain human behavior at all completely and that experience or learning or culture has important explanatory power. For example, the field of primatology investigates ape behavior to discover knowledge about human behavior and its presumed evolutionary cause. Justification for such study derives from Darwin's fallacy that human *behavior* traits have evolved from earlier species. The work of Franz de Waal, a noted primatologist, however, has begun to bend that framework (2001). For he has found that behavior learned by one ape can be *learned* from that ape by others and became typical for their entire group. He concludes that learning needs to be recognized. Jablonka and Lamb, noted authors, describe other studies showing that animals can learn behaviors that traditionally were considered as evolved (2005). Such findings have not yet been considered to indicate a grave Darwinian error even though they show that evolution does not explain at least some types of animal behavior. This recognition of learning by such scientists is new, an opening to the idea that human behavior may be caused by something other than biology, as Ehrlich has written incisively about the nature of human nature (2000).

People don't have enough genes to program all of the behaviors some evolutionary psychologists, for example, believe that genes control. . . . Human beings have something on the order of 100,000 genes and human brains have more than a *trillion* nerve cells, with about 100–1,000 trillion connections (synapses) between them. That's at least 1 billion synapses per gene, even if each and every gene did nothing but control the production of synapses (and it doesn't). Given that ratio, it would be quite a trick for genes typically to control more than the most general aspects of human behavior. Statements such as "Understanding the genetic roots of personality will help you 'find yourself' and relate better to others" are, at today's level of knowledge, frankly nonsensical. (Ehrlich 2000, 4)

One of the pillar beliefs of the biological approach holds that the human brain is prewired genetically, and that is the cause of the individual's behavior. Evidence against that view has been slowly accruing, however. The human brain is not a static organ, laid down genetically, whose development takes place through biological maturation. Even at the time I began my research program it was known that learning takes place by the formation of neural connections in the brain. As I studied human learning it became clear to me that a great amount of complex learning occurs and this must bring changes to the anatomy of the brain, that learning changes the brain, and that affects the person's behavior. This conception views the brain as *the mechanism* by which learning experiences affect behavior. *The conception makes learning, not the brain, the cause.* I first saw evidence of this over forty years ago when a study showed that Scotty dogs raised with eyes occluded never developed normal vision. The experience of visual stimuli also appeared necessary for developing their ocular organs and the relevant part of the brain. Later I ran across studies that showed clearly that experience changes rats' brains. If rats' brains are affected by the simple experiences cage-life provides, think of the large effect on the brain the hugely complex learning experience of humans would have. Generalizing from this, my conception became that "*the child's experience results in learning recorded via the formation of neural networks*. The brain's changes reflect the child's learning, not the reverse" (Staats 1996, 165).

The important point here is that biological scientists are stepping outside of the box in their work by recognizing that learning plays a role in determining human behavior. Moreover, in support of a new learning paradigm, findings are accumulating that *the human brain is plastic and changes throughout life*. "It has become clear, however, not only that [neural] synaptic organization is changed by experience, but also that the scope of factors that can do this is much more extensive than anyone had anticipated" (Kolb, Gibb, and Robinson 2004, 2). Drs. Mike Merzenich at the University of Illinois and Arthur Kramer at the University of California at San Francisco Medical School weigh in on the plasticity side by showing that elderly people, in whom the brain is supposed to lose neurons, actually grow new units and connections on the basis of experiences.

There is actually a spreading change taking place in the once-monolithic biological focus. Robert Plomin can be used as an example. He has been a leader in behavioral-genetics research that attempts to find genetic causes of behavioral traits, an approach that does not consider learning causes. He, however, has become director for interdisciplinary research on both nature and nurture. We can see the same emerging change in the National Science Foundation's establishment of four Science of Learning Centers in recent years.

Another voice indicative of the gathering change is that of Jared Diamond, a creative leader in biological science. In 1992 he published a biologically oriented book, *The Third Chimpanzee* (1992), in the tradition of works deriving from the Darwinian fallacy. Nevertheless, seven years later, he published a work that studies the different technological advancement of peoples from different geographic locations—examples ranging from Stone-Age-level cultures to modern societies (Diamond 1999). He concludes, "[T]he striking differences between the long-term histories of peoples of the different continents have been due not to innate differences in the peoples themselves but to differences in their environments" (Diamond 1999, 405). Dealing with great and important behaviors, he rejects genetics as the explanation—thereby going against widespread and long-held beliefs, in favor of environmentalism, or learning.

These are examples involving leaders within biological fields. But we can see also the same insight in scientists now coming up, who show that the views circulating through the science community are shaking belief in the Great Scientific Error paradigm. An e-mail sent to me by a graduate student in Australia shows the change.

> I have read one of your other books [Staats 1963] . . . and it made so much sense to me, so many observations I have read about became comprehensible. I have a problem with the research framework we are using [i.e., "behavior results from innate biological vulnerabilities interacting with stressful life events"] as it seems not to take these learning principles into account, which means it doesn't make sense to me—however, nobody else seems to care about this . . .—so it makes me doubt my own thoughts a little. (Prichard 2006a)

I wrote her back that I considered her views very insightful, and she expanded on her previous statement.

[W]hat amazes me is that so much information necessary for doing research in human behavior has been available for many years, I can't imagine how frustrating it must be for you to see people ignoring that information over and over again. Everyone says that the reason why we are not getting good results in this field (i.e., why it has not been possible to find genes that underlie human behavior) is because there is too much complexity, but I think the real reason is because the model of human behavior we are using is wrong, and the sad thing is that it is not wrong because there are too many unknowns, it is wrong because certain important facts are being ignored. (Prichard 2006b)

We have here a clear picture of the emerging clash between an old paradigm and a nascent challenger. Prichard is prescient in seeing the two sides, noting problems in the old that can be dealt with in the new, and noting as well the obstacles the old presents to progress.

Another example occurs in the consideration of cultural evolution. "Culture affects the success and survival of individuals and groups; as a result some cultural variants spread and others diminish, leading to evolutionary processes that are every bit as real and important as those that shape genetic variation" (Jablonka and Lamb 2005, 4). The product of this new interest is a body of thought concerned with human behavior caused by experience but described in concepts and principles of natural selection, hence the term cultural evolution. The result is an interesting mixture. "What is emerging is a new synthesis, which challenges the gene-centered version of neo-Darwinism that has dominated biological thought for the last fifty years" (Jablonka and Lamb 2005, 1). I consider this an important step despite the fact that the learning part—consisting of imitation—has not been widely or well developed (see also Richerson and Boyd 2005).

One last example to indicate the discontent with the old paradigm extends into applied-science areas. The case in point appears in letters to the editor of a national news magazine in response to a previous article that considered addictions to be determined by biochemistry and genetics.

A simple biochemical view cannot explain why 90 percent of heroin-addicted Vietnam veterans were able to quit once they returned home . . . , or why a person picks up a drink even after years of abstinence when faced with a serious loss. . . . These examples suggest an emotional origin to addiction. . . .

Addiction is hardly as simple as the biochemical reductionists would have you believe. (Dodes 2008, 18)

Stated in general terms, there are a number of voices indicating that learning needs to be included as a cause of human behavior in a way that presently doesn't occur. Some are also beginning to have reservations concerning the explanation of human behavior by genetic inheritance and inferred internal traits. For example, Malcolm Gladwell's 2008 book *Outliers* is built around the theme that not even geniuses exercise great achievements without special environmental support. Psychologist Nisbett, formerly within the old tradition studying the internal mind, switches and considers intelligence as less determined by genetic inheritance and more by experience and learning (2009). His conception actually coincides closely with the previous learning conception of intelligence (Staats 1971a, 1996; Staats and Burns 1981) and thus represents another abandonment of the old paradigm. Similarly, another new book plays down the importance of talent—as a genetically wired-in trait—and focuses on the importance of practice, as can be seen from its title, *The Talent Code: Greatness Isn't Born. It's Grown. Here's How* (Coyle 2009).

After a long period of dominance by books written within the Great Scientific Error paradigm tradition, these last books published in the same time period suggest the cat is out of the bag. Like some of the other books described, they have been highly successful in the popular media, showing the mounting need for development of the nurture view that the nature view has been blotting out.

An Opening Does Not a New Paradigm Make

An opening constitutes a potentiality, however, not an accomplishment. The "nurture" interest has always existed, but has waned for want of a fruitful learning paradigm. The environmentalist position has never produced a science tradition that explains human behavior and human nature widely and deeply in a way that produces a broad science endeavor with a framework for continuing advancement. Important as getting out of the box of biological explanation is, the new voices only constitute that first step in paradigm change. Those who lead the new learning interest in the old paradigm still focus on biological science. "However, it is a daunting task to determine if synapses have been added or lost in a particular region, given that the human brain has something like 100 billion neurons and each neuron makes an average several thousand synapses" (Kolb, Gibb, and

Robinson 2001, 1). We can't expect either a learning theory or a conception of human learning to come from the study of neurons. Those working in the old paradigm do not have the needed specialized knowledge of learning. An anthropologist can try to introduce learning into an explanation of evolution in a commonsense way by using an unspecified concept of cultural or social evolution. That still leaves the understanding of learning on a commonsense basis; it does not make the needed learning analyses of human behavior and human nature.

All this is understandable. It is too much to expect those whose backgrounds are in biological and related sciences, or in commonsense views, to provide a learning conception. There remains the formidable task of "providing the meat," that is, a new paradigm that indicates what the principles of learning are, just how learning determines important human behavior phenomena, a broad conception that provides a foundation for extended understanding, as well as programmatic advancement. The discontent that has been described, thus, represents just an opening, still inchoate, still unsure, popping up separately as important complaints. When gathered together they illuminate a weakness in the present paradigm and a readiness for change in a very important, large scientific area. They show the time is ripe for presenting a paradigm about human behavior and human nature that gives learning due place, that provides understanding, broadly gives applications in life, and engenders new science paths to take. The new paradigm calls for new science and a cork on the bottle of wasteful old science. Characterizing that new paradigm constitutes this book's central aim.

A NEW PARADIGM

Learning, as a cause of human actions, hasn't passed unnoticed. Aristotle said to the effect that the human mind is like a blank page written upon by experience. The British empiricist philosophers specified how the associations of the human mind are formed by experience. Experimental psychology later conducted many investigations to establish the principles of how human verbal associations are learned. Traditional behaviorists discovered and specified basic principles of learning for all species. Many environmentalists' studies have shown in a general way that environmental experiences can enhance children's ability. Since the 1960s our applied-behavioral fields, in many works that use the basic learning principles to treat people's problems, actually have provided a huge cache of evidence showing that learning determines human behavior. These and many other studies demonstrate that learning is very fundamental.

Still, these efforts are separated and have not coalesced into an organized unified paradigm that drives broad efforts to explain human behavior and human nature generally. The new paradigm to be presented here springs directly from my more than fifty-year program of research and theory on learning human behavior and human nature. That program involved many studies—theoretical, empirical, applied, and methodological—directed to various aspects of human behavior as well as traits of human nature. Even by my 1963 book *Complex Human Behavior*, I saw my path would become a continuing, elaborating, unified thoroughfare, presenting analytic learning conceptions of child development, the acquisition and function of language, intelligence, personality, human motivation, social interaction, educational psychology, and abnormal behavior and treatment. Many developments outside of the program were analyzed in its terms, brought into its principles and methods as a means of broadening the program's scope. Various parts of that book provided foundations for the fields of behavior modification, behavior analysis, behavior therapy, and cognitive-behavior therapy—much more than any other behavioral work of that time.

Nevertheless, the work was only a beginning, a basis for further studies. For example, my view of personality in 1963 had little specification. Later, personality was analyzed further to include the trait of intelligence, a more extensive and deeper theory indicating how learning experiences can raise intelligence (Staats 1971a). That formulation, and the findings from which it derived, were the basis for later study that demonstrated how children can gain intelligence through learning (see Staats and Burns 1981). Intelligence was analyzed into its constituents (see Staats 1996) in a manner followed later (Nisbett 2009). That was how the foundations for the new paradigm were laid, first in framework form that progressively has been enriched and deepened, the structure progressively extending and filling in.

There have not been analytic paradigms concerned with human behavior phenomena that are very broad in conception and in methodology, that have principles and concepts closely and regularly aligned, that are unified, with many avenues of advance. Charles Darwin's theory combined with Gregor Mendel's genetics composed a paradigm that saw expansion into various science areas in a way that makes those areas related, their works mutually supportive and significant. Nothing like that has emerged in the science areas concerned with human behavior that stretches broadly from sensory physiology to social science, from child development to human evolution.

Creating such a paradigm has not even been fantasized. After all, the human-behavior sciences are new on the block. At the beginning, when everything is looked at superficially, the search is for the new and different, not for how things fit together. The originator of the concept of the paradigm said early science involves a hodgepodge of conception and method. In describing the field of electrical science in the beginning, Kuhn says that "during that period [the late 1700s] there were almost as many views about the nature of electricity as there were important electrical experimenters" (1962, 13). That characterizes the young science, like those concerned with human behavior. Such science of course produces superficial and wandering knowledge. When a unified paradigm emerges, it is more profound and detailed, and it has many more implications. It deals with more complex and more profound matters, but in a simpler way than the confusion that went before.

I am going to develop a new paradigm here, in the form appropriate for a general audience. The paradigm elaborates how the human species' learning ability is gigantic, marvelous. Humans have a wonderful set of sensory organs by which to experience the environment in great complexity. And the human body is made for producing a huge variation of different, complex, coordinated, and extended kinds of behavior in a way no other animal can. In harmony with that great versatility, humans have a huge brain. The human brain and nervous system are made for connecting the complex sensory input to the complex response output such that humans can *learn* to respond with astonishing flexibility and precision to infinitely variable and complex environmental circumstances. Human behavior and human nature have been developed primarily through learning. No one can look deeply at human behavior, unless blinded by preconception, and not see that human actions vary hugely, that human behaviors constitute a most *complex* set of phenomena. No one can look at the great lines of cultural development of behavior over time and not realize that *complex* learning processes have been the cause. No one can systematically observe children develop and not see the great changes due to experience; what is the educational institution for, if not for learning? Humans have enormous learning ability for an infinite variety of behaviors—we are a species completely unique in that respect.

Are we to think that humans have this great learning ability for no reason, to no effect? That is inconceivable. Mother Nature is not a spendthrift. The brain burns up about a quarter of the energy the human consumes; why would such a powerful and costly learning organ have evolved if it did not pay its way by mightily contributing to humans' survival and reproduction? The human brain could not in major part be devoted to learning unless learning plays, and has played, a big adaptive role in human behavior and human nature. Not to recognize that is actually a rejection of evolution.

Centrally, if we are interested in human behavior and human nature, we need to know about that learning. We can't leave our understanding of learning on a commonsense level. We have to know the principles by which learning occurs in the complexity displayed by humans. We have to take that knowledge to the point where it is developed to actually deal with the behaviors that are involved in life, to solve problems, and to project paths of development.

The new paradigm brings together many varied aspects of human behavior and human nature within its unified principles, each analysis providing foundations for understanding the next. Can the various aspects of the child's behavior development be learned? Can the various aspects of personality be analyzed as learned traits? Can problems of education, of cognitive failure, like the sizeable fraction of American children who do not learn to read, be studied profitably as a breakdown in learning? Great interest exists in cognitive phenomena and the operation of the mind—could they be learned?

Could different ways of reasoning, from the precision of a logician or lawyer to the delusions of a schizophrenic, have been learned? Autistic children can learn, according to the same principles as normal children; could the disorders of language development in autistic children have involved learning or its absence? Could other abnormal behaviors that form disorders be learned? Don't we need to consider very broadly the various disorders of human behavior in terms of how learning may be involved in their origin? We already have a vast amount of evidence in the fields of behavior therapy and behavior analysis that abnormal behavior can be changed through learning. If abnormal behavior can be treated via learning experiences, doesn't that suggest their origins have involved learning?

Human origin conceptions are central in defining human nature. Clearly humans are the preeminent learners among animal species. Wouldn't it be important in considering human nature to be concerned with how that great learning ability evolved? Using its conception of human behavior and human nature, the new paradigm develops a new theory of human evolution that focuses on the role of learning in the progressive development of hominin species that preceded *Homo sapiens*. When such analyses are made within a unified learning paradigm, a large and profound picture of human behavior and of human nature—of humanness—emerges, a new picture of the human with new implications for everyone, and for science as well.

Part 2

THE HUMAN ANIMAL

Chapter 2

SUCH A BODY!
THE VERSATILE ANIMAL

Other mammals have digestive systems, respiratory systems, immune systems, and sensory systems, as do humans. Species' differences in these systems have evolved, of course, springing from different life circumstances. The immune system of a species, for example, may be different because of exposure to unusual toxins, parasites, or bacteria. Or species' digestive systems may differ because of differing diets. Yet across such differences there is general similarity, all resulting from evolution.

There are, however, two fundamental ways in which humans differ distinctly from other species. One of them is in learning ability, dependent on the huge human brain, where humans differ vastly from any other animal. The other feature consists of the human body, especially the human skeletal body; it also has fundamental differences in function from the body of any other animal. Not usually a matter of consideration, the human body plays fundamental roles in determining humanness.

THE MARVELOUS GENERALIST

Human achievements are indeed remarkable. How to explain them is the concern of this book. The focus will be on learning. However, without the human body to put things into action, learning would get the human species nowhere. Here resides a basic way in which biology and learning interact. Let's break the body into sensory organs, response organs, and the organs for connecting the two, that is, the brain and nervous system.

Human Nature in the Senses

Our general environment is made up of many specific features emitting energies—such as light waves, sound waves, mechanical pressures, chemical and temperature conditions. Humans have sense organs for some of these environmental energies, to be called stimuli. The bare human, unaided by instruments or communications from others, contacts the world and comes to know about the world only through sense organs. Some sound waves produce hearing, some light waves produce vision, some chemicals produce taste and smell, and mechanical and thermal stimuli produce tactile sensations.

There are also sense organs in the muscles and tendons of the body. When our movements stimulate those sensory nerves, that gives us information about what our bodies are doing. Other internal stimuli come from organs in the semicircular canals of the ear, providing information concerning our balance. There are also pain receptors in various organs and parts of the body that tell us when stimulation begins approaching a harmful level. It is valuable to realize that there are internal stimuli that function like those in the environment.

Other animals have sensory organs too. Some animals have sense organs for detecting stimuli we cannot, such as sharks' sensitivity to electrical fields, which we lack, and dogs' sensitivity to high notes beyond our hearing. Similarly, the sense organs of different species vary in the richness of the stimuli to which they are sensitive. For example, most animals respond visually only to light-dark differences, whereas a lesser number, including humans, respond to the different light waves as colors. The sensory organs of various species have been formed through evolution because they make the species better able to respond to the environment and thus survive and procreate.

Evolution molded our sensory equipment also, delineating what we can experience. Without some kind of aid we can respond only to those aspects of the environment that activate our sensory organs.

You Are What You Eat

The nature of the sensory organs can determine the animal's behavior and nature. As an example, let us take taste; we don't usually consider how taste organs help determine behavior and body structure. But consider giraffes, zebras, lions, and pandas. Giraffes eat leaves, zebras eat grass, lions eat animal flesh, and pandas eat bamboo shoots. Their taste buds lead them to eat that which their digestive system can handle and their actions can gather. If a lion were born with a panda's taste buds, it would eat the wrong things for its stomach to digest. Its behavior wouldn't be lionlike either. So the nature of those sensory taste buds has behavioral significance.

The animal, thus, has to be structured to "capture" and consume its food. For the aardvark that meant developing a proboscis and tongue that enable the capture of ants. For the jaguar that runs down its prey in a ferocious burst of speed, evolution took its development down another line. Humans, like chimpanzees, omnivorously enjoy the sweetness of fruits and the taste of flesh and many things in between. That means that humans, "taste-wise," can live in many places. Giraffes, pandas, aardvarks, and lions are more limited. They can live only in places where their specific food can be found. Flexibility of taste and digestive system has thus been a factor in being able to live all over the world. Health-food circles advocate that "you are what you eat," which plays out pretty well in the light of these considerations.

Sensitivity Can Be Touching

Tactual sensibility (feeling) constitutes another sense, albeit less prominently considered than sight, hearing, taste, and smell. Humans have exceptional equipment here, specialized in the hands. At the end of flexibly maneuverable arms, the pads on the palms and fingers are loaded with nerves to make them sensitive and mobile instruments for sensing by feeling the environment and providing stimuli for action. Compare human tactile sensory organs with those of four-footed animals that feel only with vibrissae, nose, tongue, lips, footpads, and body and hair. Their sensory equipment provides relatively imprecise tactile sensitivity. Humans have much greater sensitivity, for example, in establishing whether the firmness of an avocado indicates that it is not yet ripe, is ready to be eaten, or has reached the time to be tossed into the garbage. Can you imagine a doctor palpating a breast with a dog's paw, vibrissa, shoulder, or tongue in an examination for a tumor? The extent to which a species can respond to the manifold aspects of the environment depends in part on the broadness of the species' tactile sensory organs. Big brain or not, dolphins don't have the sensory organs by which to experience the world tactilely with the breadth that we do, nor can most any other animal.

**Sensory Characteristics Help Determine Our Learning and
What We Are as Humans**

By providing information about the nature of the world, sensory organs help determine behavior. That applies to *learning* behaviors. Deaf children will not learn to speak because they cannot hear the sounds their parents make. Taking away any type of sensory input will take away from human behavioral skills. Loss of smell, for example, generally takes away taste and appetite for food. It is important to realize that input from the outside is necessary for human behavior—it doesn't come from within, from the brain by itself. *Humans' sensory ability is part of human nature.*

Human Nature in Skeletal and Muscular Characteristics

Just as sense organs are determinants of human behavior and human nature, so are other characteristics of the body.

Locomotive Structure

The most striking characteristics, and most noted as human, are those responsible for fully developed upright posture. Most mammals locomote on all fours. Primates like the chimpanzee, gorilla, and orangutan are the closest to humans in body structure, and they are able to move on hind legs. Their pelvises, legs, and feet have evolved to allow them to stand and move in a semi-upright position, but not well and not for long distances. They still employ their forelimbs for locomotion, and they do not approach the human's grace, speed, stamina, power, and variability in upright movement.

In most primates, the foot is a generalized organ, employed both as a hand and as a means for locomotion. The big toe is separated from the other toes in a handlike way, serving like a thumb in grasping. In humans the foot has lost its grasping ability. In a series of evolutionary developments involving feet, ankles, knees, and hips, the human legs gained in bone and muscle structure and became more powerful. Humans' hind appendages are specialized for locomotion, making possible extended and efficient standing, walking, and running.

The change to greater specificity of the leg function opened evolution up for changes in the arms and hands. Not needed for swinging in trees, human arms and hands became free to develop for a huge variety of purposes. Thus, humans can conduct two different activities at the same time, with independence, yet in coordination. Behaviorally that must have been vital in pack hunting; signaling could occur between hunters while they continued independent actions with their legs. This trait of the human body today provides the foundation for learning many different skills, ranging from playing piano or tennis to driving a car, quarterbacking a football team, or drilling teeth.

Hands

Two stages of development have been considered in the evolution of the human hand. The first occurred with primates' adaptation to the forest environment by developing a thumb, good for grasping branches in locomotion. The second stage occurred when human progenitors adapted to life on the open savanna, where bipedalism was the means of locomotion and the hand was no longer used or adapted for tree locomotion. In this second development, the thumb became longer, became human. The joint at the base evolved such that the opposable thumb has more complex musculature, is stronger, and can rotate through 45 degrees. This evolution produced a precision grip that combines the features of great strength with great sensitivity. The grip can be tenacious, forceful as in weightlifting, or precise in manipulating small objects delicately with thumb and forefinger as in sewing, watch making, surgery, and lovemaking. Moreover, evolutionary adaptation supplied the hand with the sensory and motor nerves needed to make it the highly skilled sensory-and-response instrument it is, capable of actions ranging from crushing a kill in volleyball to constructing the *Queen Mary* or a Swiss watch.

As will be indicated, an extraordinary aspect of humanness lies in the great variety of actions people can take. Much of that ability depends on the human hand.

The Face and Articulation Organs

The head houses organs that also centrally define basic human nature. The face and articulation structures are among them.

The Face

The faces of lower organisms are mostly expressionless. The term "cold fish" characterizes the same lack of expression, whether the fish is wounded, releasing sperm, or eating. Even domestic pets like cats and dogs have almost no facial expression. They can lift their lips and open their mouths in a snarl. But it is hard to detect any other state from facial expression, notwithstanding some owners' claims that their pets can smile. I find it titillating that a dog can seek, get, and obviously love being scratched, but with a face devoid of smiling or other facial expressions a human would display when tactile stimulation is a turn-on. Chimpanzees have much greater range, and some investigators have awarded them different facial expressions for various moods such as disgust, fear, sorrow, anger, excitement, and laughter.

Humans, however, clearly differ from all animals in the variability of facial-expression responses. Perhaps chimpanzees smile. Perhaps. But humans not only smile but do so through a variety of intensities, from faint smile to broad, each smile dependent on and indicative of differing conditions. Humans facially express doubt, dissatisfaction, anger, surprise, and happiness, for example, through a number of intensities. This becomes significant when it is realized that facial expressions play an important role in social interaction. When a dog snarls, that constitutes an important stimulus for other animals. The dog's lack of facial expressions, however, makes it necessary for the animal to indicate it wants petting by body actions. The human variability of expression enables much more finely tuned actions and reactions not available to other species.

Mother Nature is not a spendthrift. There is a cost for every additional development to the body of an animal species. That includes human facial expressions. Their variability demands complex muscular and neural structure. Such equipment requires upkeep, new cost in terms of food intake. That means the equipment had to pay for itself in the form of enhanced success in surviving and reproducing. Highly variable human facial expressions would not have affected other animals, predators or prey. So the payoff must have come from improving interrelationships among individuals in the prehuman and early human groups. From facial expressions of one another, early humanoid creatures could adjust their behaviors to be more pleasing. Where a lion would launch a claw-studded slash at a mate infringing on its dinner, a human could achieve the same end by simply frowning. Harmful conflicts could be avoided. Perhaps the human facial muscles evolved to facilitate social interactions before human language developed the ability to communicate much more fully and sensitively. This previews a principle of evolutionary advancement of traits because they improve social behavior, and presumably the survival and reproduction of prehistorical humanoid groups.

The Articulation Tract and the Vocal Response

In most other animals, jaws, mouths, tongue, and teeth—and the muscles and nerves that activate them—generally are made for obtaining and consuming food as well as for self-defense, cleaning, sensing, and social-sexual contact. These organs, along with the lungs and vocal chords and the muscles and nerves that supply them, may also be constructed for making sounds, and have importance in that respect.

Being able to make controlled vocal sounds with great variability constitutes a primary part of human nature, so much more than for any other animal that it becomes a difference of kind. Dogs can bark and growl, horses can neigh, pigs can snort and grunt, lions can roar, and whales and a species of monkey can "sing." It is now being said by some researchers that dogs make a kind of laughing sound. Chimpanzees can make a number of vocal sounds that may serve as signals to other members of the group. It is clear that these animals can learn to make particular vocal responses to particular stimuli. Dolphins can click, whistle, and squeal, and use their sound-making apparatus for echolocation in such activities as group hunting of schools of fish. This vocal behavior and its use are complex, perhaps more complex than occurs with any other nonhuman animal.

But there is no evidence that any other animal can make the variety of vocal responses common to humans, put together basic sound elements to form words in complex, systematic ways, in response to highly variable, complex stimuli. In humans, the muscles of the diaphragm and ribs force air in the lungs up through the larynx, whose muscles determine the tension of vocal cords and thus the pitch of the sound. The air then passes over the tongue and through the mouth, both of which are highly maneuverable, to exit through the teeth and the lips, which are also movable. The many complex and precise speech sounds in the various human languages require that all the muscles that move these structures work precisely, variably, in marvelous coordination.

Speech itself is a very complex behavior. The body parts that are used to produce speech are intricate. And it takes a lot of brain tissue to activate those parts. Moreover, the muscular responses in vocalizations must work under the control of guidance by sensory input; the marvelous coordinations of the lips, tongue, and jaw, to illustrate, require sensory feedback as the speech actions take place. This demand requires a large contribution of nervous tissue on the sensory side. The complex vocal apparatus requires a costly upkeep that must have been paid for in terms of evolutionary coin—survival and reproduction. That tells us that the behavior they enable, language, must have contributed long ago in very important measure for survival and reproduction. Otherwise those characteristics of the body would not have evolved and been retained. Many in the field believe that language in humans evolved via genetic mutation relatively recently, one hundred thousand years or so ago. There is another possibility, with different implications, that I will develop in chapter 8.

Human Nature in the Brain

The human brain is large and has convolutions that add to that volume, it contains new parts, and it contains further developments in evolutionarily older parts. Unique among the various species, the human brain is an enormous mechanism by which the sensory input and the behavioral output are joined via learning. That enormous mechanism makes possible complex, highly variable behaviors to complex, highly variable stimulus circumstances. Animals that have a greater capacity for learned associations do so because they have bigger brains than animals with lesser performance—just as larger computers provide more possibilities than do smaller ones. Animals that can record a successful response to a situation and repeat it later in a similar situation have an adaptive advantage over animals without that learning ability. That type of performance demands nervous tissue. The demands are great for exquisitely complex, subtle, and extended stimulus-input configurations and complex response outputs, as in jazz music, science theory, and football quarterback performances. Increased development of the brain—making it possible *to sense* the hugely complex environment and to learn *to behave* in an infinity of different ways—makes possible having and recording a gigantic number of learning experiences.

Yes, but isn't this common knowledge about the brain? Of course it is. But I say it again, although differently, in the process of establishing a foundation for the new paradigm. *I am not considering the brain as containing mental modules with wired-in traits that determine the individual's behavior, with the modules composing human nature. Rather, I conceive of the human to have evolved with great ability to sense the environment, with great ability to respond sensitively, variably, and complexly, and with great ability to connect the sensory input with the response output through learning, to enable the learning of behavior that sensitively adjusts to enormous environmental complexity and variability. The brain is conceptualized as a* mechanism *that performs the learning responsible for behavior, not as the* cause *of behavior. This change—conceiving of the brain as a mechanism rather than a cause—is basic in establishing a learning conception of human nature, as will be indicated in various developments to be presented.*

The Brain and Pleasure-Pain

Emotional responding fits in as well, worked out in evolution long before hominids came on the scene. Cats, rats, dogs, and chimpanzees will avoid things like electric shock that arouse a negative emotional response, just as a human will. And they will approach things that arouse a positive emotional response, just as humans do. Humans gained emotional responsiveness from ancestor species. A unique human characteristic, however, that must have taken place in human evolution, involves the capacity for learning to respond emotionally to so many different things. That capacity is so much greater for humans than for other animals. The human brain contains evolved and enlarged and elaborated structures that are basic in emotion—the cingulate gyrus, hippocampus, thalamus, hypothalamus, basic ganglia, the midbrain, and amygdala, which constitute the limbic system. The limbic system has an association cortex and thereby connects to other parts of the brain. This enables emotional responses to be connected to all other types of stimuli—visual, auditory, tactile, taste, smell—as well as those stimuli that arise in the muscles and tendons (kinaesthetic stimuli). Such connection takes place through learning. Humans can learn positive or negative emotions to anything they can sense and, by virtue of a history of experience, learn an emotional response to a huge multitude of stimuli, a much larger number than for any other animal.

Humans, like other organisms, have unlearned, wired-in emotional responses to some stimuli, like a positive emotional response to food and a negative emotional response to a painful stimulus (see Staats, Hekmat, and Staats 1996.) However, of the many stimuli to which humans respond emotionally, most are not wired in; response to them has been learned. Humans can even learn a positive emotional response to something that naturally is negative, and vice versa.

Chimpanzees have an emotion apparatus and can also learn to respond emotionally to new stimuli. So can dogs; after all, dogs come to love their owners who feed and pet them. But neither chimpanzees nor dogs ever learn to love a hoard of things like classical music, Renoir paintings, a good poker hand, or watching a NASCAR race or a football game. Rather than carrying huge numbers of wired-in emotions, the human brain is the *mechanism* by which an emotional response can be learned to most anything, as will be further indicated later.

Human Nature in the Nervous System

As the sensory and response organs developed in the successive hominid species, evolution also increased the nervous system that ran the new equipment. Anatomical developments are only as good as the nervous system by which they work. An increased number of sensory nerve endings in the fingers would go for naught unless there were neurons to carry the sensory signals to the brain, increasing sensitivity. Receiving more complex input from sensory organs, the human brain must be more complex also. That would hold generally for enhanced sensory input.

The nervous system must also have neurons for carrying stimulation back to the response organs. As the human hand became more developed and more variable in response than a dog's paw, it took more neurons coming from the brain to activate the hand's actions, with the great variability involved. When the human facial musculature grew more complex, it took more nerve cells coming from the brain to make the various facial expressions. When the tongue gained the musculature for precisely making the complex responses involved in speech, neurons had to be added to carry signals from the brain.

Each sensory and behavioral human capability added by evolution had to include developments to the human nervous system. The sensory neurons had to be connected to the neurons that operate the responding organs.

I CAN'T DO EVERYTHING BETTER THAN YOU, BUT I KNOW A LITTLE BIT ABOUT A LOT OF THINGS

An important characteristic of a species concerns its specificity-generality. Herbivore species eat vegetation, perhaps only a restricted number of vegetable foods. Carnivores eat only meat, perhaps only a restricted number of kinds. Most species have a particular sexual behavior by which to procreate. Some animals are nocturnal, they sleep in the daytime and become active at night. Others are diurnal, active by day, saving sleep for night. In these cases there is specificity, not generality. This is limiting.

The human body has evolved so our species has perhaps the greatest generality of all the species. Take the human sensory nature. The human visual apparatus is not so acute as that of eagles and owls. In hearing the human cannot detect a school of tuna by echolocation as do dolphins. The human cannot locate mushrooms by smell like a pig or hear very faint and high sounds like a dog. But humans have fine eyesight, hearing (possibly especially for language), taste, and tactile sense organs, adding up to very good generality of sensory apparatus, thus we have a very good ability to sense the external and internal environment.

Likewise, humans as a physical animal cannot compete in skill and power with the behavior of various animals due to their specialized structure—not with a mole with its claws for digging, a jaguar that can kill with its claws and jaws, a seal with its flippers for swimming, or a horse with its hooves for running. But humans can dig, kill, swim, and run.

A reason I call *Homo sapiens* the *marvelous animal* is because the human body has such general qualities in its sensory organs and in its organs of response. Humans are not specialists but generalists. Human bodies are made to do an endless number of things that other animals are unable to do because of their specialization. That ability to do so many things, and experience so many things, is quite central to human nature. And that characteristic is in part possible because of the human body. No matter how well they could reason, communicate, and be creative, humans could have progressed very little with the body of a lion, dog, dolphin, deer, or chimpanzee. No tools or weapons could have been made, no fires constructed, no fur clothes stitched. *The human body is an essential part of human nature.* That body has enabled the human species to do all the things it has done and to create its stupendously rich and varied products.

THE BODY ALSO MAKES THE HUMAN IN OTHER WAYS

Big-Headedness Doesn't Come Free

It is interesting how intricately evolution solved complex problems that arose as the prehuman species kept getting more and more learning ability, with a concurrent increase in the size of the brain and head. That wasn't without cost, in various ways. For one thing, a large-headed fetus creates the problem of passing through the mother's birth canal. One evolutionary solution was a larger opening through the pelvis, thus the wide hips of women relative to female chimpanzees and male humans. To lessen the extent of that demand, the human fetus's head is soft at birth so that it can be squeezed and narrowed during birth. Still another aid is that the fetus is born while still very immature and small. A side effect of this, however, is great and extended helplessness. Unlike a newborn zebra, which can get to its feet and start learning to walk almost from birth, a child when born requires a year or so before taking its first steps. (A new opinion in human evolution holds that the extended time of childhood may go back perhaps two million years. Perhaps. But if one considers that children are born undeveloped to get their heads through the birth canal, that would suggest the long childhood did not arise two million years ago when hominid heads were not yet so large as now.)

As a side effect of the child's extended immaturity there is a long period when all needs must be met by caretakers. This has the effect of maximizing the learning experiences the child has with the parent, which is very important in child development. For one thing, the bonding between parent and child becomes greater in humans because of the greater length of childhood. Again, this is an example where body characteristics establish a causal chain that has important effects on human behavior and human nature. *The human child's large brain and head, and early birth and helplessness, result in long, spread out opportunities for learning in child development.*

That of course places demands on parenting, on the need for two parents devoted to child sustenance and care for a long time.

Sexual Biology and Bonding

Sexual receptivity in female mammals, including almost all the primates, typically follows the estrous cycle. The term literally means *heat*, and animals with an estrous cycle come into sexual heat periodically. Female primates are receptive during estrous and may then seek the male's response.

Importantly, humans (and bonobos) have a different pattern of sex behavior. Although women ovulate periodically (the basis for other animals' estrous periods), women do not experience estrous periods of distinctly heightened sexual interest mixed with in-between periods of sexual disinterest. Rather, women generally are sexually interested in a sustained way, relatively uninfluenced by ovulation cycles.

As with other body features, sexual biology has learning and behavioral significance. Consider for a moment some of the effects on the gender relationship. There are animals that are solitary except once a year when males and females come together to mate. Contrast this with the baboon, where there is a repeated estrous cycle during the year, resulting in multiple sexual matings. And then there is the human, where sexual activity is unlimited by reproductive cycles, the result being many sexual encounters. When evolution changes a characteristic, one can suspect there to be reasons why. The piper must be paid in this case, as in others, for sex activity takes energy, takes time away from securing food, and is sufficiently diverting of attention that those enjoying sex are more vulnerable to predators. Other things equal, the less frequent the sex, the better the survivability of the individual animal and the species.

What then could have been the advantage of the human increase in sexuality? Could the benefit lie in learning and consequent behavior that is produced, not in reproductive success? Behavioral analysis leads me to various considerations. For one thing, sexual activity involves an intensely positive emotional response. If that emotional responding occurs many times in conjunction with another person, an intense kind of emotional learning can be expected. In a word, *love relations* typically result from frequently repeated sex experiences. The strong emotion of love should enhance markedly the symbiotic relating of women and men. Would that not strengthen pairing and increase their togetherness in living? And would that not enhance human success in raising progeny? Clearly a team of two adults would be better able than one alone to feed and care for a dependent humanoid child, especially when gender differences in body size and strength make one of each pair better suited for hunting and physically fighting against predators and the other for nurturing the young. Frequent sexual sharing would provide emotional bonding, "glue," for keeping human pairs together and thus for strengthening parenting and the success of reproduction.

The suggestion is that the growth of the human brain led to the helplessness of human infants, which called for fulsome parenting. That made it advantageous in various ways for dimorphism, with women being smaller and better suited for gathering foodstuffs and providing childcare and men being larger, stronger, faster, and better suited for hunting and protecting. Pairing produced a more viable family group, and increased sexuality produced the bonding that aided close, supportive relationships. Again, evolution of body characteristics would have affected human learning and behavior.

Body, Experience, Learning, and Behavior

The common view is that the behavioral nature of mammals—aggressive, brutal, timid, affectionate, dangerous, intelligent, uncontrollable, docile, or vicious—is given by the genetically determined brain. A lion is said to be ferocious in nature, with that personality hardwired into its brain. Deer and rabbits are said to be timid because of their brains' wiring. There actually is a gene that makes houseflies aggressive. Does that lend proof to the belief that lions' ferociousness is genetic? And do the "personality" differences in mammal species justify believing that differences in human traits of behavior are genetically inherited? That is a type of question that will be answered in various ways as we go on.

In beginning, body differences among the species are genetic. But could "personality" differences in animals stem not from genetics but rather from those body differences? And could those body differences have their effects via learning, not as part of genetic inheritance? Let me explain. If a fawn had the body of a lion, would that have any influence on the fawn's behavior? Would the fawn grow up to behave like a lion, through learning? Could what has always been taken to reside in the psychological nature (brain wiring) of a species at least in part reside in the physical structure of the species and the experiences and learning that this structure "arranges"? An important point of this chapter is to suggest that *physical differences can produce different experiences and, hence, different learned natures.* There are various ways that characteristics of the body produce different experiences for the animal or human.

Physical Features as Environmental Causes

Elephants are not carnivorous and are not made to be hunters and do not display the behaviors of hunters. Yet they are not patsies. Predator animals do not attack elephants (unless the elephants are vulnerable in some way). Why? Because elephants are so big, strong, and formidable. Do those features constitute a stimulus even ferocious predators avoid? More broadly, do the physical features of an animal serve as stimuli to other animals? If that occurs, then an *animal's physical features help determine what the animal's experiences will be and thus what the animal becomes behaviorally.* That principle is not generally understood or considered. It should be realized that just because the whole species shows the same behavior does not mean that a genetically produced nature is the cause of the behavioral trait. Behavioral characteristics thought to be hardwired into the brain actually may result from learning experiences.

The lion struts in solitary glory across the savanna. Small antelope species, however, cluster in herds and then move, ever alert and ready to run. Being alone for an antelope evokes fear. Not so with the lion. Is there a difference in "mentality" here, hardwired biologically into the animals' brains by evolution? Do genes and the brain give one animal a solitary and ferocious nature and the other a gregarious and timid nature? Or does the brain's nature, and thus the animal's, result from the different experiences the animals have? Let's consider the latter. When lions walk the savanna, other animals run away. Does the lion learn something from that experience? Certainly the small antelope does not have the same experience. Other animals do not startle and run at its presence. Rather, the antelope has the experience of smelling the spoor of lions and then doing the running itself, frequently prompted by the frenzied movement of the herd.

Would lions the size of a mouse, or a house cat, strut across the savanna? My experience is that the homeless domestic "wild" cats that roam Kaka'ako Park in Honolulu do so warily, avoidant of humans, not at all in the manner of a lion. But behavior becomes more lionlike when a domestic cat faces a small rat or a bird. Perhaps its body is the root of the lion's bravado, not a trait of arrogant self-confidence wired into its brain structure genetically. Both the lion and the antelope are capable of complex learning; of course their experiences must have an effect. We know from the observations of primatologists that animals that have been defeated in contests over territory or mates become less aggressive, more timid. Why wouldn't that be at play with the "natures" of the lion and the antelope as well as the domestic cat? I suggest there is a complex interaction between body features, the experience that results from these features, and the behavioral characteristics that are learned.

Physical Features and Psychological Differences in Humans

Continuing this analysis, perhaps the same process operates with human nature. For example, men customarily are the abusers in a heterosexual pair. It is most often men who rape and sexually molest women. Men fight more (at least physically). General belief has it that this male-female behavioral difference is determined genetically, wired into gender-different brains as sex-linked mind, nature, or personality differences. However, there is a sex-linked difference in size between men and women (dimorphism). The sexes also differ in amount of muscle and fat. Even our casual observation tells us that big men, like elephants, are less the object of attack than small men. As a body feature, size thus determines differences in experience that men have. That holds also for men and women.

Dimorphism and Gender Behavioral Trait Differences

Biological differences in women's and men's brains are thought to explain other differences in the nature and behavior of the sexes. Men are said to excel in math and spatial relations; to be more logical, more promiscuous sexually, more frequently autistic, dyslexic, and criminal; to need activity more and be more distractible in academic pursuits. Women are said to be more docile, verbal, disciplined in school, social, sensitive to others' feelings, and better in fine motor skills. Sex differences in brain functioning, as measured via brain-imaging techniques, are considered firm proof that brain differences cause these gender differences in behavior, as well as gender differences on psychological tests. Could such "evidence" be wrong? Perhaps the gender behavioral differences are the result of learning, not gene-produced brain differences.

My argument here is that the *body* differences between women and men produce different experiences and different experiences produce different behavioral traits. Those different experiences also result in brain differences as well as in psychological test differences. Someone who is tall, heavily muscled, strong, and fast will have different life experiences than someone who is small, lightly muscled, weak, and slow. As soon as such body differences occur, they will produce differences in experience. The bigger, stronger child on average will display greater athletic feats and be more successful in aggression than the smaller, weaker child. As a group, smaller and weaker women (on average) are less able to get what they want through physical aggression than are larger and stronger men. Women as a group can be more successful getting what they want in other ways, such as by winning over a male rather than physically contesting him. In such a case, the two types of behavior would become typical of the group; the behaviors become expected and thus unintentionally "taught" to children on a gender basis. Gender differences in brain structure and function may not actually be revealing of biologically fixed behavioral traits and the nature of the male and female "mind."

It is important to remember that differences in experience and behavior have operated historically and prehistorically. When one half the population is bigger, stronger, and faster than the other, which half will become the hunters on the savanna and which half will do the gathering and take care of the children and home site? When it comes to planning major actions, like hunting, a member from which half will be likely to lead? Male-female differences in experience began long ago, prehistorically, when their dimorphism became definitive. Male-female differences in behavior should thus have existed so long they are taken as natural sex differences, meaning men and women would be treated differently from birth.

Physical Features Help Make the Man, and the Woman Too

An old saw states "clothes make the man." The implication is that how others respond depends on one's clothes. I amend this in saying that by affecting experience, "the body makes the man or woman." That type of causation holds beyond genderism, for example, to racism. If dark skin and light skin elicit different behavior from people in the milieu, then dark-skinned and light-skinned people will have different experiences and as a consequence learn different characteristics.

> Dozens of research studies have shown that skin tone and other racial features play powerful roles in who gets ahead and who does not. These racial factors regularly determine who gets hired, who gets convicted, and who gets elected. . . . This isn't racism per se; it's colorism, an unconscious prejudice . . . focused on . . . *blackness* itself. . . . Our brains, shaped by culture and history create intricate caste hierarchies that privilege those who are physically . . . whiter and punish those who are darker. (Vedantam 2010, D3)

As this book will show, if we want two groups that have different features to behave the same, then they must have the same experience. That means they have to be responded to in the same way.

The central point here is that genes can act as the determinants of behavior in an indirect, not biological, manner. Genes can produce physical traits that determine the environment that people will experience. Size, strength, and speed differences act in that way. So do beauty and other physical-appearance differences as well as differences in health and resistance to disease. A sickly child has different social experiences than a robust child. That includes being categorized as psychologically "sick" when developmentally delayed, low in intelligence, autistic, or uncontrolled.

As indicated in the preceding chapter, Darwin did not distinguish a physical feature from a behavioral feature. Both were considered to be determined by evolved biological characteristics. Rather, we have to be sophisticated about what may be wired into the species' brains genetically and what may enter the species' brains because of experience and learning. Common experience can account for behavior that is common to a species. Whether a behavioral trait is genetic or learned can be known only by direct proof, by isolating the genes or the learning experiences responsible.

MARVELOUS, INDEED

Humans are a marvelous species. The marvelous human body can do so many different things, from swinging on bars and rings with primate skill, to removing a brain tumor, to playing a violin, to flying a hang glider, to driving a car. Humans can smile, laugh, scowl, shout, sing, and whisper. On the sensory side, humans can hear music, ticking watches, and complex explanations; see printed letters, sunsets, unhappiness, and 90 mph pitches; and feel breast tumors, orthopedic discrepancies, the ripeness of fruit, and the loveliness of the human body. The marvelous human sensory-and-response apparatus has no peer.

Humans' life experience also is more varied and complex than that of any other animal. Humans have experience with environments that range from the frozen lands of high latitudes and high altitudes to the steamy jungles of the equator. Humans experience millions of words and myriads of musical notes and movies. Humans experience the complexities of human behavior, in parenting, with relatives, with co-workers, with competitors and supporters, with friends and foes.

Behaving well to this complexity of life can be achieved only by a being with marvelous sensory versatility and marvelous response versatility. That demands a marvelous sensory-and-response nervous system. And that also demands a marvelous brain to connect the systems. Establishing those connections constitutes learning. The complexity on both sides means that learning will be very complex. It is no wonder that the human brain has so many billions of neurons and can make trillions of connections between them. That brain indeed is a marvel of the marvelous body.

The marvelous human body, with all its potential for wonderfully complex actions, is marvelous because of the gigantic human ability to learn those actions. The two must go together—physical body and learning ability—to achieve humanness. Without the human body, learning would get humans nowhere. It goes both ways, however, for without learning the species would not be human.

Human learning has been neglected and has not been analyzed systematically in a way that represents its generality and importance. That deficit needs remedy. Commonsense consideration is not enough. An understanding of the principles of learning that came to humans via evolution is basic. For those are the fundamental principles by which individual *Homo sapiens* creatures, with their marvelous bodies, through their learning experiences become human beings.

VERSATILITY AND LEARNING ABILITY

An obstacle to advancement has been the nature-versus-nurture opposition. The new paradigm is "bipartisan," integrating learning and biology, as exemplified here. The human body is an instrument of learning by giving the species great versatility in sensing and responding to the environment, the greatest versatility of any animal. With human hips, legs, knees, ankles, and feet that enable walking, running, jumping, lifting, and twisting; with shoulders, elbows, wrists, and fingers that enable countless actions; with jaws, lips, tongue, respiratory system, and larynx that enable speaking; evolution has made humans capable of lifting four-hundred-pound weights and repairing wristwatches, playing a violin and making crushing tackles as a linebacker in the NFL, whispering to a companion and singing opera.

The brain usually stars in accounting for human nature, but what is a great brain without such a versatile body? Take dolphins. This species, too, has a huge brain and great learning ability. But dolphins lack body versatility. The animal can swim like crazy, with great skill and variation. But dolphins cannot build an Atlantis, even under the sea. Their body limits them to swimming, doing spins and flips and jumps, and group herding/harvesting schools of tuna. Comparatively humans swim clumsily, but humans also can do thousands of other things, on land, in the sea, in the air, and in space. Oh yes, humans do a lot of things, but gaining hugely versatile skills demands a mechanism for acquiring the behaviors involved. There is no way that evolution could have produced an animal, no matter the greatness of versatility of body, that had built-in behavioral skills for being a virtuoso violinist, brain surgeon, boxing champion, ballerina, opera star, or elementary schoolteacher. So let's talk of that huge human brain and the learning it makes possible, because that is the way all kinds of behavioral skills are acquired, from going potty to being a genius physicist.

Chapter 3

SUCH A BRAIN! 100,000,000,000 NEURONS BUY A LOT OF LEARNING

So humans have a marvelous body that comes from evolution. That means we have a sensory system that is responsive to environmental stimuli and sends a vivid picture to the brain, along with a brain that sends complex instructions to the body's response organs. The brain does not just relay the messages from the sensory system to the response system; it works into the messages a relevant organization the brain has learned from past experiences.

This schema began setting learning within a framework that connects the process of learning to biology. The nature-nurture, biology-versus-learning schism has been a traditional burden, one of the obstacles to advancing a good paradigm for understanding humanness. Biological sciences deal with natural phenomena. The process of learning also deals with natural phenomena. Mother Nature is not divided against herself. There is no natural schism between biology and learning. Learning and biology, nature and nurture, are sibling sciences, complementary when considered correctly. Actually the nature-nurture schism is a concocted opposition, divided by the *views* that are held, not by the phenomena themselves. Learning is a biological process carried out by biological structures.

But knowledge of learning is to be found in studying learning phenomena, not in studying biological phenomena such as natural selection, genes, sexual reproduction, or pharmacology. Frequently some change is considered as evolved when actually it has been learned. The principles of evolution simply do not apply to learning. It takes many millions of years for a species to evolve from swimming in the sea to a species that can live on land, or vice versa, just as it takes many millions of years for a land-bound species to become one that can fly. Such behavioral developments involve genetic changes that produce body changes. They are slow. Take *Homo sapiens*, however. In less than thirty thousand years humans advanced from making stone tools and running after game to zipping along at 70 mph on land, 600 mph in an airliner, and hurtling thousands of miles per hour in a spacecraft. Even before that, hominid creatures advanced in a few million years from verbal skills not much better than a chimpanzee's to the complexities of normal language use as well as special displays like the language skills of Shakespeare or an astrophysicist elaborating the Big Bang theory.

Evolution takes a long time. *Homo sapiens'* advancement has taken the blink of an eye. The facts tell us that the great behavioral change took place in humans without any biological change, when the brain and body remained the same. Something other than evolution has produced the wonders of human behavior and human nature. That something is learning. If that is the case then if we want to explain human behavior, and do something about that behavior, we have to know about how learning works. While knowledge of evolution has vast importance for knowing about a lot of things, it won't tell us at all about learning and thus about how to understand and deal with people's behavior that is learned.

Yes, but isn't the importance of learning for human behavior already well understood? Aristotle said the mind is a blank page that is written on by experience. Sigmund Freud theorized that parental treatment made deep and lasting impressions on the child's mind that determined later behavior. The common language also expresses many principles of how experience affects behavior, for example, "spare the rod and spoil the child." Almost universal today is the use of time-out by parents to decrease objectionable behavior of their child. Our construction of school systems shows how important the belief is that learning is a cause of behavior.

Nevertheless, despite general knowledge that learning experiences are very powerful, just what are the human behaviors that are learned? And what are the principles by which they are learned? Is the child's language ability learned, and if so, how? Is the child's bonding to the parent learned, and if so, how? Does learning affect intelligence, and if so, how? Are motor skills—like those needed for being an athlete, a brain surgeon, or a violinist—learned, and if so, how? Yes, everyone knows about learning, knows it is important, and recognizes that many things are learned. But that knowledge is on a commonsense level. There is knowing and KNOWING. Everyone knew about gravity before Newton, but they couldn't calculate a spacecraft's trajectory to the moon. He turned commonsense knowledge of gravity into systematically constructed knowledge that was refined, detailed, and universally applicable.

That is a common way that science functions. First there is commonsense knowledge based on casual observation. Then some interested individuals make

more special observations of the phenomena involved and find lawful principles. That provides the foundation for more systematic study and conceptualization. The knowledge produced deepens, becomes interrelated, clarified, general, and useful. Knowledge of human behavior has begun the ascent from the
commonsense level, but it hasn't adopted and advanced the principles of
learning needed for jumping to the Newtonian level.

Many years ago a television commercial featured the slogan "we make
money the old-fashioned way." Well, this book intends to make a conception the
old-fashioned scientific way, a systematic, building way. Newton didn't watch
apples fall from a tree, or dishes crash on the kitchen floor, or feathers swoop
into the sky in conjuring up a principle by which to precisely describe gravita-
tional phenomena. He had Galileo's principles of how balls rolled down an
inclined plane, Kepler's calculations concerning the movements of stellar
bodies, as well as knowledge of oceanic tides as related to the location of the
moon. Based on such stipulated principles, Newton constructed a general theory
of gravitation that had valuable uses for a universe of phenomena.

Similarly, constructing a broad understanding of how human behavior is
learned requires the use of scientifically established principles of learning, not
commonsense knowledge. Back at the beginning of the twentieth century, Ivan
Pavlov and Edward Thorndike each discovered one of the two basic learning
principles with animal subjects, opening the way to thousands of additional
studies. A few of us in the mid-1950s showed the principles could be used in
treating problems of human behavior. In addition, various conceptions of
learning have arisen in fields like educational psychology and cognitive psychology. The result has been a very mixed bag that does not add up to an understandable conception that people can widely use. Actually it is necessary to cut
through the cross-purposes and over-extensive detail to pick out and create a
simpler set of principles that function generally and powerfully. These principles provide a science foundation for understanding different human behaviors
as well as the phenomena called human nature—in works that are specific and
useful and are in play universally, in the way that biological and physical laws
are. Explained by the same basic principles, the many different behavior phenomena become unified—ranging from child development through normal and
abnormal personality traits and even into the explanation of human evolution.
As each area of behavior is explained by the same growing conception, the
framework increases in value for generally explaining all human behavioral
phenomena.

At the bottom of this conception lie the basic principles of animal learning that function for all mammals, added to mightily by the discovery of special human learning principles.

BIOLOGY AND LEARNING

Early on, when biology was not yet cleanly separated from theology, there were so-called vitalists who brought their religious beliefs into their science. They believed that a life force was breathed into all animals at the moment of conception. The certainty of that belief was shaken by the new, all-science biology that asserted life originally arose when a conglomeration of physical elements had come together under the right set of conditions. No divine creation was recognized by these mechanists. Rather, the new conception was that, poof, the right material constituents in the right conditions yielded a living, reproducing organism. Ever after, the conception said, biological reproduction passed life on. Those different views posed the contest. The movie about the Frankenstein monster provides an example of the issue; the village people with religious views consider Dr. Frankenstein, the scientist, to be usurping God's powers by bringing life to his monster, thus it was an evil act. He, with his science view, puts the various human parts together and by a lightning rod's electricity instills them with life. Of course the crimes of his monster placed Dr. Frankenstein in the wrong in the movie, subtly acceding to the nonscience view.

In the real science contest, the mechanists won. Life became a natural phenomenon without alluding to concepts of divine creation; there are no vitalists in biology today. Now it is perfectly acceptable to attempt to create life-forms from nonliving elements (see Barry 2008), albeit in far simpler organisms than the Frankenstein monster. Although offensive to some, creatures are also generated by cloning. Other scientists study the advancement from single-celled animals to those with multiple cells (Maxen 2008)—interest in the evolution of life-forms goes back to the single-cell organisms at the very beginning. Various organs and functions have been traced evolutionarily. For example, below a certain level of animal development, there is no heart to pump blood to the body's cells. Then a heart with one pumping chamber appeared. By the evolution of reptiles, the heart had developed two chambers. Birds, as a more advanced genre, have a more advanced, three-chambered organ. Mammals, from mice to elephants, and including humans, have a four-chambered heart, a much more effective pumping organ than the original one-chambered instrument. Digestive systems also evolved, as did brains, immune systems, glands, legs, hips, knees, vertebrae, and shoulders. They came down, in an advanced form, to the human species via evolution.

Animals vary in physical characteristics. They differ also in behavioral characteristics. A simple amoeba contacted by a particle of food responds differently than when encountering the point of a sharp instrument. With the food there is the rudimentary "approach" response of surrounding the particle and then absorbing it. In contrast, with the sharp instrument there is a rudimentary escape response, a "flowing" away from this offensive part of the environment.

Clearly those responses are adaptive and help those one-celled creatures to survive and reproduce. Their behavior of course is simple, not as simple as that of a bacterium, but very limited, not quick and precise. With more complex animals, different organs develop that permit more precise, quicker, more complex, and variable behavior in response to more complex features of the environment. Organs with which to sense the environment develop in particular ways, as well as organs that specialize in making responses to the environment, as has been described. The beginning hominid species were gifted with millions of years of prior evolution to have a body with very advanced sensory and response organs.

In addition to the evolution of sensory and response organs that provide for increasingly precise and variable behavior, there is another very central characteristic that develops through evolution. Going back to the amoeba, the organism responds to the environment, yes, but without anticipation and without variation as a consequence of previous experience. For example, let's say a pricking object is pressed against the amoeba intermittently a number of times, preceded on each occasion by presentation of a certain type of food particle. Despite the repetitious experience of food followed by the prick, the amoeba would never make anticipatory escape movements that would completely avoid the pricking object that was to come. The amoeba would not *learn*.

That is quite different than the behavior of advanced animals. Learning resulting from the pairing of stimuli has been shown even with insects. Bees will fly to one of two colored liquids after they have experienced one liquid has sugar in it and the other does not. The behavior has been learned because of the color/sugar pairing. So even insects can learn; they have neurological tissue by which to connect the response to the sensory input received. Bees can learn better than amoebas, but still in a very limited way. Other species have progressed much more in precision, in sensing the environment, in the versatility of responding, and in recording (learning) the experience.

There has to be neural tissue by which to connect sensory neurons to response neurons. Without the connecting neural mechanism, the action of the sensory organs would produce no action of the response organs. *A dimension of evolutionary progress lies in advancement in (1) the organs sensitive to the environment, (2) the organs by which to respond, and (3) the organs that connect the two.* When connections are established through experience, it is called learning. Beginning with little ability to sense the environment, to vary behaviorally, and to learn from experience, the evolutionary parade of species advanced. One result was an increase in the *number* of learnings different species could retain, aided by developing structures like the spinal cord and brain that connect the sensory organs to the response organs. Considered in this way we can see that the evolution of learning and learning ability was as fundamental as evolution of multi-chambered hearts, upright mobility, a grasping hand, eyes for stereoscopic vision, taste for a variety of foods, or hearing sensitive to human speech. That evolution of learning and learning ability also took place by biological evolution.

Evolution does not continue to support features that add to the cost of operation of a species unless the features make a contribution to survival and reproductive success. The apparatus needed for learning does not come cheap; sensory, motor, and connective nervous organs require food energy that has to be earned. That fact pretty much puts a stamp of approval on learning. Increase in learning ability had to pay for itself by increasing the survival and reproduction of the species involved, perhaps by increasing the species' ability to get food, or its ability to escape from predators.

Yes, learning principles are biological in nature, and systematically tracing how they arose and advanced in evolution is a joint learning-biology interest. Knowledge of the evolutionary biology of learning can help knit together nature and nurture. Relevant research does exist. Kandel and Tauc (1965) produced vital knowledge by studying single neurons of sea slugs in order to find out just what biological mechanisms underlie the principle of the association type of learning. In doing so they followed the analytic, systematic character of science. *A thorough study of the levels of learning at every step of development in its long evolution up to humans would be very important, and the Kandel and Tauc studies provide evidence of that possibility.*

The discovery of the basic principles of animal learning was momentous in science, much more so than is recognized. Learning since antiquity had been a common interest, of course. The British empiricist philosophers had theorized about association learning but had not shown experimentally that the principles were lawful and basic, extending over many animal species including humans. So discovery that animals learn lawfully—like gravity—and that the process could be stipulated in the laboratory, opened new science perspectives. A long-term storm of studies of basic learning principles ensued. Theories of this new science were constructed to organize the complex of principles and findings (e.g., Hull 1943, Skinner 1938). The theories were based on simple behaviors, like a dog learning to salivate or a rat running right or left in a simple maze. As indicated, this model led to more and more detailed study of the laws involved; very detailed in findings, methods, and theories. When the goal becomes that of laying the foundation of principles for a conception of human behavior, however, that confusion of detail becomes a problem.

The learning theory that follows has been constructed to correct those characteristics. A set of principles has been selected, organized, and further developed so it can serve as the foundation of principles for elaboration into a human learning theory.

BASIC PRINCIPLES OF LEARNING

At the turn of the century, in a physiology laboratory in Russia and a psychology laboratory in the United States, the discoveries occurred that I suggest will one day be considered first-rank scientific developments on the level of Galileo's or Kepler's contributions to physical science. Those discoveries consisted of the first experimental specification of the two basic learning principles, previously considered only in commonsense ways. The Russian discovery, although dealing first only with salivation, deals with a principle that pertains to other types of internal, physiological responses as well. This principle, it will be indicated, explains the learning of emotion responses. This learning came to be called *classical conditioning*. The American discovery dealt with the principle of learning now called *operant conditioning*. It pertains to motor responses. Every movement we make constitutes a motor response, from pressing a bar to playing a violin, to speaking and thinking and other cognitive acts. As Thorndike found, when motor responses are rewarded, they grow stronger, but not when they are not rewarded.

I will divide learning into the two types, one dealing with emotion and one dealing with motor behavior.

Classical Conditioning

The basic learning principles act widely and powerfully in everyday life. Commonsense calls classical conditioning phenomena associations. As an example, let's say two people having a torrid love affair listen repeatedly to a particular popular song as "their song." Many years later, each of them experiences a stronger emotional response when hearing that song than when hearing other songs of that vintage. The love emotion they experienced had "rubbed off on the song" from the pairing that had occurred.

It was not until the beginning of the twentieth century that Ivan Pavlov, a famous Russian physiologist, accidentally discovered how to specify experimentally, in the laboratory, the fundamental nature of establishing such associations, called classical conditioning. He demonstrated that this conditioning is not just an intriguing sometime thing but a natural law that we can see is a part of the nature of animals. This development helped make psychology into a science as well as change the conception of the causes of behavior.

Pavlov's major work actually concerned the digestive processes, for which he received the Nobel Prize. But, as part of studying digestion, he prepared dogs so the extent of their salivation could be measured precisely when they were given a measured bit of food substance. That experimental preparation serendipitously also produced the circumstances by which classical conditioning could be established as the explanation of associations. For, lo and behold, his laboratory preparation of the animals made it possible to see that the dogs began salivating just before they were fed each day when the caretaker entered the lab prior to feeding. The sound of the approaching caretaker, by itself, elicited the salivary response. Dogs salivate as a reflex to food in the mouth, like we do, as part of digestion. However, dogs salivating to sounds, with no food in their mouths? That was novel, and very informative.

Pavlov was wise enough to realize that this salivation was not without cause and, as a good experimental scientist, he took the circumstances into the lab to find out what that mystical cause was. So he presented a controlled, neutral stimulus, like the sound of a bell, and followed this by the presentation of a measured bit of food substance. The food automatically elicited salivation in the animal because of reflex neurology. But after the bell-and-food pairing, startlingly the bell *by itself* now also elicited salivation. That was exciting because it meant a stimulus that had no physiological reason for eliciting salivation would do so after it had been paired a number of times with another stimulus that physiologically did elicit the response. So here was a precise and systematic way of producing a psychological "association." The bell associated with food took on a function of the food itself, the elicitation of salivation. Here also, although not realized at the time, was a general experimental methodology for studying a type of *learning*—the pairing of two things together and producing a connection between the two, a major way that the environment can affect behavior.

The reigning belief had held that behavior was spontaneous, arising from within, according to the capricious will of the animal. In contrast, implicit in Pavlov's findings was the idea that classical conditioning was taking place because of outer, environmental conditions. The behavior wasn't spontaneous and it wasn't a result of internal, self-causation. Moreover, his findings support an idea that behavior might be lawful, just like physical laws such as gravity. Do this environmental thing to an animal and this particular response occurs—just like an electric typewriter responds when a key is pressed. That is quite different from the commonsense view. I mention this lawfulness idea because it opposes a belief that behavior can be spontaneous, like the traditional concept of "free will." We'll come back to this question further on.

In Pavlov's work the potentialities of the experimental method were apparent. Others could repeat what he did, and they did. Many researchers tested and extended the findings in a burst of research over the following years, producing principles that precisely indicated how the learning occurred. In classical conditioning, the stimulus used to first elicit the response (food) was named the *unconditioned stimulus*, and the stimulus that would come to elicit the response (bell) was called the *conditioned stimulus*. Additional research showed that (1) the more the pairings of the conditioned stimulus with the unconditioned stimulus, the stronger the conditioning; (2) the more intense the unconditioned stimulus, and thus the response, the more intense the conditioned response; and (3) the shorter the interval between the conditioned stimulus and the unconditioned stimulus (down to a half second), the greater the conditioning. The laboratory methods producing the basic principles can be further used to discover additional aspects of those principles, as can be shown with a few examples.

Extinction

Once conditioning takes place, does it last forever? Actually the association will last until some additional learning, or some other circumstance (like brain damage) makes a change. One important type of further learning is referred to as *extinction*. To illustrate, take an animal already conditioned to respond to the bell and then present the bell a number of times without the food. We will see that the animal's response to the bell will progressively diminish; there will be less and less salivation when the bell is sounded. If this treatment is continued, the animal ultimately will not salivate at all, or very little. Although the process is called extinction, a better name would be learning not to respond, or deconditioning. Knowing about extinction is important because many times we want to know how to break learned associations.

The two processes—conditioning and deconditioning, extinction—show how sensitive animals are in response to the environment. They can be conditioned to respond to a stimulus, and afterward they can be conditioned not to respond. We can see why classical conditioning is built into animals; by operating according to classical conditioning, animals adjust to environmental happenings. Through conditioning the animal can learn to respond one way, and if experience changes, the animal can learn to respond another way. So we can ask, did animals evolve to classically condition because it helped them stay attuned to the environment? Is conditioning that basic, such that evolution produced it? Did animal species able to *learn* and *unlearn* in their environment survive and reproduce better because of this flexibility? It better be so, because giving a species a nervous system that operates by classical conditioning principles creates an increased demand for food to maintain that nervous system. Again, no free lunches from Mother Nature.

Positive, Neutral, and Negative

Actually, Pavlov was not just conditioning a salivary response in his dogs. The food elicited a more general emotional response that the animals learned. He was not concerned with that. However, the animals were learning a positive emotional response to the bell. If he has asked his dogs—and there are appropriate ways to ask an animal—how they felt about the bell, they would have said they liked it. I say this because classical conditioning is so important because this is how emotion is learned, and understanding emotional learning is central in understanding human behavior and human nature. For one thing, the various stimuli in the world do not all elicit a positive emotional response. We know that from our own experience, through the many times we have experienced a negative or a positive emotional response. We also know that many stimuli elicit no emotional response in us. In sum, anticipating later discussion a bit, we know there are words like *cancer*, *rape*, and *hunger* that we don't actually like; and *the*, *from*, and *rock* that we don't have much feeling for; as well as *joy*, *vacation*, and *ice cream* that are kind of nice.

There is positive classical conditioning when the unconditioned stimulus is positive—such as food or sexual stimulation. The animal will learn to like the conditioned stimulus. But there is also negative classical conditioning when the unconditioned stimulus is negative—such as electric shock, loud noise, intense light, excessive heat, bad-tasting or bad-smelling substances, and other unpleasant or painful stimuli. Just like the already-described positive emotions of the two lovers and the song, if two people have negative emotional experiences when they are together, they will be conditioned not to like each other. Research shows that a neutral stimulus that is paired with an unconditioned stimulus like electric shock will come to elicit various physiological responses such as an increased heart rate, blood vessel responses, increased sweat-gland and adrenal-gland responses, and activity in certain parts of the brain. All of those responses are parts of a negative emotional response.

It is productive to consider that the environment for us is composed of positive, neutral, and negative emotion-eliciting stimuli. That applies to things that we are prewired to respond to as well as to things to which we have been conditioned to respond. We have here a very fundamental categorization of the stimuli in our lives. We learn copiously to respond emotionally to many thousands of things in our environment.

Deprivation-Satiation

Would it be adaptive to salivate lavishly every time food was present or we experienced something that had previously been associated with food? How about when the animal had just eaten to satiation and could not eat more? If animals salivated always, hungry or satiated, there would be a lot of wasted salivation, which of course would mean wasted energy. Wouldn't an animal be more effective if it salivated only when it was ready to go ahead and eat food?

Evolution generally makes animals more adaptive. As we might expect, evolution has made us and other animals so that we salivate only when we have been deprived of food for a while, and more so as the deprivation increases. So the extent of deprivation-satiation, of hunger, affects the extent of the classical conditioning of salivation. Deprivation-satiation also works with other positive emotional stimuli, not just food. Thus, right after experiencing sex to satiation, even a very sexy stimulus won't get much of a rise or evoke the juices. The general principle is that increased time of deprivation of a positive emotional stimulus increases the strength of emotional response to the stimulus.

That applies also to stimuli that we respond to because of previous conditioning. When the viewer is sexually satiated, pornographic pictures will be less stimulating than when deprived. Does that principle underlie the old saw "absence makes the heart grow fonder"? Classical music can serve as another example. Because of past conditioning, for some people classical music elicits a positive emotion. The power of the music to elicit that emotion will vary with deprivation, however. If the person has been thoroughly satiated through having just listened to a three-hour concert, the person will respond with more pleasure to an invitation to go have dessert than to go to another symphony. We have been able to show this effect with human subjects; people salivate more to food words—words like *bread, steak, hamburger,* and *pie*—after being deprived of food than do people who are satiated (Staats and Hammond 1972).

Further evidence that conditioning principles must have evolved because of their adaptive nature can be seen in that deprivation-satiation does not apply to negative emotional stimuli. The animal may just have received a series of electric shocks, but the next shock will still elicit much the same accelerated heart rate (and other physiological responses). Torturers have experienced this. No matter how many times they apply electric shock to their victims' genitals, the shock continues to be painful. Why? Because a stimulus that elicits a negative emotional response is a harmful stimulus, and it continues to be harmful, regardless of how many times it is presented. There is not a satiation effect because it is adaptive for there not be. Evolution of the nervous system, of course, generally follows that principle.

Let's Be Cool and Visit the Brain

In one of my early laboratory experiments, college students were conditioned to have an emotional response to a word by pairing it with a mild electric shock. We measured the emotional response by the sweat-gland activity in the palms of the subjects' hands. When the word was said, the sweat-gland response of the subjects increased, showing that conditioning had occurred. That experiment was published in 1962 (Staats, Staats, and Crawford). If I did that experiment today, I could use a method of brain imagery to measure the conditioning; for the emotional response elicited by the word would activate an area in the limbic system called the pleasure and pain center (see Staats and Eifert 1990), as shown experimentally by Helen Fisher of Rutgers University, albeit with a positive emotional response.

> She has studied love by looking at people's brains via magnetic resonance imaging machines . . . [using subjects] who were deeply in love. . . . While in the scanner [imager], they viewed "neutral" pictures of someone they knew but for whom they didn't have intense romantic feelings. Then they were shown a picture of their beloved. Compared with the neutral photos, a lover's picture triggers the dopamine system in the brain—the same system associated with pleasure and addiction. (Parker-Pope 2007, A8)

We can conclude that the subjects' experience conditioned them to a positive emotional response to the loved person, including a picture of that person. That is why their brains responded emotionally, because of their conditioning. This also provides a good example of the Great Scientific Error; for there is no mention of learning in Fisher's study. Rather, the point of the study is showing how the brain causes the phenomena of love. But brains don't create love. Only experience (emotional conditioning) creates love. Nevertheless, learning gets no respect in the experiment, no interest. Does that make sense? Why not combine the brain-imaging technology with study of the real—learning—cause of love, thus bringing together biology and learning? We need to get into a paradigm that suggests that type of complementary development.

The Emotional Response

Why consider both food and sex to be positive emotional stimuli and to elicit positive emotional responses? Sexual stimulation elicits tumescence in the sex organs. Food stimuli do not, instead eliciting the salivary response. So the two stimuli elicit different responses. Quite different. What, then, do these stimuli have in common that merits being treated equally? For one thing, both stimuli also elicit a response in the same region of the brain. So if one of two people has a song paired with sex stimulation, and the other person has the song paired with food, the two stimuli will elicit different responses indeed. Both people nevertheless will be conditioned emotionally to *like* the song, as could be shown by the activity in the same region of their brain, or by asking them how they felt.

The same is true on the negative side; different types of negative emotional stimuli elicit different peripheral responses. A powerful noxious odor will elicit holding one's nose. An electric shock will elicit withdrawal of the offended part of the body. However, if one subject has a tone paired with that odor and another subject has that same tone paired with shock, they both will learn a negative emotional response to the tone, including a neural response in the same region of the brain.

In terms of explaining behavior, the central importance of an emotion lies only partly in the specific responses elicited—such as in tumescence and salivation, or heart rate change and adrenalin secretion. That emotional importance lies in the common response made to positive emotional stimuli and the common response made to negative emotional stimuli. That emotional response lies in the brain, not in the peripheral response organs. Not making that distinction has led to inconsistent and disjointed beliefs about emotion (Staats and Eifert 1990). It can be added that this misunderstanding leads to the creation of multiple emotional terms that do not aid thinking. For example, common belief is that *anger, jealousy, negative attitude, hate, guilt, fear, anxiety,* and many other terms refer to different emotions. In some ways they do, but all refer to a common negative emotion. Understanding emotions calls for that distinction, which is not ordinarily made.

Fundamental and General?

Is Pavlov's discovery a very general, fundamental principle, or just a titillating but isolated finding? Definitely fundamental, definitely general, definitely very significant, and definitely lawful; and all kinds of animals, including rats, cats, and dogs, behave according to Pavlovian principles. Single leg responses on headless cockroaches and single nerve-cell responses in giant sea slugs have been classically conditioned (Kandel and Tauc 1965). So classical conditioning goes way back evolutionarily. Various types of physiological responses have been studied as well as various kinds of conditioned stimuli. Both positive and negative emotional responses have been investigated. Most important, humans also operate according to the principles of classical conditioning for important types of behavior (Staats 1996). The principles, thus, can be considered a deep characteristic of human nature having come down through evolution from many millions of years ago and remaining a driving principle in the explanation of much human behavior.

The classical conditioning principles have been stipulated precisely, and in detail, with animals and humans. This takes our knowledge past the commonsense notions of human emotion—where the concepts and principles are loose and unspecified—to an advanced theory of centrally important types of human behavior and traits of human nature, as we will see.

Operant Conditioning

But emotional responding is not the only type of behavior that is important. And classical conditioning is not the only type of learning. People not only feel about things in their world, they also take action with respect to those things. If anything, we are even more concerned with what people do and say rather than with only how they feel.

The discovery of the principles that govern how one behaves motorically was also serendipitous. An American psychologist, Edward Thorndike, who was interested in problem solving, selected cats as his experimental subjects and constructed an enclosure, or "problem box," into which he placed a cat. A cord hanging from the ceiling of the enclosure was the key to problem solution, a tug on it would open a door allowing the animal to escape the box as well as to receive a bit of food. As a good experimentalist, Thorndike selected a simple measure of problem solving, that is, the length of time it took the cat to get out of the box.

At the time, the turn of the nineteenth century, there was a belief that problem solving occurred in the mind. The animal confronted the problem situation and, after a time or two, the mind experienced—eureka!—insight. Following that problem solution the animal would immediately perform the solution. However, Thorndike placed his cat in the problem box repeatedly and measured the time on each trial before the animal tugged the string and got out. Usually the first time in the box the animals went through a series of unsuccessful behaviors, such as investigating the box and trying to bite and claw a way out. In the process of performing the various behaviors, however, the animal would eventually happen to claw the string, attain freedom, and eat the piece of food. No sudden insight occurred. What Thorndike's data showed instead was the cats *progressively got better at getting out of the box*, taking less and less time before making the correct response.

What Thorndike saw was very generally significant, beyond just problem solving. He observed during the repeated trials that the unsuccessful behaviors of the cat, like biting and scratching at the walls of the box, occurred less often. In contrast, the behavior of clawing the cord, which brought the reward of getting out, occurred progressively more often, and sooner. Finally, the first thing the animals did on being placed in the box was to claw the string. Thorndike astutely realized these findings experimentally revealed a basic principle of learning: behaviors are strengthened that are followed by "satisfying events" or rewards (later called *reinforcers*). Behaviors not followed by reinforcers are weakened.

The principle has been stated in different ways that have significant implications that need not concern us here. My way is the following. A behavior followed by a positive reinforcing stimulus in a particular situation will come to be elicited more strongly by that stimulus situation. In that sense, the stimulus situation comes to control or elicit the response.

Later studies provided knowledge of the subprinciples of operant conditioning, which turn out to be the same as for classical conditioning. Again, the more conditioning trials, the stronger the learning. In contrast, when a response occurs and is not reinforced, the learning is weakened—again called *extinction* and better called relearning or *deconditioning*. Bigger reinforcers lead to stronger conditioning. Shorter times between the response and the reinforcer also lead to stronger conditioning. And, for food to serve as a reinforcer, the animal must be deprived of food; the more the deprivation, the stronger food acts as a reinforcer, up to a point. The principles hold for other positive reinforcers, like water, or sex.

Thorndike's apparatus was improved upon by using T-mazes with the animals reinforced for a turn either left or right. Clark Hull used that device. Today Skinner's operant-conditioning apparatus, called the *experimental analysis of behavior methodology*, is the advanced technology used. For a rat subject the experimental chamber includes a lever rather than a cord. Depressing the lever constitutes the response and is automatically recorded. The lever is hooked up to a device that automatically activates delivery of a food pellet—thus reinforcing the response—the animal does not need to be handled by the experimenter, since it doesn't leave the box. The reinforcer is automatically recorded also, everything is automatic. The visual records produced reveal how different response-reinforcer schedules affect the rate at which the animal responds. Also a principle that operates in classical conditioning, partial or intermittent reinforcement shows animals do not need to be reinforced every time a response occurs, but only some of the times. Different "schedules" are possible that mix reinforced trials with nonreinforced trials—such as reinforcing every third response, or reinforcing according to a random mixture of ratios. Skinner was interested in the different effects different reinforcement schedules have on behavior. In the present view, the precise way that the nature of reinforcement determines the nature of the response, over the many variations of reinforcement possible, shows how finely evolution has constructed animals to *learn* to adapt to the environments' features of reinforcement. Why? It is clear that an animal that does what is necessary according to the size of the reinforcer, the extent of deprivation, the likelihood of a response being reinforced, and such, maximizes its survival and reproduction. Natural selection chooses the animal that responds in a way that gets the maximum reward for the minimum of effort over one that doesn't respond so finely. Getting the most out of the environment in terms of reinforcement must be very important for survival and reproduction. An animal that went all the time to a water hole that was frequently dry as opposed to a water hole that was always full would be wasting a lot of energy—a way of operating that wouldn't help in surviving and reproducing.

In addition, behavior can be conditioned not to just to the general situation, as an operant-conditioning chamber is to a rat, but to finer features of the environment. If the rat's lever press in a chamber is reinforced with food only when an overhead light is on, then the animal will learn to press the bar only when the light comes on. The light, a specific stimulus in the situation, then *controls* the animal's behavior. If reinforcement is only given when the light has a particular brilliance, not others, then only that particular light will evoke the response, to the limit of the animal's visual acuity. Following that principle of learning, animals, including humans, can acquire very fine discrimination of responses to stimuli. A child I know performs undesirable behaviors at home with his parents, behaviors he doesn't perform in nursery school; that says the two situations reinforce his undesirable behavior differently, and the child has learned a discrimination. The conditions of reinforcement have a very powerful effect over behavior.

Lawfulness

As I said in the context of classical conditioning, a traditional view has it that humans respond from within, determined by personal factors like internal wants, choices, their natures, and free will. But Pavlov's emotional responding was determined by the action of the environment. How about with motor behavior? Does it occur capriciously, determined by personal factors, in what the person wills? Or does past experience—learning—work its effects systematically here also, according to lawful principle?

Many do not quite believe the lawfulness of this kind of learning also, as they believe in physical principles like gravity. They see behavior as free, spontaneous, dependent on will. I have been told more than once that "you can reward my cat [or child] for doing something, but if she doesn't want to do it, she won't" and "my cat just doesn't respond to rewards like others do." Such statements assert a nonlawfulness of the operation of the reinforcement on behavior—sometimes it works and sometimes it doesn't. Or it may work with some animals, but not mine? What is overlooked is that the lawfulness of physical principles can be seen only where interfering conditions are blocked out. You can't see that gravity operates every time, lawfully, if conditions are changed from time to time. A feather with an updraft will soar upward. To see gravity operate on a feather, it must be put into a vacuum, taking out any effect of wind currents. In a vacuum, without any air resistance, the feather will fall like a stone. The same is true of laws of behavior, like reinforcement. You can't test how lawful the reinforcement principle is using food as the reward when food deprivation is not held constant, if sometimes the animal is food deprived and sometimes not. The person may have a cat that doesn't operate lawfully because food is used as the reinforcer and the animal has unrestricted access to food and is never deprived. What should be understood is that the principle of reinforcement works always, with every cat and every child, every time. However, in life more than one reinforcer may be at work, and the person making the observations may not know what the other reinforcers are, or what responses they follow. Lacking that knowledge, the person is not able to account for the behavior. When reinforcement sources are known and controlled, there are no exceptions concerning their effects on behavior.

Is Operant Conditioning Important?

Many of our studies have shown that operant conditioning principles are not of narrow importance, restricted to only a few behaviors. The principles pertain to all kinds of motor responses, including speech and "thoughts." Moreover, the principle is not restricted to cats, or rats, or pigeons. The principle pertains to all kinds of animals. Cockroaches will learn to go in a particular direction when food has been found there. Bees will learn to approach a receptacle of a certain color if it holds a sweeter liquid than others. The same basic principles of learning pertain also to elephants, chimpanzees, adult humans, children, psychotics, and those with low IQs.

As I indicated, we could expect that on evolutionary grounds. After all, an animal that *learns* behaviors that result in getting food will survive and reproduce better than an animal that doesn't learn that way. So this principle of behavior came down to humans through eons of time, carried by preceding ancestral species. We are talking about the long-term evolution of a very fundamental principle.

Describing operant conditioning is not just an exercise to please the academic "soul." All behaviors that involve the muscles of the body, the muscles that operate on the world—hence the name operant—are learned via the principles of reinforcement. If you want a child who has temper tantrums, provide some positive reinforcing stimulus when the child behaves in that way. Your child will learn temper tantrums. If you want a six-month-old infant who vocalizes a lot, then make sure you start reinforcing the child's vocal responses (not the crying, however). For a child to read avidly, she or he has to have had a long history of being reinforced for learning to read and then for reading. Lots of behaviors are involved in such an endeavor, and they have to be sustained through reinforcement. If a child is to become a successful athlete, singer, musician, or salesperson, a long operant-conditioning learning history will be requisite. Little muscle twitches of which the person is unaware can be operantly conditioned. Large, drawn-out performances also require operant conditioning. Verbal responses are included. So operant conditioning principles apply to just about all behaviors that are not emotional in nature.

Operant conditioning principles important? Oh yes, if one has interest in human behavior and human nature.

IS THERE ANY RELATIONSHIP BETWEEN EMOTION AND BEHAVIOR?

Pavlov employed his classical-conditioning methodology and established a tradition of research concerned with the conditioning of physiological emotional responses. Thorndike discovered a laboratory way of studying operant conditioning and established a tradition of research concerned with the conditioning of motor behaviors using reinforcers. The apparatus was different, the procedures were different, and different scientific literatures emerged in each case.

How to consider the two different types of conditioning? The second-generation behaviorists saw this and came up with different answers. Guthrie thought all learning occurred through association, classical conditioning (1935). Hull, too, posited one type of conditioning, but considered all learning to occur via the reinforcement of behavior (1943). Watson didn't differentiate the principles by which emotional and motor responses are learned (1930). Skinner, however, treated them as different types of learning. That was an important step forward. Unfortunately, he assumed that emotion had no effect on operant behavior. He assumed that classical conditioning and operant conditioning were separate, independent, and he never explained what the function is of emotion or emotional conditioning. So emotion and its conditioning were not considered important (Skinner 1975). The result has been that those who follow his approach have never studied classical conditioning systematically.

These differences are easily summarized, but my account lacks all the fire and brimstone involved. For those differences in theory were a major point in establishing different schools of behaviorism that operated in isolation from one another. Experiments that actually involved the same principles were considered separately because they were done within one or the other camp. Works done using Skinnerian experimental methodology were not accepted for publication in an experimental-psychology journal. So the Skinnerian group launched its own journal, which also published only its type of works. Skinnerians to this day do not study classical conditioning and do not use the principles in theorizing or in applications. Actually, all the behaviorists' theories suffered from not being able to put the traditions of Pavlov and Thorndike together.

All unfortunate. There is an intimate and important relationship between classical and operant conditioning, so centrally important it is apparent on a commonsense level. Everyone realizes that the positive emotion of love motivates much behavior. That is true on the other side also; everyone sees that people do many things motivated by hate. From an evolutionary perspective, how could emotional learning and motor learning be independent when both evolved because they aided animals' survival and reproduction? They both produce adaptive learning and are united in that effort. The next section will indicate on a basic level how emotion and emotional conditioning are related to motor behavior and operant conditioning.

Emotional, Reinforcement, and Incentive Functions

Emotional Stimuli Are Also Reinforcing Stimuli

What is extraordinary to me is how the most obvious relationship was missed—that is, *the same stimulus is central in both types of conditioning*. For example, food that serves in classical conditioning to elicit an emotional response is the same stimulus that serves as the reinforcer in operant conditioning. It is very important in constructing the learning theory to label this: the same stimulus, food, has two functions. Food is both *emotion-eliciting* and *reinforcing*, each function central in its type of conditioning. Sex elicits an emotional response, and it is also a very strong reward. Doesn't that suggest generally that there is a relationship between the two functions and hence the two types of conditioning? Absolutely!

Learning and the Two Functions

Food elicits an emotional response and acts also as a reinforcer, without any learning, as built into animals. Sexual stimulation elicits a strong emotional response and also will act as a reinforcer and strengthen any behavior it follows. Food and sex stimulation are examples of primary stimuli, however. How about stimuli that are learned?

That brings in classical conditioning. We know that a bell that is paired with food a number of times will also come to elicit the emotional response that food elicits, as measured by salivation. How about the other function that emotional stimuli have, the power to reinforce the operant behaviors they follow? When a stimulus comes to elicit a positive emotional response through classical conditioning, does that also make the stimulus a positive reinforcer, a reward? Would a dog that had been trained to salivate to a bell also be reinforced by the sound of the bell? Would the dog learn to press a bar if a bell was sounded every time he pressed the bar? A study has already been conducted that shows that principle (Zimmerman 1957). With rats, first a buzzer was paired with food. We can see that would make the buzzer elicit a positive emotional response, the rats would like the buzzer. Centrally, those rats later learned to press a bar because that motor response was followed by the sound of the buzzer. That sound had become a reward.

I was moved to show the relationship with human subjects, which we did in several studies. In everyday life people are conditioned to respond emotionally to food words because those words occur frequently when they are eating. Because of this classical conditioning we should be conditioned to salivate to food words. We found that to be the case in a study (Staats and Hammond 1972). Accordingly, food words, as they become positive emotional stimuli, should also acquire positive reinforcement value, that is, become small rewards, at least when one is hungry. We tested that with two groups of subjects (Harms and Staats 1978). Each subject faced a task where each time a light came on they had to make a right or a left response. If the subject chose the correct side (which varied for subjects) a food word was pronounced. One of the groups had not had food for fifteen hours, and the other had just eaten. As expected, the presentation of food words operantly conditioned the subjects who were hungry—they learned to make the response followed by the words—but not the subjects who had recently eaten. Food words acted as rewards for hungry people.

Besides confirming the general principles, the study also showed that words can be rewards and affect learning, a power of language. We actually knew that from before (Finley and Staats 1967). In that study, first we selected a group of words that people rated as pleasant—like *vacation*, *happy*, and *love*—and another group of words they rated as unpleasant—like *sick*, *hungry*, and *war*. These words thus elicited positive or negative emotional responses. The subjects then participated in a task where they had to make a left or a right response when a light came on. In one group, each time they made a left response, a positive emotional word was presented out loud, but when they made a right response, a neutral word was presented. These subjects learned to make a left response. In another group, a negative emotional word was presented following the left response. These subjects learned not to make a left response. A third group got neutral words for both responses. These subjects didn't learn to make the left response, alternating between right and left responses.

It all ties together: a stimulus that elicits an emotional response will also function as a reinforcer. *So classical conditioning of emotional responses determines to a large extent what will be emotional as well what will be rewarding and punishing. And what is rewarding and punishing for the individual plays a great role in determining what the individual's behavior will be.* A person who feels a positive emotion when seeing others in pain has the stage set for learning behaviors that put others in pain. A child who has a positive emotional response to signs of parental annoyance will learn behaviors that annoy the parent. A child who feels a negative emotional response to a teacher's compliments will not learn behaviors that the teacher compliments. A child who feels a positive emotional response when reading books will read books. This formulation of principles makes it very important to know what are the emotional stimuli that function for individuals and groups, for that will tell much about their behavior. The conception also indicates that it is important to know how to create an emotional response in another person, for example, a good set of emotional responses in a child. For that will help determine how that other person will behave. As we will see, these are central principles for understanding human behavior and human nature.

Emotional Stimuli Are Also Incentives: The Third Function

Emotion determines what life events will be rewards and what will be punishments, and in that way it helps determine the individual's behavior. When a stimulus comes to elicit an emotional response, however, this does not have an immediate effect on behavior. It only sets up conditions by which the individual can learn new behaviors, that is, determines whether or not a stimulus will have a reinforcing effect on behavior.

However, there is another principle that makes it so that as soon as a stimulus comes to elicit an emotional response, positive or negative, it will have immediate effects upon behavior. Simply put, *early in life everyone learns to approach stimuli that elicit a positive emotional response; such stimuli "attract" our behavior. And we all learn to avoid stimuli that elicit a negative emotional response in us; so such stimuli repel us.* We learn to approach stimuli that elicit a positive emotion in us because we are rewarded when we do so. We learn to avoid stimuli that elicit a negative emotional response in us because we "feel better," we are rewarded, when we get away from them.

If we have learned a positive emotional responses to classical music, we will approach it, that is, listen to classical music on radio, on television, with MP3 players, on CDs, and at concerts. If what we read and hear conditions us to have a positive emotional response to a political candidate, we will contribute money to, work for, and vote for that candidate. We (Staats and Burns 1981) confirmed this principle in a laboratory study with groups of religious and nonreligious people. Religious people—who had stronger positive emotional responses to religious stimuli than nonreligious people—had stronger approach responses to religious stimuli in an experimental situation than did nonreligious people. That shows how our emotions determine our behavior.

Increasing the extent to which a stimulus elicits an emotional response will affect the extent to which the stimulus will affect either approach or avoidance behavior. We demonstrated the principles by using deprivation as the means of varying the strength of the subjects' emotional response. Volunteer subjects went without food for an average of fifteen hours, and another group of subjects had just eaten. Then the subjects were engaged in a task of pulling toward or pushing away from themselves a different word, with a number of different words used just once (Staats and Warren 1974). Food-deprived subjects more rapidly pulled in food words and pushed them away more slowly than nondeprived subjects. To them, food words elicited a stronger emotional response and thus more strongly evoked approach behavior. That explains why a hungry person turns off the freeway after seeing a sign that says "Restaurant" when a more recently fed individual drives right by. Both would say the reason for their behavior was a mental decision. But actually the reason one person made the decision to turn off was determined by these conditions: (1) having previously learned a positive emotional response to the word *restaurant*, (2) not having eaten, so the word elicited the emotional response, and (3) long ago having learned to approach things that elicit a positive emotion.

Done with laboratory precision, these studies show clearly that the basic animal-learning principles apply to humans. We can also see how the principles enable us to understand human behavior. They explain, for example, why the advertiser pairs his product with handsome, happy people on television. That conditions viewers to have a positive emotional response to his product, raising the likelihood they will "approach" (buy) it. That is why political-campaign managers say derogatory things about the other candidate, for that will condition some people to have a negative emotional response to the other candidate and thus vote against her or him. That is one reason why governments strive to influence or control the sources of public information for their own population and do the same for the populations of other countries. When sources of information are controlled, the population's emotions can be manipulated, and thus the behavior of the population can be manipulated also.

The Emotion-Reinforcer-Incentive Functions

In summary, then, emotional stimuli are rewarding or punishing depending on their positive or negative emotional value. In addition, emotional stimuli have approach-incentive or avoidance-incentive value—that is, will attract or repel behavior—depending on whether the emotion is positive or negative. In these ways emotion is a broadly acting, heavy determinant of human behavior. That is what makes the emotions we learn so crucial and knowledge of the principles involved so important. A great weakness of Skinner's theory and the theories of the other classical behaviorists is that they do not deal with emotion and the emotion/behavior relationship.

LEARNING IS A BIOLOGICAL PROCESS, EVOLVED BY NATURAL SELECTION

We can see how the principle of operant conditioning is adaptive for the animal. It is adaptive for the rat to press the bar when that response is followed by food, just as it is adaptive for the springbok to learn to follow a path to a water hole because its behavior ends up by getting water. Imagine an animal that didn't learn in the ways that have been described. What if the animal didn't learn the behaviors that got it food and got it away from danger? If reinforcement did not affect the behaviors of the animal in those ways, the animal would be far less likely to survive and reproduce, and thus perpetuate its kind.

The same is true for stimuli that elicit an emotional response. Animals learn to avoid stimuli that elicit a negative emotional response and approach those that elicit a positive emotional response. That is the case because negative emotional stimuli are generally biologically harmful. Stimuli that elicit a positive emotional response are generally biologically beneficial. An animal species that did not approach the one and avoid the other would face extinction very soon.

The evolution of the biological structures that make the conditioning principles possible was gradual. With the first, simplest organisms, there are no neural structures. With more complex organisms, such as some jellyfish, there is a connected net of nerves; stimulation at one spot sets off an overall response. Advancement in the nervous system yields chordates and then vertebrates in the direction of increasingly refined sensing of and response to the environment. The increased ability to learn from the environment continues through the vertebrates. At some point in the progression, neural mechanisms by which to effect classical and operant conditioning became fully developed, and species that arose farther up the evolutionary scale had those neural mechanisms and behaved according to the principles of learning. By the time humanoid species came on the scene, the learning-behavior principles had long ago been fully evolved and came down as an evolutionary gift from antecedent species.

In terms of evolution, thus, there has been great time and "effort" expended "in pursuit" of the basic learning principles. Gradually established, the principles advanced in development. The end result of the evolutionary process was the creation of complex sensory and response structures and connective neural mechanisms. Critical to note, those structures are energy consuming. They involve a cost for the animal. The great evolutionary effort that went into the creation of the learning principles tells us how important and fundamental they are, as does the fact that those principles have persisted through the evolutionary creation of many species over a huge expanse of time. No nature-nurture conflict here when we discuss the evolved basic learning principles.

The Learning of Humans and Animals: Equal?

Humans, as the beneficiaries of evolution, inherit the learning principles that took billions of years to evolve. Those learning principles constitute a very fundamental part of human nature, but a part that is shared with many other animal species. Sharing of characteristics has been taken to show that humans are not unique in behavior and nature. That is one of the purposes behind the studies demonstrating that apes can learn language. The assumption has been made consequently that only a quantitative difference exists between humans and other primates, not a qualitative difference in kind. Additional confirmation is drawn from recent findings that chimpanzees use and modify tools more than had been thought and orangutans signal with gestures. But compare the learned achievements of humans with the learned achievements, or culture, of any other animal. The quantitative difference is immense, so great that it becomes a qualitative difference, making it hard to believe that both chimps and humans learn by the same principles *and no others.*

Doesn't a conception of human behavior and human nature have to confront the question of cause: what makes humans *the marvelous learning animal,* so much above all other animals? Do human neurons conduct more rapidly than do the neurons of other animals? Does basic human classical and operant conditioning occur more quickly, taking many fewer conditioning trials? Apparently not, disappointingly: "[A]ll species, regardless of the size of their association cortex, form . . . simple associations at about the same rate" (Diamond and Hall 1969, 252). Thus, unlike popular belief, humans are not superior to other mammals in such fundamental respects. The biological mechanisms by which basic learning takes place were evolved long before the hominid species arose. That creates a conundrum. What is it that makes humans superior to other animals in learning? One answer can be quantity, the sheer number of "learnings" that can be recorded.

Humans face the richest, most complicated, learning environment in the world. Just think of the child learning to speak. There is no systematically designed learning program for teaching that. Parents and others talk in the child's presence. That means the child hears a huge mixture of sounds, for there are thousands of words in a language, and those words are put into an unlimited number of combinations. The child hears a cacophony of sounds. Yet out of that complex the child learns consistencies. An American child after about six months of exposure begins babbling sounds that are in the American language. The Chinese child babbles the sounds to which she or he has been exposed. It takes a huge brain and great learning capacity to do that.

It is suggested that once the basic learning principles had evolved fully, another way of continued evolution of learning was in the increase of the number of neurons animals had for learning. Advancement took place in the *capacity* for learning; for example, in terms of the types of stimuli to which the organism is sensitive, the types of response the organism can make, and especially the number of learned connections that can be formed. There is no difference in basic learning principles for a dog, a baboon, a chimpanzee, and a human. But there is great difference in *learning capacity*.

Humans have a huge brain, by weight more than four times the size of a chimpanzee's. That human brain contains more than one hundred billion neurons. The number of learning experiences that an animal can acquire depends upon the number of neurons of its brain, and the human brain is huge, spectacularly so. The human species is a wonderful learning animal because of structure. And, as we shall see, that great innate, naturally selected learning capacity of the human brain actually opens an avenue for wondrous additional learning ability that humans have and no other animal does.

LEARNING AND HUMAN NATURE

Conventional wisdom has it that humans have a huge brain to house the many wired-in aspects of human nature. Supposedly there are emotional bonds between mother and child, wired into each of their brains. There are parts of the wiring of the brain that make individuals intelligent, responsible, honest, trustworthy, kind, sympathetic, ambitious, and also impulsive, selfish, lazy, disorganized, warlike, and cruel. This wiring takes place via biology, so goes belief. The child inherits the wiring through the genes, as a tendency, and biological maturation largely carries out the wired-in traits. Learning is not denied but is commonly believed to "top off" what biology has created, leaving biology as the major cause. That makes study of the brain, genes, hormones, and such central and the study of learning peripheral. Based on that, most of the science monies spent in the study of human behavior go to the science areas that study biological causes or things that are considered to be biological in nature.

But what if all those traits, from intelligence to cruelty, and all those in between, do not have biological causes? What if the biological structures of the human, especially the brain, instead are *the mechanisms* through which the *causes* of human behavior act, rather than the causes themselves? What if the sensory organs transmit representations of the world brilliantly, and the brain records these vivid pictures in addition to recording what behaviors worked best in those world situations? That makes the brain and the sensory and response systems the means by which humans learn to respond to the environment. Learning then is the cause of the individual's behavior and the many differences that occur between individuals in their behavior. If that is the case, a drastic change is called for in what we do, for example, in terms of what we study and how we approach problems of human behavior and how we think about ourselves and others.

An important fraction of that study, at the fundamental level, has already taken place in the discovery of the animal learning principles. Those same principles operate with humans. Those interested in human behavior and human nature need to know this fundamental information, to a useable extent. Learning should not be left on a commonsense level, as if we all know what learning is through our life experience. That commonsense knowledge of learning is not adequate. It contains errant beliefs, and it lacks necessary knowledge. Learning involves a number of principles that can be concisely stated and have been precisely shown in laboratory experiments with animal and human subjects. That study has established principles that show just how finely the individual animal or human can adjust to the environment. A subject can be conditioned to make a response to stimuli in any of the senses, to patterns of stimuli, to intensities of stimuli, to delays of stimuli, to combinations of stimuli, to the absence of stimuli, to the withdrawal of stimuli. Responding can vary in many ways, in many combinations, sensitively adjusted.

This tells us that adjusting to the environment via learning has been a very important ability in an evolutionary sense. That tells us that we have been shaped by evolution to learn extensively. Humans encounter the richest of all learning environments. Systematic consideration indicates that our huge brains have been developed to learn the very complex behaviors needed to adjust to that complex environment. Evolution constructed us to do that according to learning principles. Huge learning ability constitutes an essential, fundamental, deep, extensive, and valuable part of human nature. The human is a learning animal, better than any other, by a huge difference. The great significance of this has not been understood or dealt with. Moreover, that understanding has only begun; knowing the basic principles only gives us the vehicle for a much longer, much more picturesque journey.

Part 3

CHILD DEVELOPMENT AND THE MISSING LINK

Chapter 4

LEARNING CHILD DEVELOPMENT

"I Was Not a Lab Rat" runs the title of an article in the *Guardian: G2* (March 12, 2004). Then follows the refutation by Deborah Skinner Buzan of the charge that her father, the noted behaviorist B. F. Skinner, had used her in experiments to test his theories. Allegedly she had been placed in a laboratory box where all her needs were shaped and controlled. Her denial was vigorous, saying in rebuttal that her father did not conduct research with her as the subject. She was only put in a crib called an air crib, which her father had invented as a means of automatically controlling temperature, humidity, and hygiene for an infant, making clothing or blankets unnecessary.

Actually, although there has been much said to the effect that Skinner did research with children, there are no publications indicating that he employed operant conditioning to train his or other children. His students did use his experimental analysis of behavior methodology to show that reinforcement principles do apply to normal children, mentally challenged children, autistic children, and psychotic adults. But his projected program was to extend his animal research method to studying human behavior, which meant employing simple motor behaviors like pulling a plunger, not dealing with or studying the behaviors of real life.

The "crime" of employing learning conditions in raising one's children was committed, however, by a psychologist. And I confess, I was that psychologist father. And yes, this did involve testing my principles and methods, established in prior research. In pleading my case, let me say, however, that "it was my learning background that made me do it." For not far along my research odyssey I became convinced that human behavior is learned and that my behavior-analysis methods would be valuable in dealing with actual behavior. That meant that employing those methods would be good for raising children, a conclusion that made it mandatory for me to use my knowledge with my own children.

The fact is that much of what was considered knowledge in the field of child development seemed more than shaky to me, rather quite wrong. I didn't believe that children tend to develop behaviors about a certain time because the cause, biological maturation, occurs at about that time. I also didn't believe that the different times at which children develop a type of behavior depends on their differences in biological maturation. I believed in both cases that children *learn* their behavior developments, that similarity in time of development actually indicates similarity in learning experiences, and that differences in children's behavioral development comes from differences in learning experiences. Child-development norms that child psychiatrists and psychologists use are thus determined by learning, not by biology.

Parents, depending on their own learning experiences, present different learning experiences to their children, and that produces different learned developments in their children. Parents' beliefs about the child also play a large role in how effective they will be in training their children. Parents who believe deeply and widely that children develop through biological maturation are less likely to become engaged in training the child than a parent who recognizes more the importance of learning. I was sure I could use my scientific knowledge to improve greatly on what children usually experience in the home so that our children would develop advantageous behaviors and not problem behaviors. This could profit the children greatly. I also thought extending my behavior analysis in a broad sense would help raise happy, well-rounded children. Much of what I did—in emotion, sensory-motor, and language training—I just worked into the conditions of a normal home. With some behaviors, when a child's behavior is advanced on the good side and lacks problem behaviors, relationships with other family members go smoothly.

With our two children, Jennifer and Peter, I began my behavior-analytic parenting/researching really when they were born. The behaviors I addressed were significant in child development, valuable for everyday life. The purposeful training of skills was not intrusive or forced and each child was positively reinforced. That made it possible for their participation to always be by choice. Our two children were the first, and to my knowledge the only children, reared comprehensively in a systematic learning theory framework and its behavior-analysis methods. Let me emphasize our parent-child interaction was a two-way street. When I was training one of our children using my principles, I allowed myself to be guided by the child. To illustrate, when I was teaching reading, as I will describe, the enthusiasm of the child's attention guided me in determining features of the training. A few times I tried to extend too long a training session, and I saw that the child's attention flagged, performance dropped. That taught me that training sessions had to be appropriate in length—at first, for example, in learning a new word, just a trial or two. The reinforcement had to be adjusted also. The less the child's experience in the task, the more effortful the task is for the child, the shorter the session must be. Also, the more effortful the training, the greater the reinforcement must be. The children taught me that the good trainer has to know such principles and be sensitive to the child's attentiveness and level of contentment. If learning something was difficult for the child and training was not succeeding, the problem was mine, not the child's. The problem could be that the child was not prepared well for the task, not given the preparatory skills, or not sufficiently reinforced. The children taught me much more, such as what superb learners children are and how love-laden a relationship will grow when two people engage in an activity that contains positive experiences for each. I also learned that both parties, parent and child, teach each other. As they learned from me, I learned from them—about children, about child development, and about the role of the parent—things I could not learn elsewhere.

There was no reason for the children to experience any feeling of having been pushed or used for the sake of experimentation, and they never gave any sign of having such feeling. Like most every parent, I wanted our children to love me. And, first and foremost, my love for them guided my actions. They in turn gave me a great deal of knowledge about child development and, it turns out, a great deal about the nature of humanness.

These findings indicate that rather than being reprehensible, systematically employing learning principles and procedures in raising children can be productive and valuable for them. But how about doing so with research also in mind, interested in value to science? Is that reprehensible? Not if there is good reason to believe that what is done will be to the benefit of the child. Let me add briefly here, moreover, such research is essential. The traditional field focuses on trying to establish the characteristics of children of different ages. The study is usually short-term, is set in artificial circumstances rather than comprised of real-life behaviors over real periods of time, and does not deal directly with the causes of the behavior development that occurs. There is no substitute for studying the specifiable causes of children's behavior development. If that development is largely determined by learning, the development can only become known by manipulating and studying that learning.

In terms of the larger concerns of this book, if human behavior and human nature are learned, then it makes sense to begin study with children. That is, when learning begins, the behaviors still are simple, capable of analysis and explicit understanding. With knowledge of the phenomena at this level, there will be a foundation for understanding the more complex developments that later occur. Human behavior and human nature are complex, and the strategy is to approach it in successive levels, building upward from below, from the simple to the complex.

As a further way of handling complexity, I divide the panorama of humanness into three areas of behavior—emotion-motivation, sensory-motor, and language-cognitive. This does not mean that these are different in principle in any way, or unconnected—they are all interrelated—but only that their division makes them easier to understand.

EMOTIONAL-MOTIVATIONAL DEVELOPMENT

Everyone has felt emotion, ranging from the misery of unrequited love to the joy of attaining some strongly desired end. Such happenings seem straightforward. It is easy to believe that we feel these things because we are human; that's the way we are built. That is true only to the extent that we—along with other animals—have been constructed to manifest different physiological responses to different kinds of stimulation. We respond in a certain way to negative stimulation and in a different way to positive stimulation.

But understanding human emotion—why humans feel positively or negatively about such a huge number of things—constitutes a complex topic. There are many aspects to study, and emotion has been considered in various different theories. Some emotion facets are considered in one theory and other facets are considered in others. For example, conceptualized in a Darwinian context, human emotions are commonly considered to be inherited genetically, having come down through natural selection because they aided survival in human and animal evolution.

Emotion also has been studied in animals in various ways to establish the physiological mechanisms involved. Motivation in animals has been studied in terms of the deprivation-satiation of food and other biologically significant needs. These studies link motivation with emotion. The effects on behavior from deficits in the experience of emotion have also been studied with animals. For example, Harlow and Harlow (1966) found that baby macaque monkeys raised without the emotional experience of having a mother do not become emotionally normal, as evidenced by their social, sexual, and parenting behavior.

In other studies, emotion centers in the brain have been found. Electrical stimulation of one center can induce rage behavior; stimulation of another can induce a positive emotion. Such stimulation also will act as negative or positive reinforcement, as would be expected from the basic principles presented in the last chapter. Other biological studies show that during emotional responding the brain produces chemical products (such as serotonin).

Various conceptions of emotion and motivation have also emerged from clinical experience in treating clients for problems of differing types. Freud theorized that children go through stages of emotional development and that the way parents handle those stages determines the child's later personality characteristics. Similarly other concepts of emotion have been constructed in the theoretical context of children and parents. The child psychologist John Bowlby (1969) very influentially proposed that the infant from birth has an evolved need for attachment to a mother because that increases the child's chance to survive. Since everyone has experiences of emotion and motivation, our common language contains various emotional-motivational concepts and explanations— need, desire, ambition, interest, drive, anxiety, depression, jealousy, and on and on. Many commonsense beliefs about human emotion and human motivation are expressed in considering and raising children.

Human behavior has various aspects that are studied in various circumstances, with different explanations. A systematic conception is needed that can integrate and relate the various types of knowledge of emotion and motivation. The principles of the preceding chapter—describing the learning of the emotion-eliciting, reinforcing, and incentive functions of some stimuli—provide a framework for establishing that conception. Human emotion-motivation characteristics begin to be learned in child development.

Learning and Emotion

In brief, we respond emotionally to certain stimuli because of the way we are built biologically. This includes a large but finite number of stimuli that elicit emotional responses in us, positive or negative. We have positive emotional responses to foods, for example, and negative emotional responses to painful stimuli (Staats, Hekmat, and Staats 1996).

On the other hand, humans will *learn* to respond to an unlimited number of stimuli. When a stimulus that already elicits an emotional response is paired with a neutral stimulus, the latter will also come to elicit that emotional response. The human being has enormous learning capacity. And during the extended childhood period there will be innumerable experiences where an emotion-eliciting stimulus occurs in association with a neutral stimulus, changing that neutral stimulus to one that elicits an emotional response. In that way the child learns an emotional response to a gigantic number of things, to parents and family, to their looks and opinions and behaviors, to other people, to work stimuli, to political stimuli, to language stimuli, to educational stimuli, to religious stimuli, to recreational stimuli, to books, to music, to sports, to parties, to jokes, to movies, to television programs, to jewelry, to cars, to houses, to clothes, to foods, to multitudes of words, and on and on. Anything—animal, mineral, or vegetable, or any other social or nonsocial event, or even the absence of something, like someone supposed to be where he or she is not—can elicit an emotional response. The number of things, events, and activities that through learning will come to elicit an emotional response in a human being is incredibly large.

Emotional responding is important in and of itself. A child frequently experiencing negative emotional responses learns to respond to many things with negative emotion, thus having a life much less happy than a child who experiences less negative emotion learning. On the other hand, a person who experiences more positive emotional experiences than another through conditioning will be happier (see Rose and Staats 1988). That is understood on a commonsense level, and most parents attempt to spare their children negative emotional experiences and maximize positive emotional experience. Most parents (because of what they have learned) love to see their children happy.

But emotional responding has significance in addition to happiness; for anything that elicits an emotional response will also function as a reinforcer, either as a reward or as a punishment. A child who learns to love religion-related experiences, and a child who lacks such learning but has learned to love music, will have significantly different reward (reinforcer) systems. The behavior of the one child will be reinforced and learned if followed by musical stimuli, the behavior of the other child will be affected by religious stimuli.

Because the incentive (or motivational) nature of life's stimuli depends on their emotional value, those two children will also be *attracted* to different life events. Announcement of a musical program after school will draw the attendance of one child; the announcement of an afterschool lecture by a religious figure will draw the attendance of the other.

That makes the learning of emotions very important. What is positive and negative emotionally determines what will be rewarding and punishing and what will be attractive or repellant for the child. How the child behaves is largely determined by the child's learning of emotional responses to the stimuli of life.

Two children who learn different emotions will differ as a consequence because (1) life situations will elicit different emotions in them, (2) they will be rewarded-punished by different things, and (3) they will be attracted or repelled by different incentives in life. They thus will do different things and learn different things. They will end up quite different in their behaviors and in their natures. These principles can appear in commonsense notions, but hazily. Their importance requires that they be stated clearly; parents should know them.

By the time I began my parenting I had already done a good deal of research on the learning of emotion. My first federally supported research project dealt with the way an emotional response can be learned through language—if an object is paired with words that elicit an emotion, the object will come to elicit that emotion (see Staats 1968a). I call this process *language conditioning*. It is pervasive and important in life and occurs all the time, universally. If we are told in positive emotional words that a movie is good, we learn that. For the same reason we select restaurants, colleges, political candidates, international enemies, paintings, religions, and golf clubs by our emotions acquired through words.

This happens even when subjects are not aware of the conditioning. In politics, supporters describe their candidate's character, actions, proposals, and promises with positive words. Negative words are paired with the opponent's features. We will not remember the great number of times we experience emotional statements, but the need to "out advertise" their opponents drives politicians in our society to raise campaign funds. They may not understand the laws of conditioning, but they know what wins campaigns. That is why it is prudent to consider *what* we hear in terms of the motives of the source. Without such consideration, one is conditioned unknowingly by what is read or heard. Many other studies have tested or used my "language conditioning" method (see Hekmat and Vanian 1971), and the principles clearly operate with children (see Early 1968).

Knowing the principles of emotion-motivation has led me to use them in understanding and relating to other people. When we had children, I followed the principles with them also. For example, showing a youngster newspaper articles on the dangers of smoking will condition them negatively, and that will make it less likely they will seriously try smoking. What the parent *says* in the child's presence will help condition the child's emotions and motivations—such as ambitions, values, interests—and thus affect their actions.

Learning Bonding and Other Positive Emotional Responding

It would be pretty difficult to rear an infant without much positive emotional learning occurring. Just in feeding—with every mouthful—the parent can pair herself with the experience of a positive emotion. Parental feeding, or being there when feeding occurs, contains many thousands of positive emotional conditioning experiences. But there are variations even in such a straightforward thing as feeding. Some parents arrange food for the young child and then go off to do other things. Others sit with the child and actively engage with him or her. The parents who are present when the infant is feeding will create more bonding with their child than the parent who does other things while the child is feeding.

Actually knowing these things, at least in beginning form, the first training I extended to our children was in feeding. In the many times I fed our children, beginning with bottles in infancy, I used each such session as an opportunity to produce positive emotional conditioning to me. I made sure that I was visible, with a smiling responsive face, and that I cuddled the child affectionately. Frequently I baby-talked, told stories, spoke, and occasionally sang during the feeding. I wanted the children to bond to me, to have a strong positive emotional response to me, to my voice, to my presence, to my holding them, to my affectionate caressing. I knew the bonding would be important in their language learning, in my becoming a guiding figure for them, in my voice being listened to and imitated, in my facial expressions becoming signals of what I was feeling. I wanted to become a strong positive emotional stimulus because I knew that by doing so I would thereby become a stronger source of positive reinforcement for the child. I would be able to reinforce more strongly the desirable behaviors the child performed, and my withdrawal would be a stronger "punishment" for the child when she or he behaved undesirably. (My "punishing" withdrawal could be a withdrawal of attention, or a change of facial expression from positive to negative, such as in changing from smiling to frowning.) Feeding, of course, was not the only positive emotional thing I paired myself with. Playing with the child, in many different ways, elicits a positive emotion, as does holding, kissing, hugging, and such. And teaching the child skills in a way that is positively reinforced also elicits a positive emotion in the child. There are many activities the parent can arrange that condition the child to have a positive emotional response to the parent. Such experiences create a strong, valuable bond.

Research shows that the mother's face and voice become salient stimuli for her infant children. That occurs because most mothers (and many fathers), usually without planning, nevertheless provide that same type of experience. They provide positive emotional things while the child is also experiencing *them*—their looks, their voices, their caressing touch, their hugs and kisses. Wouldn't that experience produce love for the mother overall, for specific aspects of her, and for her caring behaviors? When the parent's voice elicits a positive emotional response, then the child will feel a positive emotion when making a vocal response that sounds like the parent, an important part of learning to speak. The same is true for other behaviors of the parent; the child who is hugged and kissed when experiencing a positive emotion later on will behave that way also with her or his children. The child who experiences an expressive face—including frowning as well as smiling and laughing—will be expressive also. Real differences in child behavior emerge from such experiences.

This analysis can explain Harlow and Harlow's (1966) findings that baby monkeys raised without mothers suffered various deficits in social behaviors. They did not have a positive emotional response to other monkeys. They did not display positive, affectionate behavior to them. They didn't relate well to other monkeys in a way that could lead to a sexual relationship, and they didn't approach and deal lovingly with their own babies in the usual manner of caretaking. The present analysis explains why: without mothering experience they didn't learn a love response to a mother figure or other like creatures. Harlow worked with monkeys. Wouldn't it be valuable to study children, to rate parents in terms of how much positive and negative emotional conditioning their infants receive in their parents' presence and then rate the children on the extent to which they are bonded to their parents? Wouldn't there be a relationship, the more positive the parents, the more bonding? I can tell you what the outcome would be. Parents who provide more positive emotional experience for their children would have more highly bonded children—as measured by such things as the amount of pleasure shown on seeing a parent, on being picked up, on being hugged, on being looked at, and on being played with. Everyone has seen a child who calls out to the parent, "Look at me, Dad," before doing something—that is a sign of bonding.

Much emotional bonding—such as parents for children and children for parents—is so usual, so widespread it seems it must be part of being human. But that does not indicate such love has evolved. Human love patterns, although general to the human group, are learned. Parents can love a child not yet born. Why, if the love is not inborn? That will become more clear when language is explained. Love can form through language, not only by direct experience. The human infant, however, shows no sign of love for a parent at birth but does so after experiencing numberless pairings of parents with the positive emotional things, like the food, caressing, and play that parents provide.

What is not focally realized is that parents differ widely. So when a child doesn't express much love for the parent, the problem is considered to be in the child. That can occur to an extent considered abnormal. The problem can arise in the child's learning, however. For example, the mother or father who believes the child develops on a schedule imposed by the child's biological maturation may not do much with the child that instills love. Take an advertisement I saw recently that pictured a father, a baby in his lap sucking on a bottle, while dad leaned forward to work on a computer. The infant couldn't see the father. There was no parent-child contact, no opportunity for the infant to learn bonding. Some parents even prop the bottle up, so they are not even present during feeding. I have seen parents—good, educated, loving parents—place a toddler in a high chair and just put the food down, while they busy themselves with other things. Such actions subtract a vital part of the child's experience. The child will get fed, but not experience the sight and sound of the parent paired with the food—a wasted opportunity for the child to learn an emotional attachment to the parent, as well as for learning the behaviors of a physically expressive, loving parent.

Here's another example: I played with our children from the beginning, with made-up little games, like arranging for the young infant to grasp a finger and then moving it back and forth, usually in time with a song or vocalization. When diapering one of our children, I would raise my hand high and then slowly start, and with increasing speed, to swoop it down, accompanied by an intensifying saying of "eeh," to end by tickling the infant under the arms and singing/saying "dee-dee-dee-dee-dee-dee," sometimes ending with a noisy, sloppy kiss on the neck or belly. They came to love that game and generally liked being diapered. Play and affectionate contact constitute strong bonding conditioning because they elicit positive emotion. Parents who have learned play skills will generally provide such experiences for their children. The important point here is that some parents haven't learned such play skills, so they miss many bonding opportunities.

Most parents are similar in their practices in these and other ways; but there is wide variation, with some very extreme. Children, after all, do display differences that vary from the child clinging to a mother, apparently frightened of separation, to the child who doesn't look at the parent, does not respond positively to affectionate embrace, and shows no sign of emotional attachment.

Let me repeat, for the benefit of those opposed to a psychologist systematically doing research with his or her own children, the benefit of the children was my first objective. One objective always was adding to our parent-child relationship—bonding in both is central. Let me give evidence, by way of a Hallmark® card given to me by my daughter for my birthday in 2009; it poetically describes a dad who did things like swing his daughter and chase her, and who, when catching her, would hug her. It says he loved her expressively and praised her and taught her, and in these various ways brought her great happiness. Jennifer says she is that daughter. There could be a card saying the same thing in reverse, for she behaved just as lovingly on the other side, and she taught me so much about children and human behavior generally. I loved the period that card described, and love as deeply as possible the daughter and son who provided that experience. Their lives and the relationships they have created with their children have extended that happiness. There need be no conflict in parental and research roles.

Is bonding just a joy, or is it important? Oh yes, it is very important. One of the essential reasons is because the parents' effectiveness in teaching the child and serving as a model in various ways will depend on the child's love for them. Ordinarily, the more the child has a positive emotion for the parents, the more effective the parents will be in imparting their experience and in exerting a positive influence on the child's behavior.

Negative emotion for the parent will interfere with that process. Rather than following and imitating the parent, the child will do the opposite, unless otherwise influenced. As an illustration, take religious-fundamentalist parents who raise a child "by the book," forcing the child's focused participation in religious experiences, while denying many positive experiences the child's friends have, in order that the child have a stringent religious morality. Let us also say the parents, guided by the same severe doctrine, have been harsh with the child, nonaffectionate, restrictive, and punishment-oriented. On that basis the child will learn a negative emotional response to the parents and their ways and, on that basis alone, will "negatively model" the parent, that is, do the opposite. Positive bonding has been described widely, but there can be negative "bonding" too, and that type of bonding can have as heavy an influence on the child.

Additional Emotional Learning

That by no means covers the emotional learning the child receives in the home. For example, religious parents ordinarily will introduce their children to positive

emotional experience with religious things, theological concepts, religious services, religious people, and religious values and knowledge. Parents interested in
sports introduce their children to sports-related experiences, ordinarily in a positive emotional way. Those with a love for music tend to supply special musical
experiences to their children. There are endless variations in emotional experience and in the emotional conditioning they impart.

Much of this conditioning occurs via language. I have already indicated that
as part of language development the child learns a very large number of words
that elicit a positive emotional response and a very large number that elicit a negative emotional response. Parents, by the language they use in the home, will
have a powerful effect on the emotional-motivational characteristics of their children. Parents who use positive emotional words in talking about education condition their children to have a positive emotional response to education situations,
events, and people. Education will thereby become a positive incentive for the
children, and they will do things to get an "education." Educational rewards also
will be important for them. They will learn behaviors that get them such rewards.
Much of the individual's values, interests, ambitions, goals, attitudes, and desires
—emotions—will be established through language communications that parents
and others supply. These emotions, given different names, which are carried by
words, have a great effect on us—they are part of the power of language.

Not all such experiences involve the learning of positive emotions. Parents
who are racially prejudiced ordinarily over the years of childhood will say
many things that condition their children to have a negative emotional response
to people of other races. The children will not be aware of why they have those
feelings, called attitudes, because the conditioning trials will have occurred in
everyday experiences, slipping by without notice. Much of it will have
occurred long ago. That kind of learning constitutes a powerful mechanism for
the transfer of emotional characteristics from generation to generation.

Parents need to become alert to how their actions produce emotional
learning in their child. That emotional learning will have important effects on
the parent-child relationship, on the child's learning in and out of the home, and
thus on the behavioral characteristics the child will later display.

SENSORY-MOTOR DEVELOPMENT

A very important part of child psychology consists of the study of child development by observing what children typically do at various ages, beginning at

birth. The infant at a half month of age typically lifts her head when lying on her stomach; at a month she grasps a ring touched to the fingers; she rolls over at three months, sits with support at four months, sits alone at six months, stands holding on to something at six months, and walks alone at twelve to fifteen months, as examples. Most children do exhibit behavioral developments at roughly the same time, even though much variation does exist, with some children very deviant, in both directions, early and late. Interestingly, both the commonality as well as the variation are generally explained the same, by assumed biological maturation or lack thereof. The assumed explanation covers all possibilities and resists invalidation because no evidence is introduced to prove or disprove it.

Although I had only my own work to go on at that time, when our Jennifer was born, I began observing her behavior within my own growing behavior-analysis conception. What was remarkable to me was the total lack of behavioral skill in the newborn response to environmental stimulation. The only systematic behaviors are reflexes such as a startle response to a noise or crying in response to disagreeable things, such as hunger, thirst, cold, or being held uncomfortably. The baby did not even look at things, or follow them as they moved. Her own movement of arms and legs was random. There was no noticeable affection for anyone. No evidence or suggestion that the newborn had any wired-in systematic behaviors. That is what I observed.

I immediately began to "intervene" in terms of arranging learning experiences that would bring about behavioral developments or skills. For example, as is usual, my child as a neonate did not focus on and track a moving person. Would the infant respond to learning experiences and become more skilled in tracking? Over the next couple months I began progressively to introduce "perceptual" learning to our infant daughter. This first sensory-motor training began as a primitive game of hide-and-seek. When she lay somewhere, as in her crib, and was looking at me, sometimes because I had moved myself into her line of vision, I would move a bit out of line. Her eyes would then begin to wander, and she might find me again. I made it easy at first, helping her find me by moving back into her line of vision and arranging myself so her eyes could encounter me with little movement. Seeing my smiling face and hearing my delighted sounds constituted the reinforcement of her little tracking response. (My expressions had become reinforcing because of the emotional learning experiences I described in the previous section.) Progressively, very slowly at first, she began to acquire some eye tracking skill. I did this with other objects, like showing her a rattle, shaking it slightly, and then moving it to the side a bit and rattling it again, until she moved her eyes to look at it again. She was learning to track moving auditory-visual "objects." The infant can learn eye and head movements in this way, by being reinforced for those behaviors. Progress is slow, indicating that first learning doesn't come easily. Doing that and other such things, however, was important because they showed me early on that our very young baby was a responsive, experiencing, learning organism, ready to learn appropriate skills. *A most egregious message the Great Scientific Error sends to the parent is that the newborn is a wonderful vegetative lump whose behavior development is driven by inner biological maturation, not by learning.* That belief leads parents away from being observant of how the child behaves and how the parents' responses to the child affect the child's behavior and learning. Many parents think it irrelevant to attempt to train young babies, because they do not learn. Quite the contrary, there are important learning opportunities for both infant and parent that are missed because they are not understood. It is also a joy to find out about one's child's great learning sensitivity and how child and parent can grow together.

Eye-Hand Coordination

I systematically analyzed and trained my daughter, and later my son, in other sensory-motor skills. For example, there are other basic eye-movement skills to be learned, such as those that make up eye-hand skills. To illustrate, the infant may grasp the handle of a rattle that is touched to her hand, shake it and make it sound, and hold onto it when the parent tugs it in different directions. The baby finds such stimulation reinforcing, especially if the parent coordinates the movements with rhythmic sounds and makes the activity a game. But at the beginning, the infant's grasping of an object is reflexive—not a systematic behavioral skill—so the baby may suddenly open its hand and drop the rattle. Although the infant will take the rattle again if touched to her palm, there is no ability to look for the rattle, no purposeful eye-hand coordination in reaching for, grasping, and moving it.

Experiences, however, can be introduced that will begin to teach eye-hand coordination. For example, when the infant, on grasping a rattle, randomly waves it, the parent may help guide it to the child's mouth. Mouthing things is rewarding for the young infant. She will thus learn movements that bring things to the mouth. That is why infants carry objects to their mouths and suck on them. The infant's eye-hand skills are zero at birth. The child learns positive emotional responses to many things before he or she learns the sensory-motor skills by which to reach them. The lack of skill in an infant reaching is very cute and shows even basic movements don't come wired in. At first reaching appears random and jerky. Gaining eye-hand skill takes many learning trials. The parent can hasten the learning process by showing a liked object and then maneuvering it so the infant reaches it in its random reaching. Gradually objects can be placed or held in various locations so the infant learns to reach in all directions to obtain them. The eye-hand skills of the infant, in reaching and grasping, later form the basis for learning other eye-hand skills. A parent experiencing the process can see very early how arranged learning advances the child's behavior development, a valuable example of how the child begins with no skill, just random behavior, and only slowly learns coordinated, purposeful behavior. Not only that, it is emotionally so positive for a parent to do things that affect valuable learning in the child. Babies learning are so cute.

Following the belief that sensory-motor coordination occurs through biological maturation, sports people attribute abilities—such as the exceptional "eye-hand" coordination of the great hitter in baseball—to genetic inheritance or to the grace of God. As a sports enthusiast, many times I have heard commentators explain athletic performance in terms of "being a natural athlete" or because of "God-given talent." Actually, eye-hand coordination is learned, but the Great Scientific Error stands in the way of studying the long process of learning that takes the infant from random inability to virtuoso dexterity in sport, art, music, dance, or surgical skill.

"Handedness"

In this country, about ten percent of the population is left-handed, a human condition that has attracted attention since antiquity. Just what is handedness? The Great Scientific Error considers it as a trait wired into the brain genetically, right-handers have one kind of brain, left-handers another. Notwithstanding that conviction, the efforts of geneticists have produced only a tentative finding of a gene that seems to be correlated with handedness in men, but only slightly so, and the gene is related to other things. It takes a lot of faith to consider that iffy correlation as evidence.

To be sure, there are detectible differences in the brains of right-handed and left-handed people. But why is that taken as evidence that handedness derives from the brain characteristic; the brain characteristic could derive from the handedness. The brain differences in handedness could be the result of learning rather than genetics. *That concept of learning causing brain features is of such central importance that it will recur here again and again.* Instead of that kind of concern, the traditional belief leads to circumstantial evidence to prove its case, namely that schizophrenics, autistic children, homosexuals, musicians, architects, and MIT professors are somewhat more likely to be left-handed than is usual (Goldberg 2008, A98). The implicit reasoning involves belief that traits, like schizophrenia and left-handedness, are caused by genes. The correlation of the two somehow is taken as evidence that genes cause both. But do genes produce those behavioral traits, or any others? There is a house-of-cards reasoning here—with only correlations, assumptions, and other unproved lines of causation—demonstrating again how eagerness in believing promotes poor scientific thinking. After all, if many groups are considered in terms of handedness, just by chance a few of those groups will show up as having more left-handed members. That can be a statistical artifact.

I propose another explanation entirely, with hearty evidence although lacking the technologic drama of brain imaging or gene searching. We can see the causes of handedness with our naked, low-tech eyes, evidence that hand skills, regardless of side, are learned. As an example, in becoming expert in playing the piano, the left hand learns certain skills and the right hand learns quite different skills. Skill is demanded on both sides, which of course means both hands become skilled. The brain supplying the mechanism for the left hand does not lack in learning skill. Observe also: no learning experience, no skill—with both hands. Skill with either hand requires many learning trials. A cute experiment would be to have the pianist attempt to play the right hand part with the left, and vice versa, on a piano that was also constructed in the reverse. I can tell you: even a skilled pianist would perform poorly, until the skill with each hand had been learned. The same would be true for those who type; switching the keyboard so the left-hand keys would be where the right-hand ones usually are located would take away skill from both hands. Skill in each hand would have to be relearned. The point is that there is no difference in skill level of the two hands. The difference lies in what behaviors the two hands become skilled in. I say it is silly to believe that there is a gene that determines the handedness shown in piano playing or typing, or any of the skill differences that typify handedness.

The child usually learns the common skills with the right hand because most parents are right-handed and put common objects—such as spoons, pencils, crayons, drumsticks, bottles, and glasses of milk—into the child's right hand.

This is done unthinkingly, without recording, so there is no evidence of the learning that produces handedness. How can I be so sure? My certainty rests on my findings in creating handedness in my children. I considered it to be advantageous to be right-handed—our material world is made for right-handed people. So I *arranged* our children's learning experience. When the children were *first* learning to use a spoon, I would put it in their right hands. If the spoon was dropped and picked up in the left hand, I would change it. I did the same with other objects, such as crayons in drawing and balls in throwing. The child's right hand as a result became skilled in doing the common things that right hands do. In time, the children used that hand without prompting, because it was more broadly skilled. That meant that once the preference was begun, it created the snowballing of additional learning through the children's own choices.

In taking this route of determining the child's handedness, the parent should begin the training at the start in order to avoid having to change handedness later on. A change a few years later can only produce a parent-child problem, for the child will not want to give up a skilled performance and replace it with an awkward, effortful, performance. But that does not mean, as has been believed, that the child resists because there is a fixed inner nature of the brain involved. It is just that when something has been learned, it is easier to do it that way, rather than changing hands and beginning learning all over again.

Such a good example of the Great Scientific Error: the traditional conception has consumed many scientific resources trying to find the genes involved in handedness. Where are the studies to test whether handedness is learned? It would be so easy to conduct research to end all questions and save additional wasted research. Just get a number of parents, right- and left-handed, who volunteer to follow a simple treatment with their children. Half of the parents would always place spoons, pencils, toys, crayons, knives, hammers, balls, and such in their child's right hand. The other half would do the same, but for the left hand. The children—with the exception of a few due to noncompliant parents—would learn to prefer the hand that had the preferred experience, regardless of the parents' own handedness. I'd bet on it.

Upright Locomotion, A Human Trait

Walking is a behavioral skill that constitutes a foundation for acquiring later skills such as running, dodging, and jumping, as well as for fulfilling the duties of daily life. Parents become concerned when their child's walking is delayed,

because they take it as a sign that the child is not developing normally biologically. There is no evidence to that effect with children without motor disabilities.
Let me suggest, however, that delayed walking among other behavioral delays
are indications that the child's learning opportunities have been lacking. And
delayed walking is undesirable, for that delays experiences that would produce
other types of child development.

Actually, what may seem like a simple motor act, common to all humans,
walking involves a complex of conditions: leg strength (affected by usage), balance (a complex of learnings), as well as the specific learned alternating–leg
movement skills. I began walking training for our daughter by holding my hands
under her arms and around the torso, supporting her weight as she stood facing
me with her feet on the tops of my thighs, as I was seated. Then I let her support
her weight, while holding her so I could resume support if needed. Infants will
support their weight for a time. There need be no concern with overdoing things
because the infant can, and will, stop holding its legs when it wants to. As long
as one has a firm grasp, and is prepared to hold up the infant, all is well. When
this exercise is repeated periodically, over time the infant lengthens standing
time and becomes more skillful, straightening the legs and flattening out the
feet. Reinforcement for the behavior comes from the activity itself and the fact
that it produces a face-to-face confrontation with the parent. The child's enjoy-
ment is easy to see. Through this the infant will progressively learn to stand
straighter. I have a picture (see Staats 1996, 136) of my son standing by himself
at four-and-a-half months while holding onto the slats of a crib, much ahead of
the six-month age expected by child-development norms.

I also describe and present pictures of Jennifer and Peter in different places
in my program of walking training (see Staats 1996, 135–37). With very little
time spent in that training, stretched out over six months, the child will walk
freely at nine months of age, as our children did, although traditional child
norms expect that development only by twelve to fifteen months. I presented my
analysis of walking as a learned behavior in my 1963 book and my walking
training program in my 1971 book, and they were corroborated a year later in a
formal study with a number of children (see Zelaso, Zelaso, and Kolb 1972). It
is clear that the traditional view of walking as caused by biological maturation,
and the traditional developmental-norm figure of twelve to fifteen months, are
in error. Despite this evidence, published in science works, the Great Scientific
Error belief remains that walking emerges through biological maturation as an
inborn trait of humans. Today I personally know parents of a child who isn't

walking even much past the time "normal" children do; they are afraid the child suffers some biological retardation—with no knowledge that walking is learned and the child requires the needed learning experiences, and without any notion of how to train their child.

Athletic Ability

I began training my children in various basic motor skills at an earlier-than-usual age. For example, I composed a little learning program for catching a ball, beginning with a balloon and moving progressively to smaller and faster-moving balls. Throwing was included, and it turned out that "throwing like a girl" has nothing to do with sex but only to do with learning opportunity. Girls can learn a good throwing motion as well as boys—as today's female athletes attest, disproving past belief in feminine lack of such skill. Our daughter and son became varsity athletes in tennis, both at a high level, one a state champion. I suggest that it is no genetic accident that Jimmie Connors, who began learning tennis at the age of four, and Tiger Woods, who started learning golf before he was two, became champions. Each had unusually good, long-term learning programs. By the time a young man becomes a quarterback in the National Football League, usually he will have gone through at least a fifteen-year learning process, as did Peyton and Eli Manning. Other highly refined performances—ranging from brain surgery to violin virtuosity—are usually considered different than athletic skills, but analysis would reveal they too heavily involve the learning of sensory-motor skills. I suggest if one wants a child to become a good athlete—or a good anything involving sensory-motor skill—provide good learning experiences that begin early and continue.

Having a professional level of success will also depend on the body genetically imparted by the parents. Of course there are biologically determined differences in the physical body. And those differences are indeed critical in the different sports. A short, large-boned, stocky individual will never be a champion high-jumper. Someone who is seven feet tall and built like Shaquille O'Neal will meet more success on becoming a center in basketball than a spindly five-and-a-half-footer. Speed and strength will also be in part determined by build and other biological variables. So some bodies are constructed better for certain movements. But all sports demand sensory-motor skills, and those are all learned, and there is no evidence that there is any genetic determinant of the skill itself. Whenever sensory-motor skill is involved, one has to look to learning experiences for an explanation, be it ballet, violin, billiards, golf, surgery, watch-making, or quarterbacking skill.

Toilet Training

Sigmund Freud's theory made toileting "needs" the focus of the child's second "stage" of psychosexual development. How parents responded to those "needs" was said to determine later personality traits of the child. Freudian child-development notions lacked both good evidence and theory, but toiletry does constitute an important sensory-motor skill. Retardation in this development can be a problem for many parents; strung out over several years it can become a detriment to parent-child relationships. Serious general problems for children can also result, for without a toileting skill the child may not be able to enter regular preschool and school programs and may receive a good deal of social punishment within the family and without.

The conventional wisdom in child development holds that toileting ability comes about through biological maturation. My conceptual behavior analysis indicated otherwise, that toilet ability is a sensory-motor skill that is learned through the usual principles. Problems in this area come from improper learning experiences. So in working with my daughter I developed a program for early and easy toilet training. A first step was to establish the child's usual times of bowel movement. The child was then placed on the potty-chair just before that time.

> [N]ot only should the sitting period be short at first, but some reinforcers— games with the parent, play with toys, [storytelling and "reading"] and so on . . . [can] profitably be introduced. . . . If the child sits on the toilet each day at a propitious time, a certain proportion of "successes" should occur. Specific reinforcement of this behavior by the parents would then be expected to bring the operants (such as [sitting and] straining) under the control of the situational cues. The . . . [responses] involving the lower intestines should also through classical conditioning tend to be elicited on future occasions by the stimuli of the situation. (Staats 1963, 378)

This was written more than forty-five years ago, when no one believed that such toilet skills should be taught, and no one put forth how to produce the skills. My behavior analysis was born out later by Azrin and Foxx, who did a specialized work on the behavioral method of toilet training (1974). Now, supported by evidence with various children, behavioral training is conventional wisdom for many (Gimpel and Holland 2003).

Children should generally be trained before any problem arises, but the training can be used also with children for whom toilet skills are retarded or problematic. Negative reinforcement should never be used in toilet training, not even admonition—that will create additional problems. Despite the evidence, and the preventive value of the training, the Great Scientific Error helps ensure that many parents and children still suffer toilet-training problems. The traditional view persists that toilet skills are determined by biological maturation, and parents need only to be accepting and wait until their child's time arrives. Such bad advice—following it makes the problem worse.

The same learning principles apply to bed-wetting. Ordinarily the diaper-wearing child has already had a great deal of learning experience before toilet-training is attempted. That means the child has strongly learned to urinate in diapers and in bed and to be comfortable with the result. The child needs learning to replace this. My training program was very simple. Jennifer would go to the bathroom before going to bed. Several hours later, just before my bed time, I would awaken her and help her walk to the bathroom, where she would urinate, and then I would help her to walk back to bed. Carrying the child will not do—the child has to learn the whole act. The child will learn to wake to the stimulus of a full bladder and to get up and walk to the potty. As is the case with BM training, not having on a wet or dirty diaper is more comfortable, and the increased comfort also constitutes a source of reinforcement for toilet skills.

Time-Out: Benign Discipline Training

Thus far I have considered only behaviors the child should learn. But there are also behaviors the child should not perform and learn. Moreover, I had already expounded the principles in considering abnormal behavior (Staats 1957). Not knowing those principles, and the inevitability of their action, many times the parent conditions the child to undesirable behaviors without realizing what he or she is doing. Simply put, when the parent inadvertently reinforces an undesirable behavior of the child, that behavior will be learned, even though the parent does not intend and does not want that to occur. The parent may even think he or she has shown the child disapproval for the behavior—but a parent's attention, even while lecturing the child about some undesirable behavior, may actually be rewarding the child. For example, a parent who allows his or her child, with impunity, to take a toy away from another child, even though telling the child that that is not the thing to do, is training his or her child to behave that way. Such learning experiences will train the child to act aggressively, without care for others' feelings. Parents do not want the child to do many things but may not realize that by their actions they encourage the learning of those behaviors in a way that is detrimental to both parties.

On the positive side, any behavior will be unlearned if it does not get rewarded. An undesirable behavior of any type can be removed by ensuring it is not reinforced. But, depending on how much learning has already taken place, it will take a varying number of such nonreinforced occasions before the behavior is eliminated. So getting a child not to behave undesirably, like throwing tantrums, may take a while and require considerable effort.

There are behaviors, however, that the parent wants to remove more quickly because they are dangerous or, like hitting or biting another child, are quite objectionable. A punishment, like a smack on the bottom, will have a quicker effect than nonreinforcement. The problem is that such physical punishment has unwanted side effects—punishment detracts from the child's positive bonding, for one thing. Because of the disadvantages, I tried to minimize such measures, only using such physical punishment three times with my daughter and a few more times with my son. Rather, I invented another procedure with my daughter, now widely known and used. It has a stronger and more immediate effect than nonreinforcement. I called it "time-out," and it can be effective even with young children. (Some people think I named my procedure after football's time-out, but actually I used the name of an experimental animal-reinforcement procedure that shared some features.) My time-out is quite simple. There are times when a young child of two, three, or more years of age will persist in performing an undesirable behavior, for example, crying and demanding something inappropriate. At such a point I would say something like, "Unless you stop crying, you will have to be in time-out." If our daughter continued with the behavior, I would pick her up and put her in her crib and say, "When you stop crying, you can come out." Usually, at least at the beginning, the crying will increase and persist for a while, and finally cease, at which time the parent can go in and say something like, "That's better." With a good response, the child can be brought back to where she or he was.

Time-out is a very effective method if used properly, and it has the advantage of detracting little from the child's bonding to the parent. There can be failures, however. Some parents believe they are using time-out, but actually they do things that subvert it, such as placing the child where there are interesting games, or interacting pleasantly with the child during time-out, or taking the child out of time-out before the misbehavior has ended—unlearning the objectionable behavior will not occur in these cases. (That is why having time-out limited to a set period is a mistake.) A child can even learn to do something that brings on time-out, if time-out means getting the parent's attention. Parents who do such things don't realize that what they do is not really time-out. Essentially time-out has to involve displacement from a desirable situation into a less desirable situation.

The physical situation may vary—perhaps a crib or a corner of the room—but the principle is always the same. Several times one of our children cried or created a fuss inappropriately, in a restaurant. I then instituted time-out, picking the child up and going outside to stand glumly, with no positive interaction, until the child quieted. Then I said something to the effect of, "If you are ready to behave better, we can go back in." With an appropriate answer and appropriate behavior, we did that.

After having sufficient time-outs for a particular behavior, the child will learn not to perform the behavior. The child will be "cured." Also, as the child has additional time-out experiences with various behaviors, the child will learn to stop misbehaving simply on being told that a time-out is going to occur, and he or she will immediately say, "I won't cry anymore," or whatever, thereby avoiding the time-out period. As I will indicate, crucial in training discipline is creating words for the child that will act to prevent undesirable behaviors so that there is no call for punishment of any kind, only words. A child raised consistently in this manner will simply become a well-behaved person much earlier than most other children, and without any "beaten-down" characteristics.

Consistency is essential. A parent who can't stand his child crying or being otherwise unhappy, even for a brief period, even though the child has done something that should elicit a mild, short-term unhappiness, can do many things that train the child to behaviors that are undesirable for the child and for others. What that means of course is that the child in the end will experience more unhappiness than a better-trained child. By using time-out, parents can raise children such that little or no corporal punishment is ever used, or is used only a few times during childhood. A child with that life experience will have more long-term happiness. Because time-out is simple and easy to understand and apply, and is very effective, it has been used so widely that it has become a household word (see Straus 2006).

LANGUAGE-COGNITIVE DEVELOPMENT

There is a strong belief that language is an inborn human trait, given via divine creation or via biological mutation in evolution. Many linguists, following the position of Noam Chomsky, hold the latter belief, as do developmental psycholinguists and evolutionary psychologists (Pinker 1994). Chomsky's linguistic theory has it that children have an internal language-acquisition device. Being so wired, children thus need to hear only an abbreviated sample of the language. Then they "break" its grammar, like a cryptologist would break a secret code. Henceforth, voilà, the child "knows" the essence of the language. This conception rejects the idea that language is learned. But where does Chomsky get support for his theory? Why, by never studying children in the actual process of acquiring a language. Rather, he generalizes from the linguists' method of study: they acquire knowledge of a new foreign language by interrogating adult native speakers of the language and by figuring out the grammar of the language. When they have "broken" the grammar, they consider themselves to know the language, to have a deep theory of the language. Such linguists take that as the way children come to know their language, too.

But the conception has no validity; using it, the complex phenomena of children learning language is completely avoided. It's the old "proctology effect" again; interpret new phenomena in terms of what one knows. But, just like a rectal examination does little to help the patient with a cold, the linguistic methodology gives little knowledge of how a child actually acquires language. No evidence exists of a language-acquisition device. There is actually no research showing such a mechanism in the brain that *produces* language in the child. Moreover, the linguistic approach, lacking cause-and-effect knowledge of language acquisition, cannot be of help when there is a problem of language development—as with a child who is over two years old and does not talk, with an autistic child who has no language, or with a child of fourteen years who cannot read. Nevertheless, linguistic belief is widely accepted because it ties in with the Great Scientific Error belief that human behavior is explained by inner mechanisms such as the brain, genes, and the processes of evolution.

If learning does produce language, the process of language development can be known only by studying that learning. I did that with my children and with other child studies.

Speaking Is Complex, as Is Its Learning

Although not well studied, there are large differences in children; some children of four or five months of age vocalize in a modulated way almost not at all, and some vocalize a lot. That difference is important even though those early vocalizations, babbling, are not speech. For, as indicated in chapter 2, the speech response involves a complex set of muscle groups. The child who vocalizes a lot is actually learning how to make the complex vocal response; the quiet child misses that learning experience and so will develop babbling later, and her or his first words will be later and not as well formed.

Although not revealed by linguistic theories of language, the parent's responses to the child's vocalizations plays a salient role here. Like any other sensory-motor response, if vocal responses are rewarded, they will occur more frequently. So if the parent gives attention when the infant vocalizes, these responses will be strengthened and will occur more often. A good way of doing this consists of playing a little game, by imitating the child's vocalizations in a happy way. Giving that attention constitutes a reward. Over time, the child so rewarded will vocalize more—the little game will aid speech development. Screeches, crying, complaining sounds, and other counterproductive vocal responses should be left without reinforcement as much as possible, except when they signal genuine needs. When parents are not responsive to the child's productive vocalizations, that constitutes the absence of reward. We know from basic principles that unrewarded behavior extinguishes, ceases to occur. And that happens; there are wide differences in the extent to which an infant of four months makes vocal responses. I remember a nonvocalizing infant with a very unresponsive mother. The effects of that unresponsiveness were shown in a four-year-old sibling whose lack of language was abnormal. The history of the older child was being replayed with the infant, who was also nonvocal. Children who experience fewer speech-learning experiences, including reward for their own vocal responding, will be delayed in developing speech.

Many children do not begin to say words until they are one and a half or two years of age, or more. I believe that research would show that the time of first words is correlated with parental responsiveness to infant vocalizations. Parental responsiveness would increase if parents more generally understood the important role that their responsiveness plays in the child's language development. Saying the sounds that make up our spoken language requires the child learning a set of complex, highly skilled vocal responses. Acquiring that set takes many learning trials whose occurrence depends on the infant's caretakers.

Loving the Parent's Voice: How the Child Comes to Babble

Babbling is frequently seen as the beginning of language, as a sign that the biological mechanisms responsible for language—the brain and the speech organs—are coming of age. Actually the cause of babbling is not biological maturation. Rather it is learning, especially a type of emotional learning, like the bonding already described.

Effortlessly, even without knowing the effect they have, parents generally condition their children to make vocal sounds that are like the ones they make in speaking, because those vocalizations sound good to the child. Children's babbling has been noted to occur at around six months. Some children do so earlier, some not at all. In explication, let us ask, if the infant, like one of Harry Harlow's monkeys, were raised with no parents, would the infant babble? Would an infant babble if it had parents, but they never uttered any sounds? Absolutely not, in both cases. *To babble the infant must experience two things together, many times: (1) the parent's speech, (2) at a time the infant is experiencing a positive emotional response.* That is, to babble the infant first has to *learn* to love the parent's speaking voice. How can the parent bring that about? If the parents want a child who develops babbling normally, then they want to talk to their child in giving a bottle or a breast, and more generally when playing, cuddling, bathing, or otherwise engaging in happy activities with the child. That experience, occurring as it does so many times, will create love for the parent's voice, that is, condition the child to have a positive emotional response to the voice.

Why does loving the parent's voice lead to the child's babbling? Well, the infant who babbles "Ma-ma-ma-ma-ma-ma-ma," "Da-da-da-da-da-da-da," or "be-be-be-be-be-be" is making vocalizations that sound like those the parents make. Because the infant loves what the parent says, the infant is rewarded when making a sound like the parent makes. Having a positive emotional response to parental vocalizations means those sounds will also be rewards. By making sounds like the parent, the child rewards herself or himself for making those sounds, which means the child learns to make those sounds.

When the infant makes sounds the parent does not make, on the other hand, there is no source of reinforcement. So the vocal responses of that type extinguish. The end result is that the child learns progressively to make the parent-type vocal responses and not the others. A Chinese and an American infant make the same vocal responses at birth and early on. But progressively Chinese infants more and more make sounds like those in their parents' language. So do American children, Russian children, Finnish children, and Hungarian children. Each learns to make sounds like those made by their parents. Why? Because they love those sounds, just like we learn to sing songs we like, because that is rewarding also.

As would be expected, it is not unusual for a young child to lie in the crib and make babbling sounds before falling asleep. The children are entertaining themselves, making parentlike sounds that elicit a positive emotion. The infant does so because of lawful learning principles, but isn't it endearing that the child loves the parent so much that his or her babbling is enjoyed? A loquacious parent can speed up the learning, in essence giving the child many emotional conditioning experiences. Laconic parents provide little learning experience for the child, their voices will have less emotional value for the child, and there will be little reward by which the child can learn to babble. That will slow down the child's language development and can be one of the things that results in the lack of language learning in an autistic or mentally challenged child.

Our children both babbled early and said their first words early, in part because many times I would purposely say things, preferably single words, when they were experiencing a positive emotion. For example, when I was feeding Jennifer or Peter, I would say "mmmh," "good," "eat," "more," "like," naming foods and such. Actually, I vocalized whenever we were doing things that were fun, and that involved many things. Such experience is why infants begin to make sounds like those of their parents. This could easily be precisely specified by research. Detailing the process and presenting it to parents would be important because it is basic to the child's language development, and that is basic to much other learning. The linguistic conception of a built-in language device, on the other hand, does not explain babbling, why infants make sounds like those of their parents—nor does any biological theory of language development. The Great Scientific Error conceptions in this area do not offer parents the ability to affect the babbling of their children, and thus their children's speech development.

Babbling: A Contributor to Further Speech Learning

Babbling is important in language development because by babbling, the child gets practice in vocalizing. Babbling is important also because it provides parents the opportunity to reinforce the child's vocalizations and advance the child's language learning. As soon as the child babbles "ma-ma-ma-ma-ma" or "da-da-da-da-da," the parent can enthusiastically give the child attention and repeat the sounds several times, shortening each utterance to form the words "Ma-ma" or "Da-da." With their attention and happy response, the parents become a direct source of reward for the child's babbling.

After the parent and child have gone through a number of such vocal interactions, the parent can try to initiate their little game, by saying "Da-da," or whatever, rather than waiting for the child to babble the sound. That can initiate the game—as soon as the child says his "babble," the parent can repeat what the child has said with happy words and face. The child's babbling will be strengthened, providing more good language-learning experience for him or her. This may be called the verbal imitation game, because in doing that, the child is learning an imitation response—the parent says "Da-da" and the child says something like it.

Why Can't You Say as I Say? Learning Verbal Imitation

Attention and approval of the child's vocal responding remains an important part of the process. Playing the verbal imitation game with the child begins the learning of an important language repertoire. That is, the child must learn to say all the phonemes in the parents' language, the 52 sounds in the English language. When on hearing a new word, if the child can say all of the sounds in it, the child can sound it out and produce the word. Acquiring an imitation repertoire of course takes many learning experiences. In sum, verbal-imitation skill makes up part of what it takes for the child to learn to pronounce the thousands of words of which language is composed.

First Words, and Rewarding and Punishing Speech

Another parental role that induces the child's language development consists of the parent naming the things the child experiences. When the child is fed milk, bread, juice, water, and so on, the name of each should be spoken. The child younger than one year, for example, may be asked, "Want milk? Milk?" When the child is entranced by seeing a dog, another child, a doll, a ball, whatever, the object should be named, as is the case when the child is experiencing something happily, with relevant words like "play," "fun," and "eating." When the parent comes into the presence of the child after an absence, the parent should say her or his name—"Ma-ma" or "Da-da." Since such things occur frequently, many pairings of objects and activities with their names will occur. The loquacious parent who does that will provide many learning experiences to his or her child. Speech becomes an important experience for the child. Many parents believe that the behavior of the child indicates the child's biological maturity. If the child doesn't vocalize, that means the child isn't ready yet, so the parent doesn't say anything. That belief of course robs the child of the learning opportunities needed for language development.

Just as with babbling, the parents saying words will make it rewarding for the child to say them too. When learning has progressed sufficiently, the child will begin to say individual words. At that time, the parent can be, and ordinarily is, excited by the child saying her or his first word. That response will reinforce the child for that particular word as well as for vocalizing more generally. I call words that name life stimuli "verbal-labeling responses." There are thousands of them for the child to learn.

There are two components of such learning. First, there is the complex response involved in saying the word. That must be learned. Also, there is making the response to the correct stimulus, which also must be learned. The child must learn to say "Mommy" when Mommy is there, and "Daddy" when Daddy is there. Calling the mailman "Daddy" will not get the appreciative enthusiasm as that name pronounced when Daddy is there. If the child says "milk" when she wants a cookie, what she is offered will not be a reinforcer. Natural conditions ordinarily provide different levels of reward for correct and incorrect speech responses, a difference that promotes learning correct verbal-labeling responses. The child should not experience admonition of any kind when making an error, for that would have the effect of generally punishing verbalizing. Just providing the correct word in a helping way, or a better pronunciation, as part of a good interaction, is good. Whether or not a child stutters will depend upon what is experienced in speaking.

As the child's verbal-labeling advances, the child will learn the value of words. Having necessary words will mean being able to get things, and not having the words means missing out, or delay and difficulty. Ease in getting what is wanted elicits a positive emotion. That experience makes the act of learning new words itself positive emotionally and thus rewarding. When that has been learned, the child will begin asking for the names of different things and attending to the names of things said by others. Words become "good" for the child, and a new avenue of learning opens.

Why Can't You Do as I Say? Verbal-Motor Learning

An aspect of language development concerns what I call "verbal-motor learning." Simply put, the child learns to make particular motor responses to particular words said by someone. Having learned a repertoire of such "verbal-motor" words—such as verbs—gives the child the ability to follow directions.

To begin with, the parents' words have no special power to affect the child's behavior—they are irrelevant stimuli, like a siren, the sound of a car, or a television. Words become significant stimuli through learning. I began presenting such learning experiences very early; for example, whenever I heard my daughter, Jenny, awakening from sleeping in her crib, I'd station myself out of sight and play a little game. I would say "Da-da . . . Da-da." Then I would swoop into sight and have a loving, playful interaction with her. At first my saying the word had no effect. By the time she was four or so months old, however, when I would first say "Da-da," she would stop what she was doing and listen. If she was cooing, or playing with a teddy in her crib, she would stop and cock her head. She would also look around, attempting to see where I was or where I would come from. And when I would swoop in, she would be all smiles, waiting to be picked up and hugged and noisily kissed on the neck or danced with in ways that would elicit additional peals of laughter.

In the colder language of analysis, what I have described constitutes the learning of a verbal-motor connection. Through the learning experience, the word "Da-da" came to elicit a motor response—hushing and looking around. She was learning a unit in the verbal-motor repertoire. Soon after that time, I began saying such things to Jenny as "Come to Da-da," as I held out my ring of keys to her. Lying on her stomach, she would begin a creeping, wiggling motion to reach the very proximate keys; they constituted a lovely toy. She could move them about, make them tinkle, and also mouth them lustily. They were a reinforcer. So when I said "Come to Da-da," she was reinforced when she wriggled toward me to get the keys. I taught her to respond to my saying "look" by getting her to look where I pointed and having something interesting for her to look at (like a dog, a spider, an ant, or the moon).

Importantly, she was not only learning to respond to words, she was learning that words are important stimuli in life. Words are connected to things of importance. Words are to be attended to. In language development, words have to gain that importance for them to gain any attention from the child.

A later type of training can involve saying something like "Give me the salt," while motioning toward the saltshaker, or "Give me the ball," with appropriate gestures. There are many opportunities for training the child to verbal-motor units, and parents generally teach their children to follow instructions even without realizing they are doing so.

Children generally learn many verbal-motor units; most of the verbs in the language, for example, must come to elicit a particular type of response, like *walk, sit, give, say, look,* and on and on (Staats 1971a). Later, adverbs and adjectives will be added so that the individual can follow complex instructions in a precise manner. "Please press the red button repeatedly" is an example where the verb, the noun, the adjective, and the adverb constitute a more complex language signal that calls out a complex response. Of course, one does not want to jump ahead of what the child has learned and give instructions using words the child has not learned. That will have a negative effect. It is important that the child early on receive reinforcement for following verbal instructions. If following instructions does not produce rewards, or is punished in some way, the child will cease complying. It is easy to see the differences children have in this respect. Some children do not attend to or follow their parents' instructions well at all, while other children do. We can know that the two types of children have different histories of learning that have produced those differences in behavior. In view of the place that the verbal-motor repertoire has played in the present approach to language, it is interesting to note that whether or not the child follows the parent's instruction to look at something is one of the ways of diagnosing autistic children, suggesting that a deficit in learning experiences is part of that childhood disorder (Staats 1968a).

Language Is Big, Complex

Linguists have made an analysis of the grammars of languages, including English. Their knowledge is recognizably much deeper than common knowledge. Take for example the way we add a sibilant to indicate plurality, *books* instead of *book* or *bugs* instead of *bug*. Both sibilants indicate plurality, but in one case the sibilant is the *s* sound and in the other case it is the *z* sound. We don't ordinarily care about such a distinction. But linguists do. A grammatical rule is involved, for the *s* sound is added to the end of a noun whenever the word ends in an unvoiced consonant, like *k*, *p*, *t*, and *f.* The *z* sound is added when the word ends in voiced consonant such as *d*, *g*, *m*, and *n*. Linguists have called this a mental rule, suggesting that it comes from the mind-brain. Actually, the way we pluralize consists of learned behaviors. The child learns the "mental rule," actually word associations, from the innumerable times she or he has heard the usage. I indicated just how this learning occurs (Staats 1963, 177–79), and Sailor (1971) provided supporting evidence with children.

Language has to be known in its full sense; in what it consists of as behavior, in the functions it has for the individual and the group, and in how it is learned. Reading and writing are part of language, as are number concepts, counting, mathematics, logic, and all types of knowledge. In illustration, when Jennifer was eighteen months old, I held one raisin in one hand and two raisins in the other and asked her if she wanted one raisin, showing her that hand, and then two raisins. If she said "one," I started to give her that, and when she protested I got her to say "two" and then gave her the two raisins. With such training she learned to say the numbers "one," "two," "three," and "four" correctly, with various objects.

Later, when she could unerringly do this, I introduced counting (see Staats 1968a). First she learned to count pennies placed in a row by placing her finger on each penny as she said its number. I started with two pennies, or raisins or peanuts or fingers, saying after she counted them, "Yes, you have two raisins," or whatever. After she could count a few pennies in a row, the pennies were placed in a random heap, and she then learned to pull one at a time out of the pile, saying the appropriate number each time. At the age of two years and three months, I have her on tape counting such a pile of thirteen objects. From there I went to simple addition and then to subtraction operations. Clearly the child was progressively learning language repertoires, the fact that they lead to the specialized area called mathematics does not change that, although people generally think that language and mathematics involve different mental traits.

All the knowledge areas—from archeology to zoology—are part of language. Language is a primal feature of humanness, involved in most every performance, thought, achievement, and creation, and it is essential to every aspect of culture. Language is central in advertising, propaganda, teaching, communicating—every type of human interaction. Language is a very large part of humanness; it shouldn't be broken into different scholarly fiefdoms.

Isn't Reading Part of Language?

The framework applies fundamentally to reading as a part of language. Reading ability should be understood as a continuation of language learning, dependent on the learning that has preceded its acquisition, and in turn a central determinant of much learning that follows throughout the child's life.

My view of reading was like my view of language acquisition. A very complex repertoire is involved that demands many learning trials to acquire all the responses involved. A program for learning reading that is not appropriate for the child's level of preparation will lead to a breakdown in learning, to disastrous consequences for the child and for the society. That breakdown occurs far too frequently.

With that view, I made the decision to begin teaching our children reading when they were toddlers. One of the traditional assumptions of developmental and educational psychology at that time was that four-year-olds are too biologically immature to be introduced to reading training. Not content with such assumptions, I needed to see, with explicit evidence, how young children attend, work, and learn reading skills, as a function of the learning conditions they experience. I was sure that children who have already learned language widely and well were ready to learn to read. In a first study, over fifty years ago, my team engaged six four-year-olds in a stipulated reading-learning task—words were first learned then were read in short sentences. The children's behavior in learning that material was compared during periods when they got reinforcement for their attention to task, and periods when they did not get reinforcement. I had devised a system of reinforcement based on tokens the child could exchange for toys and such, a variation of the token-reinforcement system I had developed in 1958 for work with dyslexics that I will describe later. *When the children were reinforced, they participated voluntarily and eagerly, attended well, and learned well in forty-minute training sessions.* Children that age were then thought to have an attention span of five minutes. Four-year-olds were thought unable to do just what these children did, because of a presumed biological immaturity. The findings blew away the fog of conventional knowledge.

When reinforcement was not given, on the other hand, the children's attention flagged, they began to fool around, to put off looking at the words and saying them. So they had few learning trials, and they stopped learning and soon no longer wanted to participate. They became the typical four-year-olds of conventional belief. When reinforcement was reinstated, these typical four-year-olds dramatically became super four-year-olds again, becoming attentive and hardworking, and learning well (Staats, Staats, and Crawford 1962). Surprise of surprises, the children's performance didn't depend on age, it depended on their attention to and participation in the task. And that depended on the conditions of reward. What does that tell us? Attention and participation are central to learning complex academic behaviors, and that learning operates according to the basic animal-conditioning principles.

Moreover, preschoolers have the wherewithal to learn skills like reading readily and easily. They just need to be rewarded for their work. Why would we expect otherwise; after all, do we ourselves work for nothing? When we think about it, educators' expectation that young kids should learn for its own reward when adults do not is very strange. Specifically with respect to school learning, our findings actually showed that inadequate reinforcement produces lack of attention, a formula that yields nonreading children. Effective reinforcement is a formula for producing a reading child. We supported the findings in a later study of longer duration (six weeks and thirty training sessions) (Staats et al. 1964). These were the first studies dissecting and exposing facts of reading learning, or any complex learning that is a part of child development. They were also the first studies showing that early preschool training can be very valuable for the child. These children, and those in later studies, were from economically deprived families, and the findings had important implications for helping that group, which was later targeted by Head Start preschool programs, a supportive development.

These studies yielded new types of knowledge because they dealt with real behaviors, complex repertoires, and the way they are learned. This was cause-and-effect research that told how important behaviors are learned, how to train such behaviors in children, and how to study child development and solve problems of child development. Although the studies dealt with learning in great detail, for longer periods of time than had been studied, they still did not deal with the whole span of reading learning. The preschoolers in these studies could not be said to have learned a reading repertoire. Beginning in 1962, however, I also began an extended research project with each of our children, first with Jennifer, learning to read the letters of the alphabet. Cards with a common picture on each, of such things as a doll, an orange, a dog, a glass of milk, a truck, or scissors. A card was presented in the window of a plywood apparatus, and she was asked to name what it was, receiving a token for answering. After a few trials, she was well accustomed to the procedure, and when a card was placed in the window with an *A* on it, she was told "That's the letter *A*; can you say *A*?"

After responding correctly, she then received the token reward. After that, there were more picture cards and the *A* appeared again. After a few more trials, she would say "*A*" on her own when it appeared. At that point, instead of an *A*, the letter *B* was to be presented. When the second letter is present, a child will say "*A*" because any letter is like another in comparison to a picture. Rather than allowing the error, however, I said before presenting the second letter, "The letter I am going to present now is a *B*; can you say *B*?" When she said "*B*," I said, "Here it is; can you say its name?" Additional trials with the *B* and pictures followed until she said the letter name consistently. Then both letters were mixed in with the pictures; the important thing at this point is to make sure that the child not make errors, which means giving prompting until it becomes apparent that the child knows the names of the letters. So the first time the *A* was presented again, she was told what it was beforehand.

When the child is good with two letters, a third can be introduced. As the child's letter-reading repertoire grows, she or he will become more expert in scrutinizing the letters, detecting their differences and learning their names. So the learning will progress increasingly rapidly (see Staats 1968a). Importantly, the person giving the training must be sensitive to what the child has learned and give prompting when needed so that errors do not occur. That is essential.

After Jennifer had learned about fifteen letters, I began introducing single words on cards to be learned. When she had learned a small number of words, I began making sentences out of them that she read, and later on, short stories like:

JENNIE DRIVES THE CAR.
JENNIE DRIVES TO MOMMY.
JENNIE DRIVES TO THE STORE.

Was this valuable for Jennifer? No question. Not only did she learn specifics, she learned to attend and to work and to feel positive emotionally about learning situations. She played it because she liked it; we called it the reading game. The tokens (I used pennies) were cashed in after playing the reading game for various common rewarding things—"When you get ten pennies, we can play with the doggie" (a play dog), or some such reward. Mostly the pennies earned dessert afterward. Learning new things indeed became rewarding. She was always successful and learned interesting things. When Jennifer was in the second grade of school, her reading skill was at the ninety-ninth percentile. An important by-product of the training was gaining confidence about learning situations.

I also gained confidence. The knowledge produced in this research, along with that discovered with the other children, propelled me further along my odyssey into the land of human learning. In 1965, I developed a plan for a preschool class in a public school in Madison, Wisconsin, that Karl Minke and I put into practice. Adjoining the preschool classroom was a room containing three learning stations where the children were taught basic school knowledge using my materials and methods. Each child left the class three times during the day for learning sessions that lasted five to ten minutes. Again, their participation was voluntary and their attention and participation were maintained by the token-reinforcement system. The tutors were graduate students without special knowledge of dealing with children. The teacher of the class, as well as the tutors, received behavioral training and supervision from Karl. Over the eight-month period, the four-year-old children had many learning trials and learned a great deal. They learned to read the letters of the alphabet—a skill shown to best predict later school success—as well as to write them. They also learned to read and write numbers and to count. Their school-readiness test scores went from the second to the twenty-eighth percentile (in comparison to five-year-olds) and their IQ scores advanced an average of eleven points (see Staats, Minke, and Butts 1970). There was variability, but the training benefited all the children.

It was clear by the cognitive tests that the training had advanced the children. They were mostly African Americans from poor homes with older siblings who had problems learning in school. The results had implications regarding racism; they indicated that if the learning conditions ensured good attention and participation, and the learning materials are suitably designed, African American children learn well. That suggests that the lag in educational achievement of such children is due to inadequate learning conditions, not any lack of "native" ability. (Problems can also arise of course because the child's earlier learning experiences have left gaps in his or her background and he or she needs additional remedial training—this applied especially to one child.)

Importantly, this was cause-and-effect research, unlike the traditional child development study that only observes the age at which particular behaviors appear and then assumes the cause is biological maturation. The conditions involved and the behaviors they produced were recorded specifically for each child. That exposed the nature of the learning in ways previously unknown. For instance, traditionally a feature of the dyslexic child is writing letters sideways, backward, and upside down. Traditional beliefs have taken such errors to indicate dyslexic children have a brain-induced perceptual disability. On the contrary, our study showed that children generally display this lack of perceptual ability at the beginning. As learning progresses, those types of errors drop out. It is not a "brain disability." Children simply perform that way when they are at the beginning of the learning. The extent of learning determines perceptual ability; not the brain's genetically wired-in ability. These points have been definitively shown (see Collette-Harris and Minke 1978).

The conclusion: traditional child-development and educational research, derived from the Great Scientific Error notions, give limited and erroneous knowledge of the child and of child development relative to that which can be gained from studying the child's learning directly. That knowledge of the original learning of children can be used for producing useful methods for advancing children's learning that can be further researched and developed. I have described this extended research on reading not only for its own importance but also because it indicates the type of study that needs to be conducted for all language development, actually for all human behavior that begins to be learned in childhood. Gaining knowledge of the repertoires that children need to acquire, and the learning that produces those repertoires, requires explicit study. Understanding that this is a learning process—not a process determined by genes, the brain, maturation, or affected by race—constitutes a most important step. That understanding is slowly creeping into our culture.

> [O]ne more question to consider is environment. Is anyone talking to this baby? Is something getting in the way—maybe an exceptionally chaotic household, or maybe a severely depressed parent? Speech and language development requires stimulation. Pediatricians have been faulted in the past for dragging our feet in making speech-delay diagnoses, but times have changed. (Klass 2010, D5)

But a change in interest does not mean that what has been found is being used or that what remains to be known is being sought. Huge research resources remain to be directed at language, its learning, and its central role in diverse human characteristics. That development requires use of the framework conception that begins the analysis of language in behavioral terms, that indicates its complexity and the long-term learning that produces language. Recognizing the hugeness of the learning task is essential to explaining human behavior and the nature of the human being (see Staats 1975).

THE CHILD AND THE GREAT SCIENTIFIC ERROR

A conception emerges from the developments that have just been described. It is different from the one that is generally held as part of the Great Scientific Error. Parents raise their children according to their conception of children. The conception held of the child is also basic in designing a system of education and other public policies addressed to children. So it is important to characterize that conception.

View from The Great Scientific Error

The huge gap in understanding the child's learning has left an opening for the belief—among scientists, including behaviorists, as well as laypeople—that much of the child's behavioral development is biologically determined. Common belief is that at birth the child has his or her genetically determined personality, traits, talents, emotions, and abilities wired into the brain, but in inchoate, immature form. The full appearance of those characteristics awaits the biological growth and maturation of the neural wiring that brings them forth to full adulthood. We can see an example of the belief in the words of Maryanne Wolf, head of Tufts University's Center for Reading and Language Research. She describes how recent imaging research shows that children's brains are not ready to read until around the age of five at the earliest. "To hasten that [reading] process not only makes no sense socially or emotionally, it makes no sense physiologically" (Paul 2007, 91).

That belief, strongly held, originally was based solely on the fact that children ordinarily don't learn to read before that age. Now brain-imagery research is used to back up the belief. For example, Jack M. Fletcher and his group have shown that as dyslexic children are trained to read, their imaged brains become like those of normal readers (see Simos et al. 2002). Fletcher concludes this shows that a problem brain causes dyslexia and that cure of the brain problem turns the child into a reader. Brain-imaging techniques have become so popular because they are thought to show how the brain determines how child development progresses, thus confirming what is already believed.

Sally and Bennett Shaywitz [a pediatrician and a neurologist at the Yale Medical School] found that [a particular] neural region remains inactive as [poor readers] grow up. Preliminary evidence from other researchers indicates that this [neural] structure, located near the back of the brain, fosters immediate recognition of familiar written words and is thus crucial for fluent reading. (Bower 2004, 291)

The belief is that lack of brain development causes dyslexia and treatment of the brain removes the problem: "research is beginning to reveal dyslexia's neurobiological basis. . . . Some psychologists even believe that [brain] imaging can one day help people shift the way they use their brains to boost their learning performance" (Murray 2000, 24).

These statements contain the same error, deep and extensive. The error, however, is blatantly apparent when one is not hypnotized by the belief that the gene-brain-mind determines human behavior. What is not being accepted, even though it has been asserted and proved for decades, is that learning changes the brain. Take the study of reading. Fletcher and the Shaywitzes didn't treat the brain! Rather, what they did do is train the children to read! *That is what changed the brain.* So wasn't the conclusion of the study a sort of an interpretive legerdemain, brought on by the Great Scientific Error's ignorance of learning? In the studies, wasn't the learning experience that was given to the children, not the brain, *the cause* of the reading improvement? Prior to the training the children were poor readers. These researchers didn't reach in and flip some switch in the brain, feed it some brain fertilizer, or manipulate the brain directly in some such way. Rather, they provided the necessary learning experiences. Those using brain imaging in this way are confusing *cause* with *mechanism.* The brain actually provides the mechanism by which the learning can take place, but *learning* is the cause.

Let's get this straight. The nonreaders' brains show "defective" neuronal development because they have not had the necessary learning experiences. Their brains are not defective. Give them the necessary learning experiences and their brains become like those of "normal" children, normal because they finally have had successful reading learning experience. Knowledge of the brain differences between readers and nonreaders does have value, but not for indicating the biological cause of dyslexia. Rather, the findings show that *learning changes both brain biology and behavioral development.*

It is central to understand that the reading program used in the two studies of reading was constructed without any knowledge of the brain or brain imaging. Furthermore, knowledge of how to cure a problem of behavior development does not emerge from finding out how the brains of people with the problem differ from the brains of people without the problem. Finding out how to change a nonreading brain to a reading brain does not come from brain research. Giving that impression is misleading.

Starkly put, this brain-centered, brain-imaging view is completely wrong. *It is learning that should be studied, with research to analyze the behaviors that compose reading, how those behaviors are learned, how learning conditions fail and result in dyslexia, and the ways teaching programs can remediate that learning problem. Learning is the cause, and moreover it is a cause that can be dealt with to affect both prevention and treatment.* Misunderstanding this constitutes an erroneous and costly excursion.

Well, it might be asked, what about a drug treatment like Ritalin® that produces improvement in attention deficit hyperactivity disorder (ADHD)? Doesn't this show the effectiveness of treatments that directly act on the brain? Not really, although this is generally assumed. Ritalin does not directly give the child reading or number skills or knowledge of any kind. It merely makes it more possible to *learn* such skills by calming down the hyperactive child, allowing the child to attend better in school, thus making it possible for learning to occur. No question that drugs can have a calming emotional effect, reduce anxiety, produce euphoria, and enable more effective behavior in learning situations. Drug effects on emotion have been evident since the invention of alcoholic beverages thousands of years ago. But there is no drug or other direct brain treatment that can produce walking, handedness, babbling, toilet training, speech, cooperative play skills, following instructions, writing, number concepts and counting, reading, or any of the other many behaviors the child needs to learn. What those behaviors are or how to produce them does not emerge from studies of the brain or drugs. Whether it will ever be possible to create a complex ability like reading through manipulating the brain—using drugs, brain stimulation, or genetic change—remains very questionable. Let me throw down the gauntlet: I think such behavioral developments *will never* be brought about by knowledge of the brain and by working directly on the brain. It will always be easier as well as possible to prevent and treat problems of behavior by learning procedures.

To solve problems of learning requires establishing knowledge of learning. It is essential to recognize *the secondary role of the brain and the primary role of learning.* A lot hangs on that recognition, such as how to allocate the funding for research devoted to the study and cure of dyslexia and the other problems of behavior that take place in childhood. *Brain changing does not produce learning; learning produces brain changes.*

Looking Forward in Biology

I cannot here confront all the errant findings in the study of children that derive from the Great Scientific Error conception. Generally, however, the child is not a miniature adult with genetically given behavioral characteristics that maturation develops. The child does not go through biologically set stages of development. Walking, talking, and toilet skills do not emerge when the child is biologically "ready." There is no brain-induced "stage" of the "terrible twos" that causes children to be obstinate, misbehave, and have temper tantrums. Biological "readiness" of some unspecified kind does not make it possible for the child to learn. There is no natural (biological) bonding connection between parent and child. The emotional quality of each for the other can range from murderous hate to the more usual deep love, but any biological causes for such differences have yet to be shown. The child does not have an inner nature that needs to be allowed to develop in its own way. The baby is not born as a little human being who is only waiting out the time necessary for biological maturation to occur.

Although there is great error in belief in the biological determination of human behavior, an increasing number of the leaders in biological science are coming around to recognize the importance of learning, as exemplified by the statement of Nobel Prize–winner Eric R. Kandel: "[B]ut superimposed upon [genetic capabilities] is a huge capability to modify predispositions through learning. This capability allows you to change the very architecture of your brain. Since learning leads to anatomical changes, every single brain is different from every other brain—by virtue of learning" (Kandel 2007, 8). His acceptance of the importance of learning constitutes an influential voice in the powerful biological sciences. He does not, of course, provide the needed theory of human behavior or of how human learning takes place. To be fair, however, it is not the responsibility of biological scientists to provide the knowledge of how learning produces behavior, emotion, language, and the abilities composed of them. Learning is not their field. They can recognize that learning affects the brain and has some central role. They are advanced in doing so. But a conception that deals fully with learning is needed.

SO, WHO IS THE CHILD?

The human adult, as will be elaborated greatly, is the marvelous learning animal, endowed with learning features no other animal has. Humans learn a truly fantastic amount. The infant is the young of that marvelous learning animal and thus has inherited the ability to sense the broad world finely and the ability to respond with great dexterity and complexity. Furthermore, the infant has the brain by which to connect the two. Human infants have the capacity to learn complex sensory-motor skills that range from ballet dancing to soccer to needle-work to singing in a choir to killing people in military action to constructing buildings with soaring beauty. Infants also have the capacity for learning emo-tions, from religious adoration and love of literature, fine art, and music, to racial hate, sadism, and other loathsome feelings. And the infant has the capacity to learn a full language, to reason like a human, to solve mathematical problems, to create new conceptions, to choose paths to follow, to communicate and receive communications, and to construct knowledge ad infinitum.

Aren't Newborns Helpless?

In the present conception of children's nature, learning ability demands star billing. A major reason that billing is not given derives from mistaken beliefs about the infant's nature. The infant does come forth physically quite immature, really helpless, without physical structure for performing motor skills and without having acquired those skills. This motor helplessness, however, lends itself to the misinterpretation that the infant *generally* lacks ability. Actually, the infant is far from helpless in a most critical way. For the child is born with an apparatus, the brain, of extraordinary size and quality. The newborn has over ten billion neurons in the brain. Those brain cells are capable of acquiring 100 to 1,000 trillion intercellular connections through learning. Evolution would not put so many neurons into the child, with that potential for connecting, for no reason. Consider the cost of all the changes in mother and child necessary for that newborn's large head to pass through the birth canal, including the risks to the survival of each. Those hazards of the large head/brain, and the cost of main-taining such an energy-gobbling instrument, might suggest to us that there is an evolutionary payoff for the infant to be born ready to learn, ready to make those trillions of connections.

Helpless? Yes! Unwitting? No.

I began introducing learning experiences within a few weeks after each of our children was born. And I found that they learned. Only now are studies beginning to confirm that stance. For example, in the machine-learning labora-tory at the University of California at San Diego, Javier Movellan, Ian Fassel, and Nick Butkos have constructed a camera-computer robot. As in artificial-intelligence research, the computer is taken as the model of a human brain. In their study they found their computer could learn to recognize faces in just six minutes. That surprised them and everyone holding the traditional conception that helpless little infants are not prepared to *learn* face recognition. Proof of that, by those who hold this belief, is that newborns can distinguish a parent within a short period of time after birth. To stay within their traditional belief framework they thus conclude that the newborn has wired-in neural connections already in place for face recognition, at least mother recognition. That belief was negated by the finding that robots had no wired-in computer connections for face recognition. Nevertheless, the lowly robot could *learn* to recognize faces. "'If a baby robot with the equivalent of only one million neurons in its programming can recognize human faces so quickly, imagine what a human baby with its ten billion neurons can do,' Mollevan says" (Dingfelder 2007, 42). Those neurons and their resulting learning ability, in the present view, explain how the baby can recognize its mother's face so soon after being born, through learning.

The baby has ten billion neurons. They are healthy and functioning neurons. They are not already taken, learning-wise, haven't been "used up" in any way. Why shouldn't the infant be capable of learning? The robot study dramatically refutes a biological maturation explanation of early face recognition, a behavioral phenomenon whose explanation was assumed as wired in.

Important in all of this is the present contention that right from the beginning, human infants are ready to learn. The newborn infant can begin learning to love the parents, to bond, if the learning conditions provided are appropriate. The newborn can begin sensory-motor learning, for example, how to move its eyes and soon its hands. Language learning can begin at birth, and usually does, when the parent talks as the child is fed, making parent's voice sounds emotionally positive. Such conditioning is essential, and of course the infant has to be a great learner right from the start for that learning to occur. *That makes it important to study infant learning ability, how the infant is ready for sensory-motor, language, and emotional learning right from the beginning.*

The traditional conception of the infant—as just a passive, helpless little creature that needs only love and physical care—is quite wrong. What needs to be recognized is that the infant's initial lack of physical abilities does not indicate the state of the infant's learning abilities. At the same time the infant is helpless motorically, having only a few wired-in reflexes, she or he has a great learning apparatus already in place. The infant is ready to learn at birth, and the infant will begin learning. Whatever the parents do, the child will learn. If the parents pair positive emotional happenings with themselves and their words, the infant will learn a positive emotion to the parents and to their words. The children of parents who do not present such experiences will not learn that emotional bonding. An infant who is picked up whenever she cries will learn to cry, just like a rat that is fed after pressing a bar will learn to press the bar. An infant who is put down to sleep and then cries and is picked up and soothed will learn to cry when put down to sleep. An infant not long after being fed, who cries and is then fed again, will learn to cry when her stomach empties just a little bit. How many cranky, colicky babies have been produced through learning of that type, having nothing to do with stomach problems? How many babies with sleep problems? How many children develop language and sensory-motor skills late, lacking the necessary learning experiences? How many children do not demonstrate the usual love for and attention to their parents? In all such cases there are loving parents who would do all they could to ensure the child's normal development, if they knew what to do.

I do not believe that the child can be understood on the basis of the commonsense views that exist in our common language. Nor is the child to be understood via biological knowledge, including brain-imagery observations that are related to behavioral characteristics. Attempting to establish the child's inborn nature as a function of age cannot provide explanation. Such sources of "proof" have been and are the basis for theories of the nature of the child and of child development that only give a false sense of explanation. Without dealing with learning, such theories cannot produce the needed knowledge. For the human baby is a marvelous learning animal, that is its nature. A change in conception is called for, and that means a change in research. Modeling child learning concerning face recognition was productive, but why not work with infants and directly with learning? Why avoid studying the child's learning, as involved in the acquisition of the important behaviors of childhood? Isn't it strange that research of that kind hardly is done? The present studies and their conception call for making learning central in establishing knowledge of the child's development of behavior and nature. Changing to the new conception thus brings with it a call for new types of research of real behavior development of children as a function of learning. For human children are the young of the marvelous learning animal.

THE PARENT

My research with children produced a number of widely used findings, such as time-out and the use of token reinforcement methods. Some of my procedures are fundamental, founding developments in behavior analysis and behavior-therapy treatments as well as leading to some change of conception and practice in society. Those behavioral fields continue to establish many other important developments. Some procedures are so straightforward they can be learned and applied by parents easily. But sometimes, also, that is not the case. Parents may need continued supervision and counseling. That occurs because the parents carry their own conception of the child and child development, ordinarily derived from the Great Scientific Error. The two—the behavioral methods and the parents' conception—may be in conflict. Having the traditional conception may interfere with properly using procedures derived from the learning view. Although such parents may attempt to use time-out, for example, they attempt to do so within the dictates of their own conception. They may carry out the physical part in moving the child into his or her bedroom, thinking superficially they have done what is needed, but they subvert time-out's learning aims because they also believe the child should not suffer unpleasantness. Such parents frequently try to explain to the child why they are using time-out, or how the child should behave, but long explanations actually give the child attention, which is a reward. Parents may also talk to the child during time-out and give the child access to television and toys. Or parents may relent when a child cries and then let the child do what had led to the time-out use in the first place. These procedures do not amount to time-out and will worsen the misbehavior rather than help. I describe this to indicate that *parents need more than just specific procedures. They need also a conception, a theory, of the nature of the child, of child development, and of the parents' role in child development.* That conception, a learning conception, has not been set forth in full, although there is much of that in my other works. Not only do parents need such knowledge, scientists and professionals also need the learning conception of the child's nature to guide their research and treatment activities.

What needs understanding is that the role of the parent in raising the child derives from the conception of the child that is used. That can be the problem. Today the educated middle-class parent is bombarded by advice from a mixture of sources. There are commercial materials to aid the parent: mobiles and other things to enrich the infant's experience, puzzles to stimulate cognitive ability, children's age-graded books, music stimulation, and TV shows attempting to provide various preschool and school-age skills. Parents buy such things and expect their children to profit. There are also many articles and books for parents, with a range of views. Some books set forth a behavioral position. I believe my book *Child Learning, Intelligence, and Personality* (Staats 1971a) was the first of those that address such interests. But contemporary behavioral books for parents mostly deal with specific problems of behavior and how to treat them, rather than with a general conception. Parents can read such books, use some behavior-analytic procedures, and still not know what to do when their children are delayed in language, walking, socialization, and such. There are also books that fall well within the Great Scientific Error conception, ranging up to the denial of parental influence on child development, laying it all to biology. There are other books for parents that present their authors' personal experiences and views of the child and child development in an eclectic mixture of common sense, personal belief, and trendy opinions. Confusing sources of information produce parents confused about what their role should be and what actions to take when the child begins to develop a problem. And once the child is given a clinical label—like *autism, learning disability, neurological impairment,* or *attention deficit hyperactive disorder*—the parent then begins to protect the child as ill or handicapped, lowering expectations and demands, which means a deficient learning environment for the child.

On the basis of my invention of time-out, I was considered in *Child* magazine to be one of the "20 People Who Changed Childhood" (Straus 2006). The procedure does contribute to children, but also to parents, as is the case generally for my other contributions and those of other child-behavior analysts. Nevertheless, such procedures, despite usefulness, are specific. Actually, I always considered my learning conception of child development as broader, deeper, and more important than its specific procedures—for parents, professionals, and scientists. For the conception can be employed generally for indicating ways to produce valuable behavioral developments in children and to avoid problems.

My conception pictures a human child differently than traditionally conceived and thus projects a vastly different role for the parent. The idea that the infant is ready to learn at birth and will do so, whatever conditions are presented, sets a different agenda for the parent. It makes it necessary to consider what the child should learn and how to impart that learning.

[A] learning analysis of the acquisition of behavior leads to a focus upon the parent as a *trainer.* . . . As long as the child's behavioral development consists of innumerable training experiences, many of which occur in the home, then the parent has many of the controlling variables in his hands and cannot relinquish them regardless of his philosophy of child development. (Staats 1963, 412)

All parents determine many of the behavioral traits their children display, whether or not they intend to, for they constitute a central and broad part of the child's environment, the environment that determines the child's learning. Parents play their roles by following what they themselves have learned, and that includes the theories of child development they have encountered. Most of the time, what the parent has learned works well enough. But many times it does not. And usually it could be better. "The parent needs to know how not to . . . [train] undesirable behaviors, or, when they have developed, how to decrease them benignly; and he needs to know how to train the many adjustive behaviors the child will require" (Staats 1963, 413). Very insufficiently recognized is the need for a general understanding of child development as learned and of the parents' role in providing the experiences that will produce that learning. The present conception has a view of the parent that is quite different than the role assigned by the Great Scientific Error conception.

CHILD DEVELOPMENT

What about the physical growth of the child? Doesn't that growth constitute a major part of child development? Well, yes, and no. In one sense there is no question concerning the contribution of physical growth to child development. Everyone can see that the increase in physical size and strength is important in sensory-motor skills, for example. But the growth only provides the potentiality; it is learning that provides whatever skill is attained.

In addition, however, the child's physical growth also affects the way others respond to her or him. Increasing size generally connotes greater age, and age affects the way we respond to a child. In analyzing language development, I once pointed out that the age of the child, through its effect on appearance, influences the way others speak to the child (Staats 1971b) and thus what the child learns. Later studies confirmed this effect abundantly (Ervin-Trip 1971, Rondal 1985). Moreover, the skill acquisition that growth makes possible will get other people to respond to the child differently. A child who can do things like an older child will be treated as an older child and in this way will receive the learning experiences of an older child. Research would show that the way children are treated in many ways is partly determined by physical appearance as well as emotional, motor, and cognitive skill. Gaining understanding of the child's nature calls for further study of how the child's physical growth affects the potentiality for learning as well as how physical growth affects the learning conditions the child meets.

The brain grows during childhood, with a large increase in the early years but with continued growth that goes on longer than previously thought. Tradition had it that the brain growth takes place through biological maturation. But, as I have indicated, a growing number of those in the biological sciences have begun to recognize and show that *learning* experiences produce growth in the brain. *That prompts me to ask: how much of brain growth is biologically driven, and how much of brain growth occurs from learning experiences?* Perhaps most brain growth in childhood is the result of learning, but we know that at least a good part is. In a sense, however, it does not matter. For neurons added solely by biological growth should mostly be "empty" neurons, ready for use, there to contribute to the child's learning potentiality, and ultimately so used.

Something else to ask: is there any proof at all for the belief in wired-in skills, talents, aptitudes, personality traits, or whatever, that are present at birth and blossom through biological (neurological) growth? The answer is no. Beliefs of this kind are assumptions. Although the mainstay for the field of child development, the belief should no longer be accepted; behavioral development should not be considered to result from physical, and brain, growth. It will take most people a self-conscious effort to divest themselves of that assumption, but it is central to do so. What needs to be understood is that a *biological explanation* of any child's behavioral development demands that the biological mechanism involved be shown. If there are wired-in traits, then that wiring has to be shown, and how it changes with maturation, and what that maturation consists of in the brain. If there are genes that produce that wiring, and genes that produce the wiring's maturation, then those must be shown along with what their action is. When such standards are enunciated, the absence of evidence becomes apparent. No wiring has been found that produces language, reading, bonding, handedness, walking, the terrible twos, intelligence, sociability, or any complex human behavior.

Learning to Be Human

The neonate is the young of the marvelous learning animal. Loaded with ten billion or so nerve cells in the brain, I suggest the infant has no wired-in nature, except for reflexes. That huge brain doesn't exist for nothing. It exists because it evolved to learn, to learn an individual's huge and complex slice of personal experience. It takes millions of nerve cells to learn language, more millions to learn intricate, variable, and complex sensory-motor skills, and yet more millions to learn positive and negative emotional responses in fine gradations to a huge number of stimuli. The learning task is gigantic—unimaginably so—complex, and going on minute to minute for years and years and involving uncountable learning experiences. The marvelous young learning animal begins as a human neonate, ready to learn from the start.

It is a testament to the great learning powers of the child that this learning is imparted by people who know little or nothing about what the child is as a learning animal, little about the principles of learning, and little about how to teach any of the many things the child needs to learn. Great learning powers enable children to learn in the face of such poor conditions, to pick up incredibly complex skills, like language, just from casual experiences. That's what those billions of neurons make possible. That's why humans have that big head right from the beginning. The human child also has that human body so that what is learned can be put into action. Thus, when the sound made by the parent becomes positive emotionally, the human baby, on its way to saying first words, has the sensory-motor wherewithal to learn to make sounds like the parent.

The complexity and variety in the child's learning can be seen only through working with and recording children engaged in complex learning—of sensory-motor, language-cognitive, and emotional-motivational types. The tip of the iceberg of children's learning was serendipitously revealed to me in the work that I did. However, the vastness of children's learning will not become known without multiplying that type of research vastly. The task has just begun.

These pursuits are called for in advancing the field of child development, as proposed in the new paradigm. The phenomena of child development revolve around learning. The conception is that the child develops into a full human being through learning. A child without the behaviors that humans ordinarily learn is not a human in the full sense of the word, although a full member of the *H. sapiens* species.

There is no question that in advancing through prehistory, humans have learned a tremendous amount of behaviors. Many of those behaviors, such as language, characterize the human being. Those behaviors, in all their complexity, are not given to the child as a divine gift. And they are not given to the child as a biological gift that emerged from mutations of human genes. Those behaviors are only acquired, or not acquired, by virtue of a long and complex learning experience.

That learning, and those behaviors, are part of the definition of human nature. The child learns to be a human being, that is the central nature of child development. We are concerned about what those learned behaviors are and how they determine the nature of the child. So we must study them—what they are, how they are learned, and what their functions are. Learning must become the focus of our study of the child's nature. That should include the learning principles that function for the child, the behaviors the child should learn, the behaviors the child should not learn, and the conditions by which the child learns them. The pie has been cut unrealistically, with much work on the study of the biological causation of child development and little on the study of the learning causation. It is time we give learning its due as an explanation of child development. Child development involves biological growth, true. But biological brain growth does not itself cause behavioral development. Learning experiences are the cause of child behavioral development. Any approach to child development that does not deeply and extensively treat learning remains deeply errant and terribly lacking.

Chapter 5

MARVELOUS LEARNING: THE MISSING LINK

In the 1950s, when I began the systematic study of child development in terms of the behaviors the child learns, the traditional field of behaviorism was still focused on the study of the basic learning principles. The extensions of learning principles to the explanation of human behavior were conjectural and dealt with general examples. Specific analyses indicating how specific human behaviors are learned didn't exist. The leading behaviorists did not do such research, nor did they outline prospective programs of research for that purpose or indicate the necessity of that type of research.

The traditional field of child development recorded the times when particular behaviors appeared—such as sitting, standing, walking, making vocal noises, babbling, first words, and two-word sentences. The field studied real behaviors, however. One would think there would be great mutuality here—behaviorism had knowledge of the learning principles that produce behavior, and child development had knowledge of the important behaviors children develop. But the two fields developed quite separately, actually antagonistically. Behaviorists did not accept the assumptions of internal causes that child development used as explanation, and child developmentalists did not believe the learning principles explained behavior development. So what should have been a central mutuality of interest was never advanced, and it still hasn't been. (This type of breaking apart of fields that should be connected is common in young sciences [Staats 1983].)

Unlike the traditional behaviorists, however, I was not an animal researcher. I came into psychology with a focal interest in human behavior, real human behavior. My mission, from the beginning, was to use learning principles for explaining that behavior. I studied fields in traditional psychology—such as developmental, educational, clinical, abnormal, and social psychology—and saw elements of valuable specifications of human behavior in each. Those who studied "mental illness," for example, despite having an empty theory, did study abnormal behavior, and they knew more about such behavior than a behaviorist studying a pigeon pecking a key, a rat running a maze, or a dog salivating to a bell, even though the important principles involved were based on experimental evidence. Of course the traditional field of child development, despite the empty assumption of internal maturational causes and stages, also contained more knowledge of children's behavior than did behaviorism.

My approach was based in learning principles, but its goal was to explain the facts of real, complex human behavior. At that time, Skinner's approach—the experimental analysis of behavior—was to use the same methodology he used with rats and pigeons for research on human behavior. That method involved observing a single, simple response, made many times, to see how reinforcement affected the rate of the response. Such simple responses are made in an instant. His students followed the experimental analysis method—Bijou (1957) reinforced a simple response of mentally challenged children, and Lindsley (1956) did the same with psychotic adults. Such studies added to the valuable knowledge that the basic principles of reinforcement apply to the simple behaviors of humans, even those who are mentally challenged and those who are psychotic. But such studies revealed nothing about how reinforcement principles could be used to understand and deal with either child development or abnormal behavior.

My work took a different approach, behavior focused, not method focused. Rather than applying the methods created for the basic study of reinforcement, I made what was to be studied—real human behaviors—dictate what research methods had to be constructed. I was focally interested in how those behaviors were learned. That aim started me on an endless program of study, so rich in serendipity, a career-absorbing program that progressively widened what had to be done.

Centrally, it turned out that human behaviors in great part are complex. Behaviorism had never studied complex human behaviors, so new research methods were necessary. The experimental analysis of behavior methodology—a rat pressing a bar repeatedly, a pigeon pecking a key repeatedly, or a mentally challenged child pulling a plunger repeatedly—isn't relevant for studying how a child acquires language, how to train an autistic child to speak, or the myriad of other important human behaviors.

Entering the study of complex behavior led to new discoveries. Importantly, the work even at an early time began to indicate that using the concept of *behavior* for a single response as well as a complex human "behavior" leads to error. The term *a behavior* is singular. But most *human behaviors* are actually complex assemblages of responses. Language is not a behavior. Skinner's book *Verbal Behavior* (1957) described certain types of language, for example, *texting* (reading). But his interest was in showing that such words were behaviors. Reading is not a behavior, and learning to read is complex, not to be understood simply according to the reinforcement principle Skinner studied and employed. Intelligence is not a simple behavior, being a natural athlete is not a simple behavior, being a music aficionado is not a simple behavior. Each is complex, and becoming intelligent, a natural athlete, or a music aficionado involves very complex learning.

After some time in studying such complex human behaviors, I introduced a term to name them. Reading takes place by virtue of a complex *repertoire* of different types of responses. The repertoire is different than a simple behavior. Its learning is different, and its effects are different. The word *repertoire* in our language is based on commonsense usage. The new definition of *repertoire* is systematic, based on specified study; and its learning, definition, and role continue to be elaborated. The concept of the repertoire is generally central in understanding and dealing with human behavior and human nature. In establishing this understanding, a new set of human learning principles act as *the missing link* in establishing a bridge between behaviorism and knowledge of traditional psychology and the other fields that study human actions. The basic learning principles plus the human learning principles constitute the foundation for explaining the new view of humanness that continues in the rest of the book.

HUMAN LEARNING PRINCIPLES

As part of a focus on language learning, I began my first study of reading in 1958. Its purpose was to train dyslexic children to read. The learning program I created did deal with individual responses. For example, the child was rewarded for reading every printed word introduced, whether the word was known or whether the tutor had to say its name. The word was presented in this way until learned. In each lesson-unit, words that had been learned or were known were combined into sentences that were read and reinforced until known, and these were combined into stories that when read were also reinforced. Although single words were learned, there were many of them, and they were presented in different combinations of increasing complexity.

In the first published study, of a fourteen-year-old dyslexic problem child, the words read in these various ways amounted to sixty-four thousand in one four-and-a-half-month training period (Staats and Butterfield 1965). This was long-term learning of a repertoire of many different responses. At the end of the semester, this child passed courses in school for the first time in his life and advanced more than two years on a reading achievement test. Maryanne Wolf follows the traditional child-development field's disdain for such a learning study in saying, "'Identifying a flash card at an early age isn't reading. . . . It's what researchers call paired-associate learning. That may sound impressive, but,' she says, 'a pigeon can do it'" (Paul 2007, 91–92).

That is quite wrong. The subject in the study, although learning single word responses—individually presented on a card—*learned a great many of those responses*. That enabled him to read many different printed materials, single words, sentences, and stories. Even at this level his learning could not be compared to that of a pigeon because a pigeon could not learn so many words, could not combine them in innumerable ways. What this boy learned was a large *repertoire* of words that he could read. And he was tested for, and trained in, comprehension in what he read. Many later studies using these methods with many different children have since showed the same effects.

In addition, however, an important aspect of such a word-reading repertoire is that it connects to other repertoires. For example, having learned a reading repertoire, a person can be given a note asking her or him to do something and the person will do it. Reading functions that way because the person has through previous language experiences learned to respond to instructions to do something. I described that ability to follow instructions to be composed of the verbal-motor repertoire, that is, having learned to respond, individually, to a large number of verbs and other words. After learning to read, that verbal-motor repertoire can be initiated by written words, not only by spoken words. That gives the individual a marvelous extension of the ability to learn—the individual can read about doing something and thereby learn how to do it. When getting a cable box for television, the individual reads the instructions and can connect it to his television set. Learning the reading repertoire thus opened new horizons for the boy in the study, horizons that could never be opened for a pigeon with its limited learning and its limited repertoires. Humans and pigeons learn by the same principles, yes. But humans can learn more by an unbelievably huge amount. And that learning also involves new principles.

Centrally, this research revealed that new phenomena show themselves in such complex kinds of learning. Exposing the new principles and concepts serendipitously drew me to a new way of thinking. My term for such groups of learned responses was *behavioral repertoires*. Repertoires are formed of multiple responses. Such repertoires are related to other repertoires. Sometimes a repertoire is composed of a number of other repertoires, as is the case with language. Not only how repertoires are learned needs study. Behavioral repertoires, once learned, give humans new learning abilities; new behaviors can occur and the learning of other repertoires is made possible. The concept of the repertoire, the particulars of the various types of repertoires, and the principles by which repertoires can operate constitute a nuclear part of the present understanding of human behavior and human nature.

Nonfunctional Repertoires

What is so special about the concept of the repertoire? After all, the word is in our common language. The *New Shorter Oxford English Dictionary* includes its definition as "a stock of regularly performed actions" (1993, 2548). This implies that certain actions are related in some way that makes them compose a repertoire. A musical repertoire is composed of actions—complex actions—that produce music. A poetry repertoire could be composed of poems the individual has memorized and can repeat. The dictionary definition, however, does not say much about the nature of repertoires, how they are acquired, how they are related to other repertoires, or what functions they may have in human action. The dictionary definition provides for recognition that there are different repertoires—singing, dancing, acting. That doesn't even scratch the surface, however.

There are different types of repertoires. One type has no functional value. For example, as an adolescent in school, when I was bored, one of the things I did was practice drumming my fingers on my desk, forward and backward, one hand and two hands, and in various synchronized ways. I learned a distinctive repertoire, a coordination of different responses, with the responses involved related by the fact that they were learned together, involved the same body parts, and tended to be performed together.

I am sure that not everyone learned the finger-drumming repertoire to the same skill I did. Perhaps I was outstanding in finger drumming. But the repertoire led nowhere. It had no functional value. Life situations rarely elicited it. The repertoire had no significance for me actually. Once that period in school passed, I gave up drumming my fingers and haven't exhibited that repertoire for many decades.

People can learn repertoires like that. Sometimes they are evident and characteristic of an individual, but they play no other role for the individual. Sometimes they are sufficiently noticeable and different, as with the rhythmic nonfunctional behaviors of an autistic child or catatonic adult. I saw a middle school kid waiting for the bus the other day performing a ticlike repertoire of moving his head and shoulders. The behavior was sufficiently bizarre that other waiting school kids were keeping their distance from the boy. Whether drawing that effect was the function of the repertoire wasn't clear in the brief period I had for observation. The repertoire certainly didn't appear to have a function.

Functional Behavioral Repertoires

There are, however, organized clusters of responses—repertoires—that are functional but are not widely important. Take toilet skills. They are significant in and of themselves for the individual, and they are interrelated with other repertoires, such as locomotion skills that get one to a bathroom. A child can't attend regular school without having a toilet repertoire, so it is related to relevant repertoires. One of my earliest memories was going to kindergarten when I was only three and a half years old and, on retreating to the bathroom, failing to undo my short pants because they were new and attached by buttons that I couldn't work. My buttoning repertoire was deficient, and that ended my schooling at that point. Although life demands a toilet repertoire, however, it doesn't have links to other important repertoires and thus is an isolated ability.

Frequency and breadth of functionality represent dimensions for considering behavioral repertoires. Take trigonometry. The repertoire is functional for someone who employs aspects of the repertoire in work or in further learning. For most of us, however, the trigonometry repertoire is nonfunctional. After a number of years of disuse, little of the trigonometry repertoire will remain, a story of actually wasted learning. Some would claim there are general benefits to learning such things as trig, because regardless of later use, the learning develops the mind. Perhaps this is true. I can say however that my "mind" has profited much more from my learning to type, a repertoire I have used almost every day for over fifty years.

These dimensions help distinguish types of repertoires. Some are simple, as is the toileting repertoire, and some are complex, as is the trigonometry repertoire. Some have no function; some have many functions. Some are related to many other repertoires; some are related to few or none. Some repertoires, like language, are ubiquitous and have important functions with wide connections and frequent use in life. Some are mainly significant for a particular function and occur only when the occasion for that function arises.

Language, reading, numbers and number operations—these are not behaviors. Nor are they abilities of the mind, genetically laid down, that mature as the child ages. They are not internal stages of child development. My analysis and work with children showed they are very complex, large, learned repertoires of behaviors. I call them *basic repertoires* because of the widespread and significant ways they act in determining the individual's life behavior, experience, and *nature*.

Basic Repertoires

These are repertoires, ordinarily acquired by everyone, that are central to being human. Knowledge of those repertoires is basic to knowledge of humans. The language repertoire the child learns, described in the preceding chapter, is a good example of such a repertoire. I call these *basic repertoires* (BRs); they explain so much of humanness.

What Is a Basic Repertoire?

BRs are large. They consist of many individually learned responses. Different estimates exist of the number of words in our English language, perhaps eighty thousand or more. The great size of the reading repertoire has already been indicated, and it is also an important basic repertoire. How many individually learned responses exist in the mathematics repertoire of an expert mathematician, of a college graduate in mathematics, of a high school graduate, or of a sixth-grade dropout? What are all the responses that go into being a fine athlete in any sport? Many. What are all the learned emotional responses that go into being an art or music critic? Many.

Many responses in a repertoire means a very extended, complex learning history. Every basic repertoire is learned only over a long period of time and involves many learning experiences. The linguists' view errs in the belief that the child has learned language after a few years of age. Many years go into the basic learning, and the learning continues throughout life.

Basic repertoires are called into operation frequently in the individual's life. My finger-drumming repertoire occurred only during classes that involved boring lectures. My trigonometry repertoire was only called out in that class, so it was relevant for a brief time of day and a brief period of life. The language repertoire comes into action all waking hours, every day of one's life.

Basic repertoires don't exist in vacuums; they do not exist or operate in seclusion. The language repertoire is part of the repertoire of the music critic. The language repertoire is important for the fine athlete. Part of every athlete's skill lies in being able to follow the instructions of a coach. That demands a language repertoire. A golf magazine specializes in instructional articles that demand a reading repertoire. A book in algebra is presented in the individual's language repertoire, for the individual to read.

So basic repertoires are interwoven. Although it is helpful to consider them as entities, they are not separate and independent. Categorization is valuable and has occurred on a commonsense level; that holds also for a systematic, scientific analysis. For example, many have referred to cognitive, emotional, and motor aspects of human behavior. It is productive to consider language-cognitive, emotional-motivational, and sensory-motor basic repertoires, although there is not a demarcation among them: language-cognitive basic repertoires, like language, reading, and mathematics; emotional-motivational basic repertoires, like values, interests, ambitions, and attitudes; and sensory-motor basic repertoires, like eye-hand coordination, dancing, violin playing, tennis playing, and making love. The various repertoires interact.

Cumulative Learning

Yes, basic repertoires are important for what they are, for what they allow the person to do. Very centrally, however, BRs have an enormous impact on humanness. They change the nature of the individual, especially by changing what the individual can learn and how rapidly that learning can occur.

As a simple example, take the word-imitation repertoire. In itself it is important because the child who has learned this repertoire can respond to many situations. If a father asks the child "Can you say 'ducky'?" and the child can repeat the two sounds, the child has succeeded in that situation. Thus, having a word-imitation repertoire that includes making those two sounds increases the child's learning ability. Once the child can imitate on command the two speech sounds "duh" and "key," when the parent asks the child to say the word, and the child does so, *voila!*, the child has the new word. Because of the child's word-imitation repertoire, the learning can take place in that easy way.

Actually, once the word-imitation repertoire is acquired, the child's learning *nature* with respect to language changes. Originally, learning vocal responses goes very slowly, taking about six months even for babbling to occur. After learning the verbal imitational repertoire, however, the child's learning ability, the child's learning nature, takes a leap forward. The parent need only pronounce the new word, perhaps only once, and the child can repeat it and learn it, quickly, easily. Rather than taking a year, like the first word takes, the learning takes place in a moment. Two children, one with a word-imitation repertoire and one without such a repertoire, are now fundamentally different. The word-imitation repertoire determines the child's word-learning ability. The two children have different learning natures. They will seem to others to be intrinsically different: one has an increased learning ability. That increased learning ability is crucial, for the child in every language culture needs to learn thousands of words. That is only possible when the child has become a very good learner of words.

Accelerating the learning of one repertoire, that of word imitation, accelerates the learning of another repertoire, which we call vocabulary. I have seen children of eighteen months of age or older who speak only a few single words, not well enunciated, whereas both Jenny and Peter at this age could converse in multiple-word sentences. Typically children say their first words at a year, and at one and a half to two years undergo an acceleration in the rate at which they learn new words. Children with advanced language repertoires, however, have higher learning ability. At that age, the language repertoires of Jenny and Peter were advanced enough to begin training them to number concepts, for example, and then to count, as I have recorded (see Staats 1968a). Following the number learning, it was possible for the children to advance early to writing numbers and also to adding numbers, both of which are additional behavioral repertoires basic to learning yet more repertoires. *Very large differences in learning ability are created in children through the learning of such basic repertoires.*

At the age of two, because of the accelerated language development, it was also possible to introduce reading learning, when schools ordinarily only do so after the age of four or five. By two years and three months, Jenny "had a reading vocabulary of 121 words . . . , 52 letters (upper- and lowercase) and 6 phonetic responses" (Staats 1968a, 292). (I have her on audiotape as she progressively learned a reading repertoire, including the reading of short stories.) To continue, with this expanding repertoire of skills, Jenny was able to read books on her own at an earlier-than-usual age. And this led to learning that expanded her reading repertoire and made it more skilled, less effortful. A California Reading Test given to her in a second grade class placed her at the ninety-ninth percentile. This advancement in cognitive performance continued through the awarding of a medical degree. The same learning history occurred for my son, Peter; I have him on a videotape just beginning to learn to read at the age of three. My prior experience indicated that beginning at age two provided more than the time needed to produce the level of reading skill I had as a goal. Let me add that the training I extended was by no means urgent or pressurized, usually no more than five minutes per day and always less than ten, but it began his learning of repertoires that he chose to continue in becoming a specialist in pain medicine, a founding chair of a department in that field at Johns Hopkins, and a researcher, inventor, noted author, and clinical leader.

The experiences I had with these children, especially, brought me face to face with the knowledge of the basic repertoires and how central they are in human learning. The children absorbed the learning experiences and as a consequence displayed new and unusual advancement in learning prowess. That advancement led me to introduce additional learning opportunities.

Jean Piaget said that a child could not count unarranged objects until the age of six, based on the assumption of a stage of development at that age. He did not know about human learning, however. The child's ability to learn does not depend on the age of the child, as the concept of maturation insists, but the extent to which the child has learned the basic repertoires that are the essential foundation for the learning. A child of eighteen months may have the BRs by which to learn numbers and learn to count. I showed that in the last chapter in teaching my own children. A child of seven, on the other hand, may not have the necessary repertoires.

We later showed that four-year-old children expected to do poorly in school can *very standardly* learn the counting repertoire (Staats, Brewer, and Gross 1970). The proof is there: age does not matter, no stage development is involved, nor are any brain/mind processes involved. What mattered was that certain repertoires had to have been learned before the next repertoires could be learned. In this case, the language repertoire provides a basis for learning the number and counting repertoires. These several repertoires then serve as the basis for an arithmetic repertoire. The total repertoire then is functional in learning an algebra repertoire, and on and on.

Language is a basic repertoire, so is reading, whose acquisition depends on the language repertoire. No question, learning both radically increases learning *ability*. Humans also learn repertoires of sensory-motor skills. A person who engages in playing sports develops a sensory-motor repertoire that becomes a foundation for learning the moves of other sports. A person who plays multiple sports and develops a broader and more skilled sensory-motor repertoire will have the means for generally learning new sports repertoires easily and quickly and will be considered a "natural athlete." We are often impressed by the skill with which animals do certain motor things, in comparison to human prowess. Actually that type of comparison suffers from focus. If it is broadened to consider the ability to learn new sensory-motor skills quickly and in great profusion, the superiority takes a human shift.

On the basis of one's sensory-motor repertoire and one's language repertoire, a person in our culture can quickly learn new skills that an animal cannot. Consider how long it would take a chimpanzee to learn to drive a car on a city street or become a football quarterback, compared to a person. It would be forever, for the chimpanzee doesn't have the basic repertoires on which to further learn such complex repertoires as driving or quarterbacking.

Humans also learn emotional-motivational repertoires. A musician ordinarily will learn positive emotions to many musically relevant people, instruments, musical pieces, concerts, and so on. That constitutes a repertoire that will be foundational in learning various additional musical repertoires. Similarly, a person high in religious values has complex repertoires that people who measure low do not have. My research has shown those two types of people learn differently and behave differently in certain situations. They have different natures because they have learned different basic repertoires (see Staats and Burns 1981).

Quickly, readily, *just on the basis of words*, people learn emotional responses to various things in life, ranging from political stimuli to aesthetic stimuli to racial stimuli to religious stimuli to business products. In each such case an emotional repertoire is involved. How does this learning take place? It can occur because humans learn a large repertoire of words that elicit a positive emotional response, and those words can be used to create such an emotional response to new things. The classical conditioning of an emotional response in a laboratory is a slow process. Once language has been learned as a BR, with its host of emotion words, the human being can learn emotions easily and quickly and does so in great quantity. Whole new emotional repertoires—like religious values—are largely learned on the basis of language.

Humans have extraordinary skills, as readers, cooks, drivers, golfers, accountants, lawyers, concert pianists, surgeons, inventors, scientists, NFL quarterbacks, and children learning to count. In each case the skill displayed comes about through a complex learning progression building through successive repertoires. *Developing such skill areas involves a cumulative learning of repertoires, one repertoire serving as the foundation for learning additional repertoires.* A lawyer as a child acquired an extensive language repertoire that served as the foundation for learning to read, the reading repertoire then served for the learning of additional repertoires in history, English, business, science, and the like. These various repertoires then provided a foundation for law-school study, and on and on. The repertoires play their roles in specific ways open to study. A scientist in our culture learns many repertoires in a cumulative-hierarchical process that begins with language, reading, counting, arithmetic, and mathematics, and that continues on for a professional lifetime. A successful career of that type involves an unrealized complexity of learning.

Group Cumulative Learning

That same marvelous learning occurs also for groups of people as they progress through generations. Thirty thousand years ago, individual artists painted bison and other animals on the walls of the Lascaux cave in France. Much has been said of cave art, how it has spiritual and symbolic meaning and what it indicates regarding the conceptual ability of those who performed it. In the present view, that cave art represented a learned repertoire, having had an earlier a history of group learning over generations; surely children before thirty thousand years ago drew figures in the dirt and must have progressively learned how to do so. We can see continuous learning of painting skill, for example, in creating three-dimensional representations on a two-dimensional surface. Just look at the development that occurred between 1200 BCE and the Renaissance many centuries later. Was the art of the old masters unaffected by the preceding history of art development? Ultimately, of course, the basic repertoire of artistic representational skills was taught and learned so widely that new horizons like that of French impressionism and abstract impressionism represented new variations.

As another instance of group cumulative-learning process, take sports. Stone-Age-level human groups living in the Amazon do not today have games with the complexity, precision, and social features of our football, baseball, or golf. We don't have to go back far in our own history to find an absence of such modern games. Some claim baseball had its first set of rules written in 1845, invented by Alexander Joy Cartwright, the author. Others claim Abner Doubleday invented the sport in 1839. There is also an opinion that the game was the fruit of a long-term development that was not recorded. Prior to baseball there were various kinds of games that involved a thrown ball object that was hit, sometimes by a bat, as well as running for a goal. Baseball did not spring up without previous behavioral developments. Relevant repertoires of players were learned over centuries of time before the step up to a full, formal game in its modern form. Multiple lines of learning had to occur, for a pitcher has a much different repertoire than a catcher, a first baseman, or a shortstop.

Every complex human organization shows the same type of development of behavioral repertoires: government, education, business, entertainment, military, and religious. The advance in the technology from the hand ax to the cell phone is enormous, beyond belief. But the learning involved took more than one million years to cumulate. Humans make historical progress by many participating formally or informally in learning basic repertoires. The process involves different participants who have different experiences that result in the learning of new behavioral skills that then become part of the skills of others. From such a base the next generation has a new level of skill and thus faces new environmental experiences. In that way there is cumulative learning of a group kind. That cumulative learning can go on for thousands of years, no doubt millions. Prehistoric research given impetus and direction by this conception could be expected to give us more knowledge of the process than we have now.

We do have ample history, however, that tells us, as another example, that government did not emerge full-blown. We know about the simplicity of Stone Age tribal life. An isolated Amazon tribe does not have an elected president, a congress, a supreme court, a department of justice, a secretary of commerce, or a CIA and a standing military. We can also see that there was a time when the basic repertoires and the government features we have today did not exist. As human groups advanced in size, because of agriculture and livestock domestication, new population sizes were experienced and conditions of larger group living arose. Those new conditions were responded to, creating governmental organizations. Just as it would be unrealistic to expect a technological jump from the hand ax to power saw, a jump from a tribal chief to an elected legislature would not occur. We have seen the advancement that has occurred in historical times, with movement through serfdoms, small kingdoms, a national king and royalty, a kingdom and a parliament, a parliament and a symbolic king, general democracies. Humans learned how to live together; they did so by learning complex repertoires of behavior over extended generations. There remains a great deal of learning yet to go.

Advancement in learning generally begins slowly. The child learning to write takes four times as many learning trials for the first four letters of the alphabet as are taken for letters 13 to 16. Acceleration in learning occurs in all areas in which there is great skill to be acquired. The skill of manufacturing a hand ax was a remarkable learning achievement some million years ago, and that skill was only improved on in thousands of years, until new skills were learned. As technology advanced—with agriculture and domestication of animals and larger communities—the learning accelerated. For an individual's learning could be communicated to a larger number of people who could advance the skill further. Cumulative learning applies to groups as well as to individuals, and the same acceleration is involved. Advances in technology are today occurring at lightning speed in many fields. Learning acceleration through the cumulative process for individuals and groups constitutes one of the features of human learning. The quickness of later points of development makes it seem like it really isn't learning, because the slowness and simplicity of the original learning is no longer evident.

The slow beginnings of learning repertoires tend to be missed, not recorded. Games that would have contributed elements to a foundation for the ultimate game of baseball, for example, were played hundreds, probably thousands, of years beforehand. Yet the creation of the game is attributed to the creativity of an individual person. That, however, attends to only the last phase of the extended learning process. The jump the individual makes in such a creation is very significant of course, but the original small jump from smacking a moving object with the hand to smacking it with piece of wood was also an important advance.

Great figures in science are celebrated for making such last-phase jumps. What is left out in explanation, however, is that even great discoveries and creations are always built upon basic repertoires that have been learned from others. Newton's large theoretical jump forward did not occur until Galileo's and Kepler's laws concerning moving bodies had been set forth. Einstein's theory did not arise until problems of explanation had arisen within the practice of Newtonian physics.

A common belief is that culture *evolves*. For example, the surge in culture that occurred around thirty thousand years ago is believed to have been due to a mutation process that also produced language in humans. Such an explanation totally leaves out what had to be in play, namely the enormous cumulative learning that began far back in prehistory, as will be indicated in chapter 8. Culture does not advance by evolutionary principles. That belief is detrimental; culture advances by the cumulative learning of individuals and groups of individuals. That applies to material artifacts as well as to organizations, economies, governments, religions, entertainments, sports, educational systems, literatures, legal systems, and on and on, for humans have cumulatively learned many and great repertoires. Huge as it is—and as productive and creative as it is—human cumulative learning of basic repertoires does indeed represent a most wondrous thing.

THE MIND AND THE MISSING LINK

Young sciences are chaotic sciences, disunified (Staats 1983). That holds for the sciences that study human behavior; they have innumerable lines of separation, ranging from small idiosyncratic areas of study that flare up and die away to persistently acting major divergences. One of these concerns what I call internalists, those who attempt to explain human behavior by the internal nature of humans. That includes those who believe that the genes and the brain determine the individual's mind and how the individual behaves. On the other side are the externalists, who believe that the individual's environment determines her or his behavior. Of course the man in the street usually has both beliefs in commonsense explanations, but the separation does govern a great deal of science, exemplified clearly by Pinker's *The Blank Slate* (1999).

The schism is unfortunate because what is studied in both cases is important and, despite the clash of conceptions, is complementary. What is needed is a framework that eliminates the conceptual discrepancy. The human learning concepts and principles—the basic repertoires and cumulative learning—do that, resolving the internalist-externalist, nature-versus-nurture conflict. Take the "mental" activities of reasoning and problem solving, an internalist area of study. Judson, Cofer, and Gelfand (1956) had one group learn a list of twelve words in which three words—rope, pendulum, swing—were presented in sequence. Another group learned the same twelve words, but the three words were not sequenced together. Then the subjects in both groups were given a problem whose solution demanded tying a weight to a rope and swinging it like a pendulum. The first group showed greater mental-reasoning ability in the problem than did the second group. It was the learning of the small word repertoire that made the first-group subjects superior reasoners. Problem solving and reasoning, considered to depend on the innate qualities of the mind-brain, should be explained by what has been learned.

As another example, an experiment by Brown and Lenneberg (1954) showed that the extent to which subjects could select from a chart with many colors the four that had previously been shown them depended on whether or not the subjects previously had learned the *names* for the colors. Perception and memory are usually thought of as determined by the nature of the mind-brain. But this study shows that perception of and memory for a situation or a problem actually depends upon what has been learned, in this case on a part of the verbal-labeling repertoire.

Whether or not children have an "impulsive or deliberate mind" similarly appears to depend upon the repertoires that have been learned. Palkes, Stewart, and Kahana (1968) trained some boys, who today would be considered to have attention deficit hyperactivity disorder, to say things to themselves while taking one type of intelligence test. The impulsive boys who had been taught to say such things as "Look and think before I answer," and other instructions to be deliberate in action, attained higher intelligence-test scores than such boys who hadn't learned those elements of their verbal-motor repertoire. Meichenbaum and Goodman (1971) supported and extended those findings. Again we see that mental activity is not some personal, biological quality of the individual's brain. The children's mental characteristics, and hence behavior, consisted of differences in learned repertoires.

Emotional-motivational repertoires can also affect cognitive (mental) behaviors. To illustrate, an experiment by DiVesta and Stover (1962) conditioned children to have a positive emotional response to some of the nonsense syllables they learned in a group. Later, the children learned to give names to some objects using those nonsense syllables. When the children were allowed to take one of the objects home, they chose one that had been labeled with a nonsense syllable that elicited a positive emotional response. Our choices, our goals, our purposes depend upon the emotional-motivational repertoires we have learned. We usually don't know why we make our choices and set our goals; we just consider them features of our being, our nature. If we could trace back, we would see they are the product of our learning.

I will give one further, actual example to indicate that complex creativity in life can also be considered in terms of the repertoires of responses that have been learned and the new stimulus configurations that are encountered. Early chemists were interested in knowing the fundamental nature of matter. They began finding some of the elements of matter. In describing them, they recorded their different weights (masses). Looking for the elements and indicating their weights was possible because of repertoires they had learned. The repertoires, thus, produced new knowledge. Then along came Dmitri Ivanovich Mendeleev; he had the previous findings to consider and saw the regularity of their weights, even though gaps were involved that took away from complete regularity. Mendeleev constructed the periodic table of elements' weights that made apparent the regularity and the gaps that suggested there were elements not yet found. Looking for the elements in the gaps became a goal, and that involved other repertoires the chemists had learned.

Furthermore, Mendeleev's periodic table added reasons for speculation concerning what it is in the elements that makes them increase in weight in an ordered manner. This in turn was relevant to the conception that there must be commonality in all the elements, in all matter. Perhaps the elements themselves are composed of common constituents in varying amounts. Such considerations then provided the stimulus for additional theoretical and experimental endeavors.

The principle here is that individuals with certain repertoires of behavior, faced with certain novel stimulus situations, will perform certain novel response combinations that constitute discoveries, creative findings, and original statements. These products can then serve as new stimuli that can bring on additional novel responses, producing stimuli for yet other creative behaviors. Such complex, cumulative "mental" processes can take place in one person or can involve multiple individuals interacting. A detailed history of various long-term acts of creativity, analyzed in terms of the learned skills of the people involved and the manner in which their findings served as stimuli to themselves and to others, would give a new and profound view of human creativity. Creativity, even of scientific developments, does not stem from extraordinary minds-brains, but from learning. Discoveries ordinarily come from actions made possible by the cumulative learning of complex repertoires that occur for an individual and across multiple individuals. The central point is that when (1) basic repertoires include many units, (2) there are a number of repertoires, and (3) the stimulus configurations impinging on the individual are complex and new, creative acts of combination may occur. All kinds of human creativity can be analyzed in these terms, in a manner that can serve as the foundation for various types of research and give a unified view of human behavior and human nature.

The present conception of human behavior and human nature provides principles and concepts with which to make objective the subjective mind and its actions. What we call the human mind is not a mental-brain thing itself. It is actually composed of the behavioral repertoires the individual has learned and the learned connections of the brain. The various actions of the mind are actually the workings of those repertoires in response to stimulation. The behavioral repertoires can be studied objectively. Individuals can be trained to specific repertoires, and the manner in which this changes their mental operations and the way the repertoires affect behavior can be observed. This type of study can bring mental processes into the grasp of direct observation. Moreover, the result is a unification of the traditions involved. The brain can be seen as the mechanism by which mental actions occur. The mind, however, is composed of the behavioral repertoires that have been learned. The actions of the "mind" affect the behaviors the individual displays. This approach yields an objective way of conceiving of the mind and thus is an important part of the human behavior and human nature conception.

This framework resolves the problem of separation, of the internalist mind/brain against the externalist behaviorism. Thorough knowledge of the basic repertoires will provide the means for understanding the workings of "the mind," such internal activities as consciousness, thinking, reasoning, problem solving, creativity, genius, and purpose.

COMMONSENSE AND SCIENCE

"Wait a minute," the critic might say. Everyone knows that humans learn skills that are necessary for later learning. Why would we send children to school if that weren't the case? College premedical, preengineering, and prelaw curricula are taken because they prepare students for learning the program proper. Isn't it obvious that language skill comes before reading, that reading makes possible learning in various areas. Isn't it also obvious that people learn repertoires of skills? Don't we all know that people have different groups of motor skills for tennis, football, and dancing; different groups of emotions in social attitudes, interests, and needs; and different language skills in storytelling, poetry, and logic?

Oh yes. Certainly everyone does know that humans learn from experience in a cumulative way. No question. Everyone also knows that many human skills are complex and involve a long, cumulative line of learning. But let's look at this knowledge. What level is it at—is it refined scientific knowledge that systematically analyzes those repertoires of behavior, indicates how those repertoires constitute traits of individuals, and describes how those repertoires determine the individual's life behaviors? Can that knowledge be used to advantage in raising children? Does the man in the street have the needed knowledge? Do those who consider human behavior on the scientific level have that knowledge? Or is the general state of knowledge of repertoires and cumulative learning on a commonsense level?

The fact is that with phenomena open to everyday observation there is always commonsense knowledge. Only later do systematic scientific study and theory appear. The existence of earlier commonsense knowledge does not detract from the value of the later scientific theory, for a great difference exists between commonsense and scientific knowledge. Commonsense knowing is undeveloped, usually vague, not specific, not wide-ranging, lacks clear explanation of cause, and has limited knowledge of what to do in dealing with the natural phenomena involved. In commonsense there are usually various opinions, some quite the opposite of others, and there is no way to decide which is true. Although everyone knew that objects fall, that commonsense knowledge did not compare to the knowledge of gravity gained by the principles Galileo, Kepler, and Newton set forth. Many people knew on a commonsense level that, unlike biblical description, animal and plant species had arisen and died out on Earth. But that knowledge did not compare to the great advances in science and general understanding that Darwin's and Mendel's theories made possible. Various ancient peoples knew about how the wind could drive sailing vessels, but that didn't compare to scientific knowledge of airfoils that underlies the modern design of sailboats and airplane wings. Long before physics had knowledge of aerodynamic laws, Australian bushmen threw boomerangs and Wilbur and Orville Wright succeeded in constructing an airplane. But when the forces that maintain an airplane aloft became systematically known, as well as those that impede its passage, methods of scientific study could be developed by which much more effective aircraft could be designed.

Scientific theories systematically constructed on the basis of empirical findings can constitute very original knowledge that has great power. That constitutes the type of advantage science knowledge has over commonsense knowledge. There is no systematic conception, no empirical analyses, of basic repertoires and cumulative learning in commonsense, everyday beliefs. There is a great deal of conjecture about genes, the brain, brain chemicals, the mind, human evolution, primate behavior, evolutionary psychology, and such in the Great Scientific Error as explanations of human behavior and human nature. Yet, despite their evident importance, no systematic treatment of either basic or human learning is included. There is no recognition of the difference between humans and all other animals in cumulative learning. Yet that is a difference that completely separates humans from other animals.

CONCLUSION

The development of knowledge of the basic repertoires and cumulative learning opens a new understanding of the importance of learning in the explanation of human behavior and human nature. This constitutes a new development, a theory level with which to construct a theory of any type of human behavior. Humans have astounding learning capacity in the combination of a versatile body that can respond in so many ways and do so many things, a set of sensory organs that can respond so broadly and sensitively to the stimuli of the world, a very elaborated nervous system that connects the sensory input with output to the response organs, and very centrally a huge brain that makes those connections possible through learning.

But as unimaginable as it is, that is not all that humans have. For that built-in apparatus, general to the species, enables humans to learn in a new way, an astounding way, a marvelous way. Not only do humans have a marvelous built-in learning ability, but that learning ability enables humans to multiply endlessly that endowment. Humans can build upon their learning; humans learn repertoires that make them better learners in new ways. Humans learn repertoires that allow them to learn new repertoires that provide the basis for learning still newer repertoires, and on and on, in an accelerating acquisition. The human species indeed becomes the marvelous learning animal. That is the importance of the basic repertoire and the cumulative learning of successive repertoires; they make the individual's learning inexhaustible during a lifetime and unlimited and infinite for the human species.

That is why I have called the knowledge of the basic repertoires and cumulative learning "the missing link." For these developments make it possible to understand aspects of human behavior and human nature that have not been understandable with only the basic learning principles that operate with all advanced animals.

The missing link provides the basis for a real explanatory theory of human personality. It makes possible an abnormal-behavior theory that can unify the understanding of the various behavior disorders. The missing link provides a framework for considering the mind in terms of learned repertoires. These accounts of the missing link provide the foundation for a new theory of human evolution. All these constituents sum to yield a definition of who we humans are. And the broad, multilevel conception has deep meaning for society, for science, and for individuals as they live life—for creating science as well as applications. These are thus the topics of later chapters, in which the marvelousness of human learning will be deepened and broadened.

An evolved biology accounts for an important part of humanness. But biology does not account for marvelous learning, does not make humans into absolutely extraordinary learning animals. *The marvelous learning ability of humans comes partly from past evolution and greatly from each individual's learning.*

Part 4

LEARNING HUMAN NATURE

Chapter 6

MARVELOUS LEARNING
OF PERSONALITY

Amost salient aspect of the human species consists of its variability. The variations are huge, from a macho wife abuser to a milquetoast spouse, from a primping party girl to a competitive medical student, from cautious and conventional to different and openly original, from sparklingly intellectual to stolid disinterest, from overly generous to narrow self-concern. In behavior some people are talkative, some quiet; some outgoing, some shy; some confident, some insecure; some honest, some not; some happy, some depressed; some optimistic, some pessimistic; some criminal, some law-abiding; some intelligent, some not; some athletic, some clumsy; some pugnacious, some timid; some hardworking, some lazy; some aggressive-brutal, some kind; some religious, some not; and on and on. People within the same society even eat differently and are sexually turned on by different others and engage in sex differently—behaviors traditionally considered as biologically determined. Not only can the dimensions of difference be huge, but there are many differences, so endless combinations occur, which produces great uniqueness. No other species approaches that variability—a special nature needing explanation.

Those individual differences take up much of our interest in humanness, for they have great portent, they affect what happens to the individual in life and thus what happens to everyone associated with the individual. A suicide bomber who had displayed in many ways deep religious fanaticism finally kills multiple people and causes great destruction, thus ending his own life and affecting the lives of many others—his family, friends, community, even his country and other countries. Less dramatically, variations in human behavior are important in all kinds of ways, for how humans behave toward each other has great mutual impact. Some behaviors are highly valued in a culture and others are abhorred. Some behaviors get rewarded, others get punished. A parent wants a child with traits of intelligence, talent, studiousness, and love of family. The head of a company wants an employee who is industrious, honest, loyal, and conscientious.

People desire a leader who has a fair, competent, compassionate, and forward-looking nature.

No wonder, then, that differences in individuals' behavior have attracted attention since antiquity and no doubt long before. Early peoples explained differences in individuals' behavior by internal spirits. Theological conceptions explained human behavior variations by a divinely given soul. In the fifth century BCE, Hippocrates expounded a general personality theory that human variations were due to the mixture that each person had of four "bodily humors": blood, phlegm, black bile, and yellow bile. Each type of humor produced a certain broad type of behavior—like blood inducing impassioned ways of behaving. Those with high proportions of blood behaved passionately. Here was an early natural, biological attempt at explanation. Another early Greek personality theory held that each person contained a *homunculus*, a little man inside the head who directed the person's behavior. These were early "internalist" explanations, that is, explanations that inferred things *inside* the individual to explain her or his behavior. The surge of theology turned explanation toward supernatural internal causation, such as belief in the soul.

> Each one of us has received a special gift in proportion to what Christ has given. . . . He appointed some to be apostles, others to be prophets, others to be evangelists, and still others to be pastors and teachers. (Ephesians 4:7 and 11)

In a later turn back toward nature and science, the British empiricists in the seventeenth century transformed the concept of the soul to the natural concept of the mind. Those philosophers looked for earthly determinants of the mind in the individual's experience. Later, Charles Darwin laid a basis for the scientific study of personality as a product of biology by considering behavioral traits as well as physical traits to be caused by evolution via natural selection. Lions are ferocious, deer are timid, and pit bulls are pugnacious.

Darwinians considered *human* traits of behavior in the same way. Intelligence, for example, was regarded as an evolved trait of behavior. Some individuals, or groups, were thought to be more intelligent than others by virtue of biological inheritance. Just as a breeder could produce a new type of dog, or a habitat could produce a new species, an environment that selected for more-intelligent individuals would over time raise the intelligence of the whole group. In this view, groups that show less intelligence by their lower achievement have evolved less on this trait. The belief continues to this day (see Rushton 1999);

based on it there is a social philosophy of eugenics, the position that since intelligence is inherited, intelligent members of the human species should be selected for breeding in order to weed out the unintelligent, thereby increasing the human level of intelligence.

A later scientific theory, by Sigmund Freud, proposed that humans are born with biological drives or needs, a part of personality called the *id*. In the individual's attempt to gain gratification, the id adjusts to environmental constraints and demands; and in that process, out of the id develops a new part of personality called the *ego*, which governs the individual's adaptive behaviors. The ego, in further experience with the world, grows the third part of personality, which incorporates the mores and customs of the social group, into the *superego* (conscience or morality). Freud's theory of personality thus considered both internal biological causes and external environmental causes, but conjecturally in both cases, without close definition by evidence. Various such theories followed.

Individual differences in behavior are taken to indicate that personality is more than just behavior. It is something that is inside the individual that works generally in determining the individual's characteristic ways of behaving. For example, everyday observation shows that the variety of human behaviors is not just random. Behaviors come in groups. A child who learns to write letters better than other children also learns to read better, spell better, write stories better, and better learn math and science as well. Some children not only play kickball, softball, and touch football better and earlier than other children but also gracefully and skillfully embark on acquiring any new athletic skill—gymnastics, dancing, driving, whatever. Some will behave in ways that favor themselves across any activity in which they engage with others—business, friendships, marriage, even raising children. Some individuals are very truthful and can be believed about any number of things. Some are very reasonable and logical; what they say about any topic hangs together, is clear and consistent. Others speak haltingly— with little organization—and are difficult to communicate with and understand. A person will behave shyly on meeting all people on various occasions or will be defensive in various situations involving criticism or generally will not display a sense of humor or will not usually be generous with others or will not get along with colleagues. Take Bernard Madoff, the self-confessed defrauder of others—to the tune of perhaps $56 billion. Golf certainly constitutes a very different situation than creating and running a record-setting Ponzi scheme. Yet the golf scores Madoff reported one year were more consistent than those of Tiger Woods (Wiesenthal 2008). Since consistency in golf comes with skill, Madoff's scores appeared too good to be true. This apparent carryover from his finances to golf suggests consistency of behavior across widely different situations and types of activity. That type of consistency is general. A person dishonest in one situation can be dishonest in other situations; a person who is kindly and sensitive with one person can be that way with other people. This helps give the impression there is something operating besides the acquisition of separate behaviors.

Behavioral characteristics also tend to persist. Parents frequently say that a child showed a particular type of behavior very early and that the behavior continued to increase with age. Not all of these reports are accurate, of course; anecdotal accounts are notoriously spurious and have to be questioned. But for many there is a consistency that begins in childhood and lasts a lifetime. That gives the impression that there is something within the person that is responsible.

Another feature of personality is familial similarity in behavior. While there are many variations among people, we have all heard descriptions such as, "He's stubborn like his father," "She's smart like her brothers," "I wouldn't trust her any more than her mother," and "The whole family will give you the shirt off their backs." Widely described are consistencies of behavior within families, while there are great variations over different families. That gives the impression that what operates inside is inherited.

In view of such a large set of important phenomena, there is no question attention is drawn to describe them, attempt to explain them, and attempt to devise ways of influencing them. We need to know about personality traits, what they are, how to detect them, how individuals with certain traits are going to act, how individuals with certain traits should be treated, and, especially, what the causes of personality traits are and how they can be created or prevented.

Despite all of the variations in explanations of personality, there is a commonality of belief that people carry something inside that in some way makes them behave characteristically in their own way and differently from others. One of the proven methods of science consists of measurement, even of assumed entities that cannot themselves be observed. After all, aren't black holes a proven scientific fact even though they can't be observed because they emit no light? Weren't genes a useful scientific concept long before they were observed? So personality is a very solidly accepted concept in the behavioral sciences, making it fair game for attempts at measurement. If personality can't be directly observed, perhaps it can be indirectly measured, as black holes are.

Efforts to measure personality began with the personality trait of intelligence. Measurement of intelligence had obvious value, for example, in selecting the type of education appropriate for a child. One early attempt involved measuring reaction time on the supposition that the speed of neural action underlies intelligence differences. That avenue was abandoned as futile after a time, although it springs up again from time to time.

PERSONALITY MEASUREMENT

The first successful attempt to construct a test to measure intelligence emerged from the work of Alfred Binet and Theodore Simon around the beginning of the twentieth century. They composed problems with increasing difficulty, like asking a young child to follow a simple verbal instruction, or to name objects, or to indicate the number of objects in a group. Their test successfully predicted success in school. The methodology for constructing intelligence tests became that of selecting items that predicted children's performance in school. The belief is that such tests measure the internal intelligence—brain quality—but the tests actually measure people's ability to solve problems of various kinds.

Differences in the way individuals behave also have led to the belief that there are additional personality traits, which stoked interest in measuring them. One method for doing that was established by Gordon Allport and H. S. Odbert (1936). They took eighteen thousand words out of the dictionary that name human characteristics (like *talkative, ambitious*, and *cruel*). In several steps, Allport and Odbert found redundancy in the trait names and reduced the number to 171, and finally to 35 traits of personality. Their basic data, however, is based on observations that common people made of different kinds of behavior along with the conception that something inside people is responsible for those differences.

Allport and Odbert's work was only the beginning. Many others in the field of personality use and refine their approach. One of the most active lines of work in personality today, called the *five-factor theory*, is based on their "lexical" method. With modern technology, the traits have been reduced to five: neuroticism, extraversion, agreeableness, conscientiousness, and openness. These five, however, also have facets that are trait measures, so they actually cover a much wider number of personality traits. The general measurement approach has produced many different tests whose purpose is to measure aspects of personality, tests that are used widely in cases where knowing about individuals' characteristic behaviors is important.

A PROBLEM WITH PERSONALITY

This shows that work to construct tests of personality has been valuable, but not for the reason generally believed. Tests produce no evidence of internal personality traits; they do not get within individuals. They are based on the implicit assumption that the behaviors they measure are indicative of some internal personality process. That assumption has no proof, so theories of personality based on tests cannot tell what personality is, determine the causes of the behaviors measured on personality tests, or explain the great differences in behavior that people exhibit. Neither tests nor personality theory have helped in finding genes or brain features that cause behavioral characteristics. Really the basis for a belief in such personality processes is that we all learn that belief in our common language; it is the traditional belief. "Intelligent kids do well in school." "She inherits self-confidence, that accounts for her social manner." "He's got a lot of courage to fight off that robber." We all have heard thousands of such statements, all of which indicate that there is a personality process inside that explains the external behavior. While this is the traditional belief, there is no evidence of inherent brain differences that account for different traits of behavior.

The behaviorist John Watson rejected concepts like soul, mind, or personality because they were just assumptions that were not defined by directly observable events. That position is followed by radical behaviorists to this day. The tests of personality are not systematically considered by behaviorists, which is a mistake, for it throws the baby out with the bath water; personality measurement has produced valuable knowledge.

This neglect has to be considered a profound weakness. Intelligence tests measure behaviors that predict school performance. Are intelligence-test behaviors and school-performance behaviors important? You bet. That is true of other personality tests. The field of personality attempts to specify what composes human nature. A theory of humanness has to address the behavioral phenomena involved and add understanding of them. Let's look at the present approach in this respect.

THE MISSING-LINK THEORY OF PERSONALITY

The missing-link theory takes a different approach than either traditional study or behaviorism because it is based on a different conception, different empirical evidence, and a different program for gaining knowledge of human behavior.

That program began with both the belief that learning was an important cause of human behavior and the goal of testing this empirically. This involved various types of immediate study, among them the study of the child's behavior in early development. But I found continuity in children's learning, of such repertoires as language. Children learn to babble and then advance from babbling to saying single words. Saying words brings rewards through the responses of others, and learning words becomes rewarding in itself, so the child's learning of words accelerates. As indicated, the child is rewarded for saying words that name different objects and events, nouns and verbs. I also found that learning a small language gave the child powers the child did not have before and provided the foundation for learning additional repertoires such as numbers and reading.

Fortunately, my doctoral training at UCLA included a wide number of courses, including one that focused on intelligence testing. At some point in my expanding program of research on children's learning I realized a startling overlap existed. I was studying how children learn language, on the one hand. And, on the other hand, I realized that those who constructed intelligence tests very prominently measured children's language repertoire. *That meant that as I was training children in language repertoires, I was in fact training them to be intelligent.* That had startling implications.

That wasn't all. In other studies, I investigated another aspect of language, namely the emotional properties of words. I thought that many words elicit an emotion, positive or negative, and that property comes through learning. Having a repertoire of such words gives humans a great new learning power. Humans can easily learn emotions to new things through language, no longer requiring firsthand experience with those things. When some object is described with emotional words, it will itself come to elicit that emotion. I did a number of studies (e.g., Staats and Staats 1958) that proved just that. This study showed that people would learn a positive or negative attitude toward a nationality of people if the name of that nationality was presented at the same time as positive or negative emotional words, common words like *vacation* and *food*, versus *sickness* and *lost*. Important emotional differences, like attitudes and interests, are treated as personality differences. Could these be made up of variations in the emotional responses different individuals have learned?

As indicated earlier, those in the personality-testing field never showed what their assumed personality traits were, how those traits operate to produce the individual's behavior, or what causes people to develop personality traits. Their theories totally lack explanatory value, useful explanatory value. However, what I was doing appeared to do just those things. That possibility had the potential of a momentous development. Let's consider in specific terms what personality consists of, how personality determines behavior, and what causes personality. The various traits can be considered to be of three types of repertoire —language-cognitive, emotional-motivational, and sensory-motor repertoires— for ease of classification, although personality actually consists of mixtures of basic repertoires.

Intelligence

When it is recognized that the child learns repertoires of behavior and that these repertoires straightforwardly and clearly change the nature of the child, the stage is set for a theory of personality. But the full realization of this, in the context of individual differences and personality, did not come in one fell swoop. The first of the dominoes to fall ensures the rest will follow. In constructing the theory, as I indicated, the first domino was the realization that the repertoires I was teaching my daughter were of the same type as those *measured by items on intelligence tests* (see Staats 1968a, 1971a).

To illustrate, I trained her to discriminate and name numbers of objects and then to count multiple objects. The learning conditions and the behaviors were straightforward, explicit, and could be publicly seen and recorded. But intelligence tests measured those very skills. An item at the six-year level of the Revised Stanford-Binet Test of Intelligence (Terman and Merrill 1937a, 95) shows this clearly; the examiner asks the child to "Give me . . . blocks" and the child is successively asked for three, nine, five, and seven blocks. This showed that what the child learns is what is measured on an intelligence test.

The child must have learned various aspects of the number repertoire, including counting, in order to score on that item and be measured as intelligent. There are many children who have not learned that number repertoire by the age of six and are thereby measured as less intelligent. Children are measured as more intelligent if they have the repertoire and can respond appropriately to the intelligence-test item. While the item appears at the six-year level of the test, my daughter had the necessary repertoire components before she was three years old. *Does that finding mean that my training had increased her personality trait of intelligence?*

The early number repertoire, however, is not the only one that constitutes intelligence. The preceding chapter described how I began training my infant children to respond to words with particular actions. For example, my daughter learned to approach when I said "Come to Daddy," and to look at me first and then at whatever I was pointing to when I said "Look." She learned such "word-motor units" because she was reinforced for making certain responses to certain words. The child ordinarily learns a large vocabulary of words that control actions such as *go, give, sit, run, whisper, stop, show, lie down, push,* and *turn,* to give a few examples. Adverbs also come to control variations in responding, and nouns and adjectives come to control the objects to which the response is made, as shown in responding to the request "Turn the red handle quickly."

When we get down to the level of analyzing intelligence in terms of behavioral repertoires, we can see that all the items of a children's intelligence test actually measure this verbal-motor repertoire. For each item involves the examiner giving an instruction to the child that the child has to follow if the item is to be passed. Moreover, there are many items that more specifically test for this repertoire, as the following examples demonstrate. "Show me the kitty" (among a group of miniature figures) (Terman and Merrill 1937a, 75). "See what I am making [a structure of blocks]. . . . You make one just like it" (76). Another item involves presenting different instructions that must be followed—"Give me the kitty. . . . Put the spoon in the cup" (77).

What I am saying is that the extent to which the young child has learned the verbal-motor repertoire determines the extent to which the child can follow instructions, an invaluable skill basic to much human learning. No wonder measuring that repertoire constitutes an important part of measuring intelligence. The child who has the fortune of getting good language training gets an intelligence boost; the unfortunate child does not get that training. I said this in 1971 (Staats[a]) and showed it clearly in 1981 (Staats and Burns); now there is much research showing that social conditions that produce slow language development yield children with low intelligence. A recent book now even takes that same position (Nisbett 2009).

My last example concerns the verbal-labeling repertoire. This is more commonly and vaguely referred to as the individual's vocabulary. The verbal-labeling repertoire consists of the words the child (or the adult) has learned to use in labeling the various objects, events, and people that make up her world. When my daughter was nine months of age, she had a thirteen-word speaking repertoire. She would say such things as "Moo" for *moon,* "Dah!" for *dog,* "eeh" for *food* or *eating,* and "Daw" for *doll.* If I said "What's that?" and pointed at her doll, she would say "Daw." If I said "Give me the doll," she would perform the expected action. The more words the child has learned, the more the child's behavior can be guided by language. There are many items on intelligence tests that measure the extent of the child's (and later the adult's) verbal-labeling repertoire. At the three-and-a-half-year level of the Revised Stanford-Binet Intelligence Scale (Terman and Merrill 1937b), one item involves showing the child pictures and asking "What's this?"; the child scores a point by labeling the objects correctly (Terman and Merrill 1937b, 83). As another example, another item asks the child "Which stick is longer? Put your finger on the long one" (84). Such items measure both verbal-labeling and verbal-motor repertoires. Many such items on intelligence tests guarantee that the child with a complex enough language repertoire will measure as more intelligent than the child whose repertoire is less rich (Staats 1963).

This theory of intelligence as various learned repertoires sprang directly from my research with my daughter and my son as well as my research with other children. Much additional evidence that children trained in relevant basic repertoires will test higher on relevant intelligence measures was still needed. I began providing that proof in my University of Wisconsin and University of Hawaii preschool projects that involved the training of four-year-old children. Karl Minke and I constructed procedures for training a group of culturally deprived four-year-olds based on the methods I had used with my own children. These preschoolers were expected to have problems in school because older siblings already displayed such problems. Extensive records were made of the training and of the repertoires the children acquired. The children also took several personality tests before and after training.

They attended our preschool for seven and a half months. At preschool in the morning, each child left the classroom to receive three brief training sessions in numbers, letter reading, and letter writing, for a total of about fifteen minutes. As a result of the training, the children advanced an average of eleven IQ points on an intelligence test and from the second to the twenty-eighth percentile on a test measuring children's readiness for entering school, another test of intellectual skill (Staats 1968b).

Later, Leonard Burns and I conducted several additional studies with other preschoolers. Half the children received training in basic repertoires and were compared on intelligence tests with the other children who did not. In one experiment, the children were taught to write and read the letters of the alphabet. Although their training was not broken down in that way, they actually were learning a repertoire that included the sensory-motor skills of holding a pencil properly, beginning a line drawing in a particular place, making lines in determined directions, looking at the model repeatedly during copying, drawing letters on request, saying the letter names, and so on.

I analyze the repertoire that is involved in writing letters because the several skills so described will generalize to any task that requires those skills, not just to the specific task of writing letters. On that basis, Burns and I explored the possibility that some tests of intelligence quite different than letter writing actually measure that same repertoire. These were geometric-design and maze tests on the Wechsler Preschool and Primary Scale of Intelligence (WPPSI) (Wechsler 1967). The geometric design subtest requires copying line drawings of geometric figures. The maze subtest requires tracing a path through a printed maze with a pencil without touching the lines that form the maze's sides. Both tests required that the child make purposeful lines with a pencil. Our expectation, thus, was that if we trained children to write letters, they also would be acquiring a sensory-motor repertoire that added to their ability to perform better on the other intelligence tests than a comparable group of children who had not had that training. The results were very clear. Four-year-old children who had previously learned to copy and freely write the letters of the alphabet were significantly more "intelligent" on these two subtests of the Wechsler intelligence test than children without the letter-writing repertoire (see Staats and Burns 1981). So, here, tests that were believed to measure a mental trait of intelligence were really composed of a complex, learned sensory-motor repertoire.

Those findings also answer the crucial question of the cause of intelligence differences. Some preschool children receive experiences from their parents that involve holding pencils or crayons and drawing pictures, making lines, even writing letters. Other children do not have such experience. We can conclude that a parent who trains her/his child in letter writing will be producing a more *intelligent* child. All parents who want to raise an intelligent child can use our training methods to do so.

Burns and I also trained another group of preschoolers in the numbers skills I developed very early with my children. The children in our study measured to have higher intelligence on the number-concept items used on intelligence tests, as compared to another group of such preschoolers who were not given that training. Again, it is clear that children who are fortunate enough to have had the numbers-learning experiences thereby become more intelligent on most intelligence tests.

A third study, done by Burns, dealt with more-complex language repertoires. He trained a group of four-year-olds to name the objects in several different classes. Then he trained them to name the classes themselves—for example, *pancakes*, *cheese*, and *bread* name objects in a class, and the word *foods* names the class of objects. He then presented these trained children with similarities-intelligence-test items that involved those objects. These children were more intelligent on these items than children who had not had the language training (see Staats and Burns 1981). The complex language training thus also raised the children's intelligence. Mind you, the children in these three experiments were not trained to answer items on the intelligence tests.

Nisbett, a cognitive psychologist, has recently presented a theory of intelligence that adopts the same conceptual view (2009). However, these studies and the theory on which they are based make clear that intelligence tests need to be analyzed completely for the repertoires they measure (Staats 1975). A large research area is needed to establish what the intelligence repertoires are and how they are learned. Then studies have to be conducted to show how parents and their professional consultants can generally use the learning procedures.

The conception of intelligence as an inherited quality of the brain gives the parent nothing to do in the matter of the child's intelligence. It is thus relevant to ask which approach is more useful. The answer is clear. The ordinary conception of intelligence gives the parent no role in helping the child. The learning conception gives the parent a central role. When all the basic repertoires, and how they are learned, are known, it will be possible to ensure that all children get the needed learning experience.

Why Do Intelligence Tests Predict School Performance? The Function of Intelligence

Intelligence tests have credibility because they predict school performance. The implicit idea is that if intelligence tests predict the ease and quality of children's school learning, then surely that is evidence that people differ on the extent to which they have an internal intelligence trait in their brain or mind. If that were not true, why would intelligence tests predict later performance in school?

Intelligence tests predict performance because they measure a sample of the basic repertoires that are necessary for good learning and good performance across a wide set of situations. The child with a well-developed verbal-motor repertoire follows directions well, and following directions is necessary for much school learning. The writing repertoire makes it easier to learn to write the letters and numbers. Kids with that repertoire will do better in school as a consequence.

The number repertoire, including counting, will provide a basic repertoire that will make the child a better learner of arithmetic operations than a child without the repertoire. Even more generally, the child with a well-developed language repertoire will do better on many learning tasks than a child with a less-developed repertoire. This is to say that intelligence really is important because it measures basic repertoires that determine how well the individual is able to do on various learning tasks. Intelligence tests measure important elements of the extended human cumulative learning process.

Those who constructed intelligence tests composed items intended to measure the quality of the mind-brain, believing that performance on the test revealed that intelligence quality. What they really were doing, however, was selecting items that measure basic repertoires that serve as a foundation for learning in school. Developing a conception of those basic repertoires, showing how those repertoires are learned, as well as showing how those repertoires improve performance on intelligence tests, changes the whole perspective and produces a new theory. *What intelligence-test constructers should be studying is what basic repertoires are needed for the learning tasks that children confront in life.* The present theory of intelligence opens whole new avenues of research and development not only for measuring children's potential for schooling but also for advancing all children's learning abilities. The same is true for those who construct other types of tests for children: readiness tests, achievement tests, developmental tests, and other cognitive tests.

The Theory of Intelligence

There is no quality of the brain, or mind, or soul—personality—that makes people differ in their ability to respond well to learning or problem situations. The marvelous-learning-animal theory of intelligence, as one of the central traits of personality, can be stated simply. Intelligence consists of some of the individual's basic repertoires, largely language-cognitive repertoires but also including sensory-motor and emotion-motivation repertoires. The intelligence repertoires are determinants of how well the individual can perform in many situations. Natives of isolated tribes in the jungles of the Amazon could not perform simple arithmetic tasks present on intelligence tests; their culture lacks basic number skills and children in that culture do not have the opportunity to learn them. One can only guess at a question like "A conundrum is a flower, a knife sharpener, or a riddle?" without having the word *conundrum* in one's verbal-labeling repertoire. One cannot manage a course in beginning algebra without having learned a number repertoire that includes arithmetic operations. Intelligence tests contain valuable information on repertoires necessary in our culture for successful human learning. Intelligence and intelligence tests could be so much better understood within this theory.

The Emotional-Motivational Personality Trait

There are other basic repertoires than those measured on intelligence tests. That means, as everyone knows, there are other personality traits besides intelligence. Few would disagree that some traits are emotional and motivational in nature and account for important personality differences among people. For example, a newspaper column by Nicholas Kristof (2009, A23) says liberals and conservatives differ in basic emotions and this defines them; conservatives, for example, are more disgusted by stepping with bare feet on an earthworm than are liberals. All of us recognize that people in different occupations differ emotionally on what they like and dislike occupationally—many people have a negative emotional feeling doing income taxes or other computations, but luckily accountants do not. People differ widely in values; strongly religious people have more emotionality about religious events, topics, and figures than do secular people, and they differ in their emotional responding to such topics as abortion. Very patriotic people have much stronger emotions to many things than do run-of-the-mine people. There are tests constructed to measure such emotional differences in people. Even in the case Kristof delineates, a "disgust test" has been constructed on which conservatives and liberals differ.

Emotional Learning of the Animal Learning Variety

In describing the differences in disgust, Kristof suggests this emotion is inherited: "Some evolutionary psychologists believe that disgust emerged as a protective mechanism against health risks, like feces, spoiled food, or corpses. . . . Humans appear to be the only species that registers disgust" (2009, A3). That coincides with conventional wisdom that some emotions are inborn, products of evolution.

   In the present view, humans are born with neurological sensory structures that when stimulated will produce either displeasure or pleasure. These emotions are reflexive: milk in the mouth of an infant elicits a positive emotion, the smack on the bottom by the birth physician to induce crying elicits a negative emotion. Thus there are stimuli humans experience that without any prior learning will elicit either a positive or a negative emotional response within them. Those structures came down via evolution.

But humans are the marvelous learning animal, and humans also learn emotions, as Pavlov showed. While he was interested in digestive processes, actually his dogs learned a general emotional response, not just salivating. No question, many species lower on the evolution ladder than humans learn emotional responses to new stimuli through classical conditioning. Humans also can learn emotions in that way, and without end, throughout life. However, humans are born with a brain that will record an infinity of emotion-response conditionings, either displeasure or pleasure, to innumerable life stimuli. All of the emotional responses—to food stimuli, to sexual stimuli, to caresses, rhythmic soft sounds, comfortable temperatures, potable water—can be learned to new stimuli. All the negative emotional responses—to all kinds of aversive stimuli—can be learned to new stimuli. The human species can and does learn an emotional response to a massive number of stimuli and in commonsense terms recognize them as emotional stimuli.

The child's built-in emotion mechanisms are ready for action at birth. So the environment elicits emotional responses in the child, and the child begins learning emotions to new stimuli immediately. At the very beginning, the sights, sounds, and touch of the parents coincide frequently with the presentation of food and other positive emotional things. That fulfills the requirements for conditioning, so the infant learns a strong positive emotional response to the parents. That emotional learning, positive and negative, is monstrously complex, continuing throughout life.

Even stimuli that are emotional on a biological basis also have a learned component in their character. To illustrate, the sexual organs are sensitive to touch. Tactile stimulation of them elicits a positive emotional response. That's built in. But in humans much sexual-emotional responding is learned, and that means many stimuli—other than genital touching—can come to elicit a sexual-emotional response through learning. That learned component becomes very important in eliciting sexual-emotional responding and in determining sex behavior. That is why there are such great individual differences in this area. The same is true with other biological stimuli that elicit positive or negative emotional responses through learning.

Humans' great learning ability, and the great number of emotion-eliciting stimuli in the lives of humans, makes for infinite variation in emotionality. Beginning in childhood, differences in learning experiences give people wide individual differences in emotional responses to people, objects, customs, recreations, occupations, behaviors, sexual stimuli, artworks, beliefs, political stimuli, and such that are encountered in life. The emotional responses the individual *learns* toward such categories of life stimuli, along with those that are inborn, constitute a central personality trait. Much of the learning of this personality trait occurs through language, a very special human learning mechanism.

Learning Emotion through Language:
The Marvelous Learning Animal Is Emotional

The basic animal-learning principle of classical conditioning produces much learning of emotions in humans. Language, however, is an emotion-learning machine. Through language, humans learn emotions to a wide range of things in an indirect way, without ever having direct contact with those things themselves. This needs to be elaborated upon in considering the emotional personality trait.

While still a graduate student, I read a study by Osgood and Suci (1955) in which they found three types of words, each having a different type of meaning. One type they called "evaluative meaning." Words such as *food*, *vacation*, and *happy* have evaluative meaning on the positive side, as *ugly*, *sick*, and *war* do on the negative side. I wasn't into the concept of evaluative meaning; rather, I saw that the words Osgood and Suci called evaluative really were those that elicited an emotional response. The words had come to do this through classical conditioning. Each such word had been paired with, or named something, that elicited an emotional response. The word *good*, for example, is paired with many different things that elicit a positive emotional response. The word *ice cream* names a tasty food and occurs many times in the presence of that food, usually coming thereby to elicit a positive emotional response. The word *sex* is paired with events or words of a sexual nature that are typically very emotionally positive. Through conditioning, the words *good* and *ice cream* and *sex* ordinarily elicit a positive emotional response, as do all the words that were rated as evaluative by Osgood and Suci.

My analysis posited that there are thousands of words that elicit an emotional response. That had many implications. If many words elicit an emotional response, for example, then it should be possible to produce new emotional learning through language, without having to present the actual things themselves. Just pairing emotional words with something should make that something emotional also. This hugely important learning mechanism is unique to humans because only humans have language.

When a freshman in college, I wrote a term paper on propaganda that in essence described how words could be used to change opinions and attitudes. Words are powerful enough to get people to support a war that might kill them, as well as to get them to do mightily beneficial works that take time and effort. The behavior analysis of emotion motivation showed how in propaganda, advertising, or just plain conversation, emotional words could pass on their emotional character to anything with which they are paired. Tell a friend that a movie "stinks," and he will have a less positive feeling about it than before. Tell an audience that abortion is murder, and many who hear that will be conditioned to a strong negative emotional response to the word. Actual murder, of a doctor who performs abortions, has occurred because someone hated him enough to kill him and go to jail for life for doing so. How, other than through words, could the killer have learned that hate toward someone unknown? Language functions this way generally, not only in dramatic cases; we manipulate others constantly through language conditioning, and we too are manipulated in the same way. That's a reason governments seek the opportunity to send programs to other countries by radio, in order to influence the emotions of their people. Emotional conditioning through language is easy and effortless, and the conditioning experiences may be spread out in time and go unnoticed. Yet the effects may be very powerful.

I began formally studying emotional learning through language when I began my first academic job, supported by the Office of Naval Research. I called my method for creating new emotions *language conditioning*. Just what I expected happened: a neutral stimulus, a nonsense word like YOF—presented visually multiple times, each time along with a positive emotional word—came to be positively emotional itself. However, with different college-student subjects, when the nonsense word *YOF* was presented with negative emotional words, they came to dislike the word. For this language-conditioning study I selected the positive emotional words *beauty, win, gift, sweet, honest, smart, rich, sacred, friend, valuable, steak, happy, pretty, health, success, money, vacation, love.* (Actually, can't you feel the emotional response just in reading such words?) I also selected the negative emotional words *thief, bitter, ugly, sad, worthless, sour, enemy, cruel, dirty, evil, sick, stupid, failure, disgusting, agony, fear, insane,* and *poison.* Each word elicits other responses (meanings) as well. But all in each group are the same in terms of the emotional response they elicit.

Just like Pavlov's dogs learned an emotional response to a bell because it was paired with food, human subjects learned an emotional response to a word that was paired with emotional words, either positive or negative (Staats and Staats 1957). Importantly, the subjects were not even aware that they had been conditioned. Life is complex, and humans have vast experience that produces emotional learning. One can hear or read about places, things, events, and activities in great abundance—impossible to experience personally—and learn emotional responses to them. One can, through language, learn emotional responses to things that can never be directly experienced, like God, framers of the Constitution, foreign political leaders, sexual encounters, government, and terrorists. We all learn likes and dislikes from what our country's media tell us. The media in other countries may tell their people different "truths." Thus the two peoples are conditioned to feel emotionally opposite on important matters.

Language is full of emotional words; that constitutes the mechanism. Those words are important for creating new emotional learning for humans. No other species has such an engine for learning emotion. Those emotional aspects of language contain central explanations of human behavior and human nature involving aesthetic values, religious values, political values, tastes, interests, needs, attitudes, and preferences, as well as fears, anxieties, depression, moods, dislikes, and prejudices. Centrally, the emotional-motivational aspects of personality are to a great extent produced through language.

Emotion-Motivation and Personality

Thus far I have described how humans learn an extremely large repertoire of emotional stimuli. It should be understood that the learning experiences are huge in number and are very different for different people. That means their emotional repertoires differ greatly. But what is the importance of this? The answer is that those differences in emotional repertoires are personality differences; they are explanations for the wide differences in the behavior that people display.

To understand this, we have to go back to the basic principles described in chapter 3. First of all, the way we feel about the stimuli in life determines our general feeling state. A person who has positive emotional responses to most of the stimuli encountered in life is happier than the person who encounters many negative emotional stimuli (Rose and Staats 1988). That difference is recognized as a personality difference, and it will be considered again in the next chapter, on abnormal personality.

Happiness or sadness, however, is not the only personality effect of the emotional repertoire. Traditionally, personality traits are conceived of as causes of the individual's behavior, and that is indeed part of the emotional repertoire as a personality trait. Again, we have to consider the basic principles of learning to understand the explanatory actions of the emotional repertoire, for one thing, that positive emotional stimuli act as reinforcers, rewards, and negative emotional stimuli act as punishments. That means a person's emotional repertoire determines what will be rewarding for the individual and what will be punishing. And that means also that the individual's behavior will be determined by the nature of her or his emotional personality repertoire. We learn to do things that are rewarding; we learn not to do things that are punishing. That determines a great deal of the individual's behavior. A person who finds heterosexual relations rewarding, other things equal, will engage in heterosexual behavior more frequently than a person who finds such relations punishing. A person who finds playing a sport rewarding, other things equal, will spend time playing that sport. A person who finds reading rewarding will read more than a person who does not, other things equal.

But that is not the only principle that makes the emotional repertoire a personality trait. Another principle is that we are attracted to positive emotional stimuli and we avoid negative emotional stimuli. If opposite-sex individuals elicit a positive sexual-emotional response in an individual, he or she will approach members of the opposite sex in the various ways the individual has learned to do so. If children elicit that emotional response, they will be approached, at least in imagination. If same-sex individuals elicit a sexual-emotional response, they will be approached. Of course, the behavior of the individual will depend upon other things. Not every individual who has a sexual-emotional response to children will act upon that, and that holds for every type of behavior. This third effect of the individual's emotional repertoire, thus, helps determine the individual's incentives and the behaviors that incentives induce. That is why I refer to the emotional repertoire as the emotional-motivational repertoire.

Emotional-Motivational Personality Tests

The hugeness of the emotional-motivational repertoire can be seen from the many personality tests that measure emotions. Sometimes the tests measure the like/dislike (emotional) aspect, sometimes the reward-punishment aspect, and sometimes the approach-avoidance (incentive) aspect. Some tests will have items of these several types. The tests of attitudes, values, interests, and needs will be used to characterize the emotional-motivational repertoire.

The tests do not really differ basically, even though they are considered different, for each measures emotional responding, and thus how individuals behave. Their differences lie in *what* elicits the emotion and thus in what area of life the test can predict behavior. Attitude tests usually measure emotional responding to classes of people. Interest tests measure emotional responding to things that differentiate the members of different occupations. Values tests, on the other hand, measure emotional responding to aspects and issues of society, like religion, economics, political affairs, and education. There are tests of negative emotions also, like tests of fears, anxiety, and depression that also have some differences.

I am thus saying that a number of psychological tests are related; they involve the same principles, and they should be considered as related. When that is done, the knowledge gained of one type of test can be generalized to others; the knowledge of the lot can provide a broader, related view of human emotion and human behavior. Several test areas to be described may be considered to measure aspects of the more general emotional-motivational personality trait.

Attitude Testing

In the language-conditioning study described above, positive and negative emotional responses were learned to nonsense words. Such studies indicate basically that humans learn emotion via language and that language conditioning applies widely to life. When a child grows up in a home where ethnic, religious, political, feminine, national, or homosexual group names are paired with negative emotional (derogatory) words, the child will learn negative attitudes (emotional responding) toward the group names involved as well as toward the people who bear those names. When a joke is told about a group using words that indicate that they are not bright or are oversexed, dishonest, clannish, and such, that experience will produce negative emotional conditioning. The opportunities for such emotional conditioning to occur are vast in number. That means deeply held attitudes can be produced in the everyday experiences in life. More prejudice is created through words than through direct experience; that is the nature of human experience.

When there is an emotional response to a group of people, it is referred to as an attitude. We showed in two studies using my language-conditioning method that a positive or a negative emotional response can be conditioned to a national group—like the *Swedish* or *Dutch*—by pairing positive or negative emotional words with the name of that group (see Staats and Staats 1958).

An attitude-test item with respect to nationalities could ask that the nationalities be rated on a one-to-seven scale of like/dislike. Or many different characteristics could be listed separately, such as honesty, intelligence, friendliness, clannishness, generosity, selfishness, and cowardliness, and groups could be rated on those characteristics. Like or dislike can be measured with a variety of items that indicate attitude, strength of emotion, to various kinds of groups and other things.

An attitude toward a group of people is important because the attitude has motivational properties, a determinant of how those people will be treated. For example, Early (1968), using my language-conditioning, positively changed the attitudes of the students in a class toward the name of a "loner" student in class. Those students then changed their behavior to the loner, interacting more with him. There are many studies in social psychology that show that attitude toward people determines whether they are approached or avoided, supported or opposed, helped or hindered. There are various tests and methods of testing to measure attitudes because they can reveal a personality trait that determines social behavior.

A dramatic case from everyday life can be seen in the murder of Dr. George Tiller on Sunday, May 31, 2009. He conducted a clinic and gave late-term abortions to women with legal medical need. Abortion is called "murder" by some people, and this was applied to Dr. Tiller—he was called "America's Doctor of Death." Unlike the old saying "sticks and stones can break your bones but words can never hurt you," words can hurt you. Words can hurt you because they can create an emotional response in others that results in actions that hurt you, as in the case of Dr. Tiller. Attitudes caused by words are very generally applied by nations to their people to create negative attitudes necessary for going to war.

Interests

I had occasion to observe the complex learning of a set of interests that importantly determined the nature of a young man's life. When he was still a toddler, his father died and the boy's mother considered him to be inherently like the man she loved. The mother frequently voiced a resemblance, for example, in describing how the boy was interested in mechanical things, "just like his father."

Mechanical activities and things came to elicit positive emotional responses in the boy and he sought them out, taking a preengineering curriculum in high school, completing university training in engineering, and becoming an engineer.

I am suggesting that interests are not biologically inherited personality traits; as many laypeople and professional people think they are. Rather, they are learned. The tests that measure them are measuring emotional responses, the extent of liking and disliking, for a wide number of vocationally relevant things. The *Strong Interest Inventory* (Strong, Hansen, and Campbell 1985) has a large number of items that groups successful in different occupations have rated for like/dislike (positive or negative emotion). Such groups like and dislike different vocationally relevant things, events, and activities. While lawyers, librarians, and lithographers may not differ in how they respond to the item "I like picnics," lawyers may distinguish themselves in items showing that they do not dislike confrontational interactions, librarians by items showing that they like contemplative activities, and lithographers by items showing that they like visual aesthetic pursuits. That is what the interest test does: it matches the likes and dislikes of individuals to those "interests" of people in occupations. The individual should enjoy things that people with shared interests enjoy. A student with interests like a radiologist should enjoy working as a radiologist. If she has the intelligence repertoire sufficient for medical school, radiology would be a good career goal. Interest tests actually do have that value in vocational counseling; they characterize part of individuals' emotion-motivation repertoire that is relevant for occupation. An individual who has positive emotional experiences on the job will be rewarded by job-related behavior and will be attracted by job-related activities, such as reading about the job, attending lectures about the job, and associating with colleagues from the job. Such behaviors lead to success on the job.

This conception called for research to show that interest tests actually measure emotional responses that determine whether job-related stimuli elicit emotion, have reward value, and act as incentives. College-student subjects took an interest test in class. The emotional value of interest items was measured in one study, and it was found that the items did elicit emotion differently depending on the vocational interests of subjects. On a second study it was found that the reward value of the items on the test differed depending on the vocational interest of the subjects; the subjects learned a response that was followed by the auditory presentation of an interest item that was in their occupational interest.

In a third study, the incentive value of interests was measured by presenting subjects with a choice: they could choose to read an article that was in their occupational interest or an article in a different occupational interest. One group of subjects had "art" interests on the interest test given to them, and the other group had "chemistry" interests. The members of the two groups were given the choice of reading an article from a pile labeled "art" or one from a pile labeled "chemistry." The results showed that the differences in the emotion-motivation repertoires of the two groups perfectly controlled their choice behavior. All those with art "interests" selected art articles to read; those with chemistry "interests" selected the chemistry articles. The predictability of behavior was remarkable. Their personalities, with respect to the emotional quality of vocation-related stimuli, strongly controlled their choice behavior (see Staats 1996).

Further, regardless of the way the stacks of articles were labeled, one half of the subjects in each group were given an article to read that was opposite their interest. The other subjects were given the article that was the same as their interests. We found that those who read an article in harmony with their interests enjoyed the article more. We also measured how much they learned from what they read and found that they learned more when the article was harmonious with their interests than when the article was not (Staats 1996). These findings show why interest tests are predictive of both job satisfaction and job success. When the job coincides with one's interest, one will do more things—such as reading about the job—that advance one's job-related knowledge, and learn better while doing so. These findings indicate very well the importance of individual differences in the emotional-motivational repertoire. This personality trait determines individual differences in behavior.

Values

One's values also constitute an aspect of the emotion-motivation repertoire. People feel positively or negatively about different things of issue. They have different values, and that affects how they behave. The test named "The Study of Values" (Allport, Vernon, and Lindzey 1951) is constructed of items such as number 9, "Which of these character traits do you consider the more desirable? (a) high ideals and reverence; (b) unselfishness and sympathy." Number 29 reads as follows: "In a paper such as the *New York Times*, are you more likely to read (a) the real estate sections and the account of the stock market; (b) the section on picture galleries and exhibitions?" Number 4 states, "Assuming that you have sufficient ability, would you prefer to be (a) a banker; (b) a politician?" Just these items indicate that the test concerns political, social, aesthetic, economic, and religious values.

My view is that the individual learns emotional responses to a wide variety of stimuli, some of which have a thematic, or topical, quality considered to reflect values. Religious values, for example, affect behavior, as shown by participation in religious activities, rituals, and social events, and by having religious beliefs, opinions, and knowledge. Religiousness, thus, depends on having learned emotional responses to certain stimuli that will determine in important measure certain behaviors. Leonard Burns and I found this to be the case in a study in which we first selected subjects from a class in which they had been given the values test. Later their behavior was measured. Those with high religious values made a quicker "pulling toward them" approach response to religious words than subjects with low religious values. The nonreligious subjects responded more quickly than the religious subjects to religious words when the response they had to make was pushing away the word (Staats and Burns 1981).

A person's values indicate how the individual will behave on a wide variety of issues, from political liberalism-versus-conservatism to parents' materialism-versus-idealism.

Needs

One theory of personality revolves around the concept of *needs* (Murray 1938). People are considered to have different needs, like the need for cleanliness or sociability or achievement. Those with different needs are considered to behave differently. For example, Murray defined the need for achievement:

> To accomplish something difficult. To master, manipulate, or organize physical objects, human beings, or ideas. To do this as rapidly and independently as possible. To overcome high obstacles and attain a high standard. To excel oneself. To rival and surpass others. To increase self-regard by the successful exercise of talent. (1938, 154)

Edwards (1953) composed a test to measure people's needs along the lines indicated by Murray. In one example item, the person picks the statement liked best, as in "I like telling amusing stories and jokes at parties" versus "I would like to write a great novel or play." We can see that someone selecting the writing attraction would score higher in the need for achievement than a person who likes to be entertaining at parties.

The Tests of the Emotion-Motivation Repertoire

What shines through here is that there are various tests independently developed, with different conceptions. The composers believed they were measuring different personality entities. But all these different tests measured the same behavioral mechanism, the emotion-motivation repertoire, just different aspects of that repertoire. It is valuable to understand them in a unified way, to realize the ubiquity of the repertoire in affecting individual differences in a wide range of behaviors. The wide-ranging personality structure involved operates according to lawful, set principles for all the tests. Understanding those principles, the way the personality repertoire is learned, and the important effects that repertoire has in determining individual differences in behavior, provides a framework for understanding human behavior on the one hand and for understanding this type of psychological test on the other. This framework would be of high value in the further construction of the field of personality measurement to advance its general meaningfulness, to eradicate redundancy, and to simplify its conceptual basis.

Raising a child to have emotional feelings determines the child's later behavior. This applies to interests, values, needs, attitudes, preferences, and various other aspects of personality. The emotion-motivation repertoire involves a huge number of stimuli that occur in life. The fact that there are so many personality tests that measure the emotion-motivation repertoire attests to the broadness of this repertoire and the many situations in which the repertoire acts on behavior.

Sensory-Motor Personality Traits

Ordinarily people think of personality as a causative mental process that lies within, a cause of overt behavior. Behavior is external, what we see, and what we are concerned about, the thing one wishes to be able to explain and predict. It is the internal personality—supposedly determined by our genes and wired into our brains—that is believed to be the cause of that external behavior. Behavior becomes the effect.

Individual Differences in Motor Ability

That presents a problem, however. For there are large individual differences in sensory-motor skills, and those are important. One seven-year-old skillfully dribbles, passes, and kicks for the goal in age-group soccer. This child goes to the right place, makes the right choices in what to do, and attends closely not only to playing but also to instruction. This child is called upon to play much of the time. Another child hangs back, is awkward, seemingly at a loss concerning where to go and what to do, and does not attend focally to coaching or concentrate when playing. Little playing time is awarded to this child.

Looking back we see that the first child spent time with an athletic father playing, throwing and kicking balls, running around while being chased, and having fun in various physical activities. Enrolled in a soccer league for young children, the child then goes on to play various sports, spending time learning, becoming better in each, and learning more rapidly as this is extended.

The other child has no physical interactions with a parent and is awkward when engaging in games. He does not stay in soccer very long or enthusiastically; it takes him away from more-liked activities. The same choice leads him to avoid playing other sports as much as possible. After a time his awkwardness compared to others becomes more pronounced, lacking as he is in athletic skills.

The two have become quite different. In gym class, one does well in the various sports introduced, the other does poorly. These individual differences extend widely, even when both have not had any experience with the athletic skill being introduced. One picks up the skill quickly and gracefully, the other fumbles along.

The first child is called a "natural athlete." Natural athlete? That's a personality concept. Natural athleticism is believed to be internal, inherited, residing in the body and the brain and the genes. That is thought to explain athletic skill. So individual differences in sensory-motor skill are considered personality differences.

The Sensory-Motor Repertoire

But we have a different explanation in the marvelous-learning-animal conception. There are huge differences in sensory-motor skills among populations of people. These differences arise from different learning experiences, according to the same principles as those for the language-cognitive and emotional-motivational aspects of personality. The sensory-motor aspect of personality consists of repertoires that demand an extensive number of training trials.

Before a child can learn to play soccer well she must have learned to run, kick, follow a ball, and follow instructions regarding sensory-motor actions. A child who has learned to throw a ball will more quickly learn to serve in tennis. That is one example. New skills are built on the repertoires already learned, and this can go on in extensive, successive, cumulative learning of repertoire.

The individual learns a large number of sensory-motor skills, ranging from eye-hand coordination, sitting, and standing in balance (as described in chapter 4) to driving, writing, singing, pouring a glass of orange juice, and making love. Each individual learns a unique, huge sensory-motor repertoire. There may be commonalities in elements of two individuals' repertoire, but there are also differences. Carpenters, auto mechanics, neurosurgeons, and concert pianists have commonalities with those in their specialty, yet even then there will be differences.

Sensory-Motor Repertoires, Part of Personality

Sensory-motor repertoires result from cumulative learning. With the sensory-motor repertoire learned from being a defensive lineman in football, the large and muscular man can more easily learn to be a professional wrestler than a person with the same physical structure who has spent his work life as a horticulturist. Common knowledge attests to that, but much remains to understand the extent of this process and how it constitutes an important part of personality. Sensory-motor repertoires not only result from cumulative learning but also determine the individual's behavior and can serve as the basis for later sensory-motor learning. A child who has learned to run, to skip sideways, to throw overhand and underhand, will learn tennis better than a child who does not have that sensory-motor repertoire.

Sensory-motor repertoires also act as causes of the individual's behavior in another way. For example, certain skills are valued by human societies, such as having an especially good memory, being quick-witted, and having special knowledge or academic degrees. The same is true with skilled sensory-motor repertoires, such as being an exceptional Major League Baseball pitcher, the winner of a major tournament in tennis or golf, an all-state cornerback, or just being a good athlete at the lay level. A person who excels in athletics receives esteem from others, and that conditions the individual to have a positive emotional response to herself or himself. That comes across to others as self-confidence, self-assuredness in action. One's sensory-motor prowess contributes to one's the repertoire that composes one's self-esteem.

The individual's sensory-motor repertoires also have that effect on the individual's personality by acting on the experience the individual receives—experience that acts back on the individual. For example, the child who demonstrates the sensory-motor skills of speaking some words will be treated as an older child than one of the same age who has not yet become verbal. In the same way, the NFL defensive end who is a winner of a Super Bowl ring will have different social experiences than a layperson with the same inherent physical structure. The defensive end gains social attention and adulation.

Ballet Dancers, Linebackers, and Surgeons—Peas in a Pod?

Different specialized developments of the sensory-motor repertoire—as in medicine, art, music, sports, and such—are generally considered to spring from different internal talents. But in the present view they have common origins. The weightlifter, pathologist, ballet dancer, violinist, and NFL quarterback have all *learned* their special repertoires via the same principles. And for each that learning was long and cumulative, building repertoires of skill on the basis of already-learned repertoires. The NFL quarterback's learning may be traced to Pop Warner youth football, but actually it began in infancy in learning basic eye-hand coordination, standing, walking, running, throwing, and catching. The others' backgrounds go back to those same roots.

There are differences, of course, in basic features in the different types of performance. For example, sometimes the skill lies in the sensory part, sometimes in the motor part of the sensory-motor repertoire. Thus the pathologist's special repertoire—as in perceiving differences between normal and abnormal cells—involves specially learned visual-discrimination skills. Motor developments are less important. The opposite is true of the weightlifter where, in addition to great strength of large muscles, precisely sequenced movements depend on learning—internal stimuli are important in the motor learning, but the discrimination of external stimuli is not. The watchmaker, in another contrast, involves both fine visual skills combined with fine, delicate movements of the hands and fingers. A running back in football or a basketball player needs to make complex, powerful, yet precise motor responses to a quickly changing array of environmental stimuli. The stimuli involved in various sensory-motor skills will vary for the artist, the musician, or the wine taster. On the motor side, some skills involve the fine muscles, some involve the large muscles, some require quickness, and some require power.

Probably a sensory-focused sensory-motor repertoire skill (like that of a radiologist) would show up differently on brain-scan imagery than would a motor-focused sensory-motor repertoire skill (like that of a weightlifter). It would be interesting to see if brain injury to sensory parts of the brain would affect the radiologist more than the weightlifter, and if the opposite would occur when the injury is to the motor parts of the brain. The marvelous-learning conception calls for new types of research.

How about Body Differences in Motor Performance?

Thus far the discussion of sensory-motor repertoire skills has centered on learning. The possibility of differences in talent hasn't been mentioned (and the question of God-given talent not entertained and that of gene-given talent not considered). Is this to say that the biology of individuals plays no role in determining their sensory-motor characteristics? Not so, biology plays a powerful role. A five-foot-three-inch young man will never become a National Basketball Association player today, let alone a power forward; a large-boned, huge, three-hundred-pound person will never become an Olympic-class high-jumper.

Long fingers are an advantage in becoming a concert pianist. Slim, athletic, not-too-large builds are more propitious for ballet dancing than large-boned, large, and heavy builds. Female Olympic gymnasts usually are not tall and do not have broad and heavy hips. No question that individual biology impacts sensory-motor performances. Different body characteristics are quite central in developing different kinds of sensory-motor skills. Differences in musculature, joints, bones, tendons, and ligaments may all play a role.

That does not extend to the belief in sensory-motor connections prewired into the brain that compose inherited talents. There is no evidence to support that belief, although the belief is strong. I suggest that the human brain is blank in that respect, that there is no proclivity in the brain, no talent, for developing skill in any sensory-motor repertoire. Behavior skills are learned—that applies to the whole sensory-motor basic repertoire. This does not mean the brain does not have a role. For the learning of that huge sensory-motor repertoire occurs by virtue of the brain's neural associations. Motor skills are indeed internal, recorded in the brain. But that doesn't mean those skills were prewired by the genes.

There are great individual differences in sensory-motor repertoires. That great human variability distinguishes humans from other animals. No question other animals can become highly skilled in sensory-motor behavior. A lion chasing, leaping on, clawing, and biting into a running blue wildebeest, dragging it down and killing and eating it, constitutes a most virtuoso performance. A border collie performs skilled behaviors under the control of different instructions. However, no other species has shown the great variety of sensory-motor skills that humans exhibit. It takes the human brain and the human body, and the missing-link learning principles to be able to produce that kind of sensory-motor ability. Many human skills take years of repertoires learned on top of earlier learned repertoires

Personality Tests of Motor Skills

As we might expect from this conception, there are personality tests that measure individual differences in the sensory-motor repertoire. Some tests measure sensory-motor skills in order to judge the child's developmental progress. For example, developmental tests address sensory-motor behaviors like sitting, walking, balance, throwing, and catching. Another includes tests for running speed and agility, bilateral coordination, strength, upper-limb coordination, response speed, upper-limb speed, dexterity, and visual-motor control. There are tests that select children for special education programs that assess everyday life skills, such as eating, dressing, toileting, and performing household chores, by the ratings of a parent or a teacher. Tests have been constructed that measure occupational behavioral skills, such as a typing test or a hearing test for selecting candidates to operate underwater sound-detecting apparatus. There are also items on intelligence tests that measure aspects of the sensory-motor repertoire. The maze and geometric-design tests on the WPPSI (Wechsler 1967) measure the basic sensory-motor repertoire described earlier. Child-development norms, which indicate the types of sensory-motor skills developed by children of different ages, are also used as tests by pediatricians and child clinical psychologists. A pediatrician may also simply observe whether the child's behavioral development meets that indicated by the norms.

Tests, however, do not begin to assess the great variety of ways humans differ from one another in terms of sensory-motor repertoires. Over all peoples, the number of different sensory-motor skills is uncountable; one individual can learn only a small portion of them. Contrast that with wild animals. They learn the same sensory-motor skills of their species, although there will be individual differences in those skills. It is interesting to consider, as I will in chapter 8, that when *Homo sapiens sapiens* first appeared in full form perhaps one hundred thousand years (or so) ago, the vast human basic repertoires of sensory-motor skills was tiny in comparison to what it is today. The central point is that humans have the ability to learn an astonishingly huge number of sensory-motor repertoires, which helps account for the great individual differences in personality that humans display.

WHAT REALLY IS PERSONALITY?

What does this all say? First, it is clear this constitutes a 180-degree turn away from the traditional view of personality, away from the usual inference of mental-brain-gene types of internal causes of people's behavioral differences. Personality is indeed an internal cause of the individual's behavior, that is, learning as recorded in the brain. However, those personality causes are composed of learned repertoires of behavior. The preceding developments begin to specify personality in terms of such repertoires.

This knowledge of personality commences with the specification of the relevant events. Those are the basic repertoires the child begins to learn in infancy. They can be observed, along with the learning experiences that produce them. The way the child learns a language repertoire can be observed, and has been, even though much more study is needed. As an outgrowth of that observation, it becomes clear that as the child learns the elements of a language repertoire, the child's nature—personality—changes. The child's behavior, then, can be dealt with by instructions; the child can also, via instructions, learn more easily and rapidly. That constitutes a big change. The child becomes more intelligent. That is shown by the central role of language measurement on intelligence tests for young children.

In childhood the repertoires that constitute personality are open to observation, and not only with respect to intelligence. A child who is "spoiled," that is, gets his own way by cajoling, crying, "temper-tantruming," and demanding; who dominates actions in the household; who is protected when he mistreats other children; and so on, will develop a different personality—different basic repertoires—than a child raised to learn more socialized behavior that respects others' feelings. That window of observation of such early child learning needs to be enlarged immeasurably, but we can see such effects through our own observations.

As the child's chronology increases, her repertoires grow more complex. At age two, all of the child's word repertoire may be observable and known to parents. Not so by age four. That makes a general point. Human learning quickly becomes complex and evanescent. The learning causes of repertoire expansion can no longer be traced easily. That holds for the repertoires themselves; they no longer are evident, observable, to a great extent on display in their completeness. Soon they become so complex, only sparse elements of the internally recorded repertoires are sampled in the various life situations encountered. And the behavior elements that are elicited by the individual's life situations are usually mixtures from various repertoires. The repertoires that can be distinguished separately in their beginnings lose individual identity. For example, one can consider that the child's learning to say words and her learning to respond to words can be studied separately. However, an item on an intelligence test may require the child to have both repertoires well developed. A person who reads a lot will acquire a language repertoire, a vocabulary, that includes elements that will never be said, never be exhibited. A vocabulary test, thus, can provide better knowledge of the repertoire involved than day-to-day observations of what the individual says. This provides a good reason why traditional behaviorists should become interested in psychological tests.

Personality is formed of learned repertoires that become increasingly complex and increasingly intermixed. The individual's behavior consists of elements drawn from those already-learned repertoires that are assembled in new combinations continually by life's situations. The life situations that individuals experience can be new, never experienced before, and yet be responded to in an effective way because elements can be put together from different repertoires that make up a new whole behavior. The young child may have the words *man* and *running* in his repertoire, see a man running and thus say "man running," despite having never said that before. That happens continuously, most frequently in mundane ways such that, although the individual's behavior is actually creative, it is nevertheless commonplace. Individuals continually face situations never before experienced and behave in new, creative ways because of the repertoires they have learned. Sometimes such acts are recognized as creative and new because the individual has repertoires that no one else does, or because the individual has encountered situations that no one else has. Einstein, for example, not only had training in physics, like many other physicists, but he had also worked two years in a patent office where he had to decide whether a new work in its basic sense was like a previously patented work. Did that work enable him to cut to the heart of complex developments so he could compare them in basic ways, gaining skills valuable for cutting to the heart of developments in physics?

Only in early childhood are the personality repertoires and their learning readily available for observation. Those repertoires are learned via basic learning principles. But, even more so, they are learned by using previously learned repertoires, such as learning through language in newspapers, movies, television, books, magazines, and stories and accounts from others. These learning experiences are ephemeral, not noted and recorded, untraceable. That is why the traditional concept of personality has remained an inference attributed to invisible causes like genes, brain, and mind. The real causes, evanescent and complex, have never been realized.

Aren't Personality Traits Prewired in the Brain?

Many would agree that parts of traditional personality traits are learned. And yet, in the face of such evidence, the widest belief is still that personality is an inborn quality of people, prewired into the brain. For example, researchers at UCLA's Ahmanson Lovelace Brain Mapping Center measured brain-activity differences between Republican and Democratic subjects when shown political stimuli (Tierney 2004, C6). The idea that political personality traits are caused by inborn differences in the brain illustrates the Great Scientific Error's neglect of learning. This study ignores how learning could have operated in its results. When an individual learns a political orientation, this ordinarily involves beliefs that elicit strong emotional responses. A liberal learns positive emotional responses to some things and negative emotional responses to other things in a different way than a conservative does. Those differences will be recorded in the brain. Brain scans may very well pick up the differences in these two people. But it wasn't their brains that produced those differences; it was their learning.

The same is true of familial similarities in personality traits. While it is true that family members share genes, as shown many times in shared physical characteristics, family members ordinarily also share many learning experiences. Two children raised with a father who has a prejudiced view of a minority group will likely hear many derogatory things about that minority group and be conditioned also to come to have a negative emotional response to that group. Children of an athlete likely will get sensory-motor learning experiences that give them athletic ability. Children of a salesperson are likely to get particular learning experiences, and so on. The various phenomena—used to validate the belief in an internal, inherited, brain-determined personality that gives each individual her or his personal characteristics—are explained by a personality composed of learned behavioral repertoires. All repertoires are potentially observable; none have the insubstantial, subjective aspects of the soul or the mind given in some popular views of personality.

HOW ABOUT BIOLOGICAL CAUSES OF PERSONALITY?

Am I saying that the individual's biological characteristics have no effect on the individual's personality? Well, yes and no. On the negative side, the individual's personality is composed of learned repertoires, and that leaves no place for genetic-brain causation.

What about the traditional belief that people with certain biological traits have certain personality traits, like fat people being jolly, shifty-eyed people being dishonest, limp handshakers being indecisive, and so on? To that I respond that people with different physiognomies have different life experiences. A tall, muscular boy has different experiences than a plump, short boy. As a consequence, the two boys will learn different basic repertoires in certain areas. That different life experience may of course undergo further learning if the plump boy goes on to become a doctor and the muscular boy becomes a personal trainer. A beautiful, shapely girl throughout childhood and as an adult will learn different repertoires than a much less attractive girl. There are many studies that show that attractive people have different social experiences than less attractive people. Differences in experience will create differences in the repertoires later learned, thus constituting different personalities.

So, yes, an individual's biological makeup does affect personality. But that occurs through the way that makeup determines the experiences the individual has and the way this affects the personality repertoires that are learned.

PERSONALITY, A CAUSE AS WELL AS AN EFFECT

Thus far the emphasis has been on defining personality and indicating that it is learned in a cumulative way. However, personality is not actually known by what it is and by what causes personality. Personality is known by what it affects, by how it is a cause of how the individual feels, thinks, and acts. Children high in intelligence tend to do well in school. Children who lack intelligence tend not to do well. Intelligence as a personality trait is of interest because it is viewed as a cause of performance in school and other situations.

Explanation of intelligence, thus, demands stipulating how intelligence determines how people behave. That involves showing how having a basic number repertoire (of stating how many objects are presented and of counting objects) will be an advantage to the young child in kindergarten. Such a child learns better because she has the foundation for learning additional number-conception skills. On a higher level, it is important to understand why having a high score on the Miller Analogies Test is an advantage to graduate students studying a behavioral science. The reason is that it is a test of the individual's language repertoire. The more developed the repertoire, the better the individual will do in all the additional language learning involved in such graduate study. The quality of one's language repertoire becomes a cause of further language-learning ability.

Establishing how a basic repertoire affects positively or negatively the learning of additional repertoires is part of the explanatory task. Why is intelligence important in terms of affecting the way the individual behaves? Why are the different aspects of the emotion-motivation repertoire important? Why is the sensory-motor repertoire important? All the learned personality repertoires have to be studied so that the ways they affect human behavior can become known.

USEFUL EXPLANATION VERSUS PIE-IN-THE-SKY EXPLANATION

Beginning with Francis Galton, Darwin's cousin, there has been a concerted attempt to construct a conception that heredity produces individual differences in personality. Galton (1869) calculated how many eminent personages arose from a general group of people versus a family of eminent people. A much greater fraction came from eminent families than from people in general. His conclusion, of course, was that heredity was the major cause of high ability, of genius.

Proof of that belief has been the goal of scientific studies and theories since that time. The twin studies are one example; generally the intelligence test scores of identical twins are more closely related than the test scores of those less closely genetically related. The weaknesses of such studies have been indicated variously (see Kamin 1974, Staats 1971a), but the field of behavioral genetics still goes on with studies that attempt to show behavioral similarities based on genetic inheritance.

Gordon Allport considered the personality traits he distinguished through his test construction to be "cortical, subcortical, or postural" in nature (1966, 3). The advancement of biological science has brought more direct ways of seeking proof that gene, brain, and brain chemistry are the major causes of personality differences. Today the attempt is to show that individual differences in behavior in people are determined by their brain differences, using brain-scanning technologies to show those biological differences.

Belief in such an explanation has existed for a very long time. Scientific investigation has sought to support that belief for a long time. Sophisticated genetic science has been directed to provide the explanation for some time. Yet the status is still pie-in-the-sky, expected, projected success in the future. The Haskin research proves only that the brains of readers and dyslexics differ. But differences in learning experience create differences in the brain. The marvelous-learning-animal conception says that those with different personality traits have had different learning experiences, and that is what has created their different brain characteristics.

If one wants a child to develop language, reading, knowledge of history, physics, or biology, there is only one way to achieve that goal: by providing successful learning experiences. If one wants a child to be intelligent, no way has been found to manipulate genes or the brain to achieve that goal. If one wants a child not to have temper tantrums, to not go into the terrible twos developmental stage, and other aspects of personality, nothing biological is available for that purpose. However, a child can become intelligent by having learning experiences that produce a good language repertoire, a good number repertoire, and other repertoires. A child with certain learning experiences will not go into the terrible twos stage of behavior. A person with certain learning experiences will be a good mechanic or orthopedic surgeon. Knowledge of the *learning experiences* necessary to produce certain types of personality and to avoid others is better than pie-in-the-sky. What has already been achieved holds hope that if that type of scientific-applied learning development is pursued with the effort now being put into genetic and brain research, that much more "here and now" knowledge will be gained.

LINKING LEARNING AND THE BRAIN

There have been two approaches, widely separated. One seeks to explain individual differences in behavior by personality conceived of as a brain characteristic, wired-in genetically. The other approach rejects the concept of an internal personality that determines behavior. Personality is rejected on the methodological grounds that it is not observed, just made up of inferential names for certain types of behavior—someone behaves intelligently, so it is inferred that she or he has a high intelligence inside. So behaviorism has said there is no personality, just behavior. There is no communication between the two approaches.

The marvelous-learning-animal approach rejects the separation that has been imposed on the interest in individual differences in human behavior. Everyone would agree that personality is affected by learning. That means that those who specialize in the study of learning should be addressing their expertise to the study of individual differences in behavior. In doing so, the knowledge that has already been gained by the study of individual differences in behavior should be used.

The missing-link principles provide the framework by which to bring together these separate fields. That framework says that human learning involves learning large, complex basic repertoires that determine how the individual will behave in the later life situations that are encountered. Those basic repertoires are a result of very complex learning experiences that have occurred over years of time. Such a great quantity of learning and such complex repertoires and so many life situations guarantees that there will be huge individual differences in behavior.

This understanding of human learning provides a new framework for understanding personality. It recognizes that personality is inside, is in the brain. That is where the learned repertoires are recorded, and that changes the brain's features. The framework recognizes that limiting study to just the observed behaviors that occur in life situations is insufficient. The example given is that the extent of the individual's vocabulary repertoire can't be observed that way; people ordinarily know words they will not express in a lifetime.

The conception of personality becomes objectively, scientifically, defined when it is realized that personality consists of learned basic repertoires. That opens the way for the study of personality traits as causal mechanisms that produce individual differences of behavior. It opens the way for the study of how personality traits are learned. It opens the way for the study of brain difference—in structure and function—of different personality traits. It opens the way for the construction of personality tests and for advancing the field of personality testing. Finally, it opens the way to new applications, for example, how parents can affect their children's personality development—as in increasing their intelligence.

CONCLUSION

The missing-link theory has traditional natural-science characteristics. It indicates causes and provides avenues for dealing with those causes. It also indicates effects, the phenomena that are to be explained. And it indicates the principles by which particular causes produce those effects. The theory should be compared to the other approaches: to psychological personality theories, to theories that the characteristics of the brain and its chemistry constitute personality, and to the search for genes and drugs that determine personality traits. How do the various attempts serve as classic theories of personality? To what extent do they inform us about the nature of personality in specific terms? To what extent do they provide means for shaping personality development? To what extent do they offer means of dealing with problems of personality? To what extent can the approaches unify the separated behaviorism and personality-testing fields and increase their productivity?

At present, almost no money, time, and effort have been spent in the attempt to establish what personality is behaviorally, how it is determined by learning, or what the processes are by which personality affects the individual's behavior. The present conception, in contrast, calls for a very broad range of research in addition to biological and pharmaceutical research.

A conception of human behavior and human nature must deal with the phenomena of child development. It must also deal with phenomena of individual differences in behavior, that is, personality. This applies to normal personality. However, some of the individual differences in behavior that humans display is considered abnormal, harmful to the individual or to others. Abnormal behavior must also be explained, which is the concern of the next chapter.

Chapter 7

MARVELOUS LEARNING OF ABNORMAL PERSONALITY

There has been a strong belief for a very long time that mental illness is inherited. But the evidence to substantiate that belief has not been found, despite great efforts with all the methods of science.

> So scientists are turning their focus to an emerging field; epigenetics, the study of how people's experience and environment affect the function of their genes. . . .
>
> In studies of rats, researchers have shown that affectionate mothering alters the expression of genes, allowing them to dampen their physiological response to stress. These biological buffers are then passed on to the next generation: rodents and nonhuman primates biologically primed to handle stress tend to be more nurturing to their own offspring, and the system is thought to work similarly in humans. . . .
>
> [Perhaps such epigenetic markers affect the way humans inherit mental illness.]
>
> The National Institutes of Health is sponsoring about 100 studies looking at the relationship between epigenetic markers and behavior problems. . . . [S]uch studies are expensive, and . . . the findings may be as difficult to decipher as studies of the genes themselves. (Carey 2010, D7)

The article from which this quotation is taken illustrates dramatically that the contemporary conception of abnormal behavior, and the Great Scientific Error involved, have been a dismal failure. Evidence that abnormal behavior is caused by some sort of abnormal genetics that acts on the brain is lacking. Nevertheless, the National Institutes of Health (NIH) is described as pouring money into furthering the same type of search with the added interest in epigenetics. Nowhere in the NIH's studies does much interest in learning appear, despite the great evidence that environmental happenings produce learning that greatly affects people's behavior.

Strange. And our behavior-analysis movement makes the lack of interest in learning even stranger. Since their beginnings more than fifty years ago, our fields of behavior analysis and behavior therapy have shown time and time again that a wide range of problems can be dealt with by using learning principles and procedures to treat many aberrant behaviors. The same principles and methods that can be used to affect the behavior development with normal children (see Staats 1963, 1968a) can be used to correct behavior development with abnormal children (see Lovaas 1977, Staats and Butterfield 1965); the advances since that time show growing verification and growing applicability.

What makes this not actually strange is that the behavioral movement has focused on correcting abnormal behavior, not on how learning may produce that abnormal behavior. In the same way as the concept of mental illness established a framework with fundamental error, the contemporary behavioral movement, which bases itself on a behaviorism that focuses on behavior and ignores learning (see Skinner 1957), also has need for fundamental development with respect to understanding abnormal behavior in terms of its causes and thus its prevention.

THE MARVELOUS-LEARNING-ANIMAL CONCEPTION OF MENTAL ILLNESS

Ancient belief held spirits as the internal cause of both physical illnesses and abnormal behaviors. Advancement to religious thought changed that belief to possession by the devil.

A naturalistic medicine began with the classification of the various diseases by the symptoms displayed. At a later time, the causes of some of the diseases, such as the spirochete causing syphilis, were found. Medicine that identified diseases and found their causes became progressively advanced following that method. It made sense that those trained in medicine would generalize that method to abnormal behavior. Different types of abnormal behavior were considered as different mental illnesses. The first task was that of identifying the various mental illnesses by the specific symptoms—abnormal behaviors—that patients displayed. The symptom behaviors were not the illness, only the signs of the illness. The illness lay within. The task was to discover the illness itself.

Based on this belief, diagnostic manuals that list the groups of behaviors that compose the "symptoms" of the various mental disorders have been established. Just as personality determines normal behavior, abnormal personality (mental illness) determines abnormal behavior.

The present approach considers the identification of the cluster of behaviors that identify a behavior disorder as a powerful source of knowledge of human behavior and human nature. The diagnostic manuals that describe the different behavior disorders provide such knowledge. Those manuals are based on the experience of clinicians who work with individuals with behavior problems, who observe the types of behaviors they display and also use other types of knowledge, such as information supplied by family members. Tests that attempt to diagnose individuals who have those disorders are also used.

The present approach considers abnormal behavior in the same conceptual framework as normal behavior. Why would it be assumed they are different, explained by normal internal processes or structures in one case and by abnormal ones in the other? After all, normal and abnormal behaviors involve the same response organs. Learning principles were developed in evolution from single-celled organisms through the many advancements to and through the mammals prior to *Homo sapiens* because they aided survival and reproduction. Evolution would not have supported the development of principles of learning and behavior that diminished survival and reproduction. It wouldn't make sense for humans to have diverged into two learning types, some members of the species learning through normal principles and others learning through abnormal principles. Other wild animals do not demonstrate principles of abnormal learning. Why should that occur within the human species?

There is no evidence that people with behavior disorders learn differently than normally behaving people. It is sensible to assume that people with behavior disorders learn according to normal learning principles. Well, then, how could they have learned abnormal behavior repertoires? That could occur if their learning experiences had been "abnormal," if their environments had involved circumstances that led to the learning of abnormal behavior repertoires.

This chapter presents the theory of abnormal behavior that is part of the marvelous-learning-animal conception, beginning with some of its basic principles.

Abnormal Environments Produce Abnormal Behaviors

The environment acts as an essential cause of abnormal behavior. I first pictured that in my analysis of a psychotic with "opposite speech" (Staats 1957). He said the opposite of what was called for because he was rewarded for doing so by his doctors. For example, one doctor asked him if he wanted a cigarette and, when the patient said no, gave him the cigarette anyway, which he then smoked (see

Laffal, Lenkoski, and Ameen 1956). That was an abnormal environment that would condition abnormal language behavior. The doctors, because of their psychoanalytic theory, focused on their patients' subconscious motivations, missed realizing the abnormality of the behavior itself, how that behavior affected the patient's life adjustment, as well as the abnormal learning experiences that caused the abnormal behavior. They did not consider that what they did was an important part of the patient's environment. But it was.

Another example involves a more complex abnormal behavior, borderline personality disorder, which includes unstable and disturbed interpersonal relationships, impulsivity (for example, in unwise spending, engaging in sex, taking drugs, reckless driving, and binge eating), frequent displays of temper, and constant anger (DSM-IV 1994, 654).

> People with the disorder . . . often behave like two-year-olds, throwing tantrums when some innocent word, gesture, facial expression or action by others sets off an emotional storm they cannot control. The attacks can be brutal, pushing away those they care about most. . . . *In an effort to maintain calm, families often struggle to avoid situations that can set off another outburst. They walk on eggshells, a doomed effort because it is not possible to predict what will prompt an outburst.* (Brody 2009, D7, ital. added)

Intended as a description of a behavior disorder, the account also reveals an abnormal learning environment. The patient displays behaviors that are offensive to others, without receiving the usual responses of opposition. The family is reinforcing abnormal behaviors. They have the best of intentions, but their society has given them an egregiously incorrect explanation of human behavior, its concept of "mental illness," which misleads them in dealing with the patient.

Behavior disorders are in large part created by abnormal learning conditions.

"Mental Illness" Suggests the Person Is Sick

Once it is believed that a child is abnormal, the child is treated differently than the normal child. When the "ill" child does not learn normally, for example, the parent tries to protect the child from the experience of failure in learning. What needs understanding is that this constitutes an abnormal environment. That will not produce normal behavior, it will exacerbate the child's abnormality.

Similarly, children diagnosed as having ADHD, autism, or other disorders, are considered ill and treated specially. Again, such treatment constitutes an abnormal environment. That does not mean there has been any malign motivation involved, no intent to do other than love and support the child. It only means the environment was different than what normally occurs.

Abnormal Environments Go Unrecognized

The widespread lack of knowledge of learning principles and learning conditions ensures that the "abnormality" of the environment for the child goes without recognition. Parents do experience their special efforts to help the child. They do not realize that what they do may be creating an abnormal learning environment that produces more deeply and broadly abnormal repertoires.

There are various kinds of abnormal environments. Early in my career, living in a faculty housing development, I had the opportunity to see temper tantrums develop in a young child by virtue of an abnormal environment provided by a mother who didn't realize that when she gave in to the child's demands for something the child should not have, she was training her child to behave inappropriately. She loved her child deeply and did not want the child to experience harmful life conditions. At the beginning, the child was learning to cry with increasing intensity and duration when something wanted was not given. By the time my observation began, the child already cried more frequently and more uncontrollably than is customary. On realizing this, the parents, actually the father, decided not to coddle the child when she cried. For the mother that rule had a proviso, however: no coddling unless the child was seriously affected and needed her support. When the child started a complaint, the mother would at first delay, but as the child's unhappiness continued, the mother would first try to verbally make her feel better. That of course encouraged the child to continue, and eventually the mother gave the child what she wanted. The problem was that the mother judged what was serious by how long and intensely the child cried, which meant the child would not be rewarded until she cried with sufficient vehemence and duration. The level of crying demanded, thus, was progressively raised by the pulling and tugging of the two parents; he wanted not to give in to the child, the child raised the crying level, and then the mother couldn't stand the child's unhappiness. The result was an abnormal environment that rewarded the child progressively to cry with more and more vehemence and duration. Over time, the child learned full-force tantrums and she learned how she could use the tantrums to get what she wanted. I observed in a supermarket how she demanded something inappropriate and got it as the crying began to slide into a tantrum (called a "melt-down" by many today, which also calls for rewarding the child).

This child's behavior was the result of an abnormal environment. An additional aspect of such an environment is that it is encouraged by public response. Attempts to resist rewarding the child or to press the child to stop crying will draw social disproval, another pressure on the parent to continue the abnormal environment. The cure for such behavior is not rewarding the child, which takes longer, or calling a time-out. That would be a normal environment for a child with that behavior problem.

Two Environments

Why would a child learn disadvantageous behavior? The basic learning theory states that behaviors that are reinforced are learned, not those that are not rewarded or are even punished. So why would anyone learn behavior that actually results in no reward or even punishment? One answer is given by David Brooks, a political columnist for the *New York Times*, who considers psychological findings in his analyses; he gives an example with college students that contains one of the principles in the present approach. "College students are raised in an environment that demands one set of navigational skills, and they are then cast out into a different environment requiring a different set of skills, which they have to figure out on their own" (Brooks 2011, A23).

Brooks's analysis is not addressed to behavior disorders, but he nevertheless illustrates an important principle in producing abnormal behavior. The principle is that children can face two different lives, two environments, not just one consistent environment. They can learn repertoires at home that gain them reward but that will be a detriment to them in the life situations outside the home. For example, getting what one wants by inappropriate behavior can be reinforced at home. But that type of behavior in later life will put off most people, causing the individual social difficulties and isolation. That difference in learning situations—home versus life, college versus work, military versus nonmilitary—can be huge and is widespread as a cause of the behavior disorders.

Additional Considerations

In the present theory *deficits in behaviors* may be abnormal. Mental retardation, autism, schizophrenia, and antisocial personality disorder involve deficits in repertoires necessary for normal life adjustment. Deficit behavior is explained by deficits in learning experiences, deficits in environment.

In addition, some behavior disorders will involve learned *inappropriate behaviors*, behaviors that interfere with getting along well in life. Temper tantrums are inappropriate, violence is inappropriate, the suspicions of the paranoid personality disorder are inappropriate, as are the fears of those with a phobia. Inappropriate behavior is the result of inappropriate learning experiences, inappropriate environments.

There is also interaction between the two types of abnormality. For example, the ADHD child has the problem of hyperactivity. That behavior interferes with learning, creating deficits in needed repertoires. The development, moreover, can be the other way around. The dyslexic child with a deficit in reading may as a consequence learn a very negative attitude toward school and therefore avoid going. The child may then join other children with problems and in that group learn additional inappropriate attitudes and behavior.

A behavior analysis of a behavior disorder constitutes a theory of the behavior disorder that should reveal what the disorder is in terms of the behavioral repertoires involved, what behaviors are deficit or inappropriate, how those behavior problems detract from life adjustment, as well as how these behaviors affect the environment in ways that produce additional abnormal learning. A behavior analysis should focus on how the behavior disorder has been learned. Such a theory should stimulate research to augment knowledge of the disorder as well as suggest a therapy to treat and prevent the disorder.

With these principles in hand, I will consider first behavior disorders that prominently involve abnormality in the language-cognitive repertoire. Then disorders in which the emotion-motivation personality repertoire plays a focal role will be exemplified, and finally disorders involving the sensory-motor repertoire. *The Diagnostic and Statistical Manual of Mental Disorders* (1994) published by the American Psychiatric Association will be used to behaviorally define the disorders treated. In each case the disorder will be considered in terms of conventional wisdom and then in terms of the marvelous-learning approach.

LANGUAGE-COGNITIVE DISORDERS

Various behavior disorders focally or heavily involve abnormalities in the individual's language-cognitive repertoires. Dyslexia is one of those disorders.

Dyslexia

> The essential feature of dyslexia, or Reading Disorder, is reading achievement (i.e., reading accuracy, speed or comprehension . . .) that falls substantially below that expected from the individual's chronological age, measured intelligence, and age-appropriate education. The disturbance in reading significantly interferes with academic achievement or with activities of daily living that require reading skills. (DSM-IV 1994, 48)

The believed explanation of dyslexia comes from the Great Scientific Error. "Reading disorder aggregates familially and is more prevalent among first-degree biological relatives of individuals with Learning Disorders" (DSM-IV 1994, 49). Originally the belief in this type of mental illness was supported by such familial data. Today more sophisticated evidence is used for support. For example, in one study, dyslexic kids were given training in reading skills and a functional-magnetic-resonance-imaging (fMRI) scanner found that their brains had an increased blood flow to the expected parts of the brain as they identified letters, so reported Elise Temple of Cornell University and her coworkers (see Bower 2003b, 173). They concluded that these results show how the brain determines reading skill, lending credibility to the belief that the brain should be studied in order to understand behavior disorders.

> [T]he researchers at Haskins . . . find that the brain of someone with dyslexia functions differently from a typical brain as it processes phonemes—the "c," "a," and "t" that come together to form "cat.". . . [T]his research is beginning to reveal dyslexia's neurobiological basis. . . . Some psychologists even believe that [brain] imaging can one day help people shift the way they use their brains to boost their learning performance. (Murray 2000, 23–24)

Such studies, however, are not evidence that back up the interpretations made. Becoming a reader involves learning very complex repertoires that take thousands and thousands of learning trials. That is plain to see—no extensive learning experience, no reading. That a dyslexic child sits in a classroom for years doing different things when other children are having those thousands of reading-learning experiences guarantees that their brains will be different.

Many children arrive at school without having learned the skills of attending and working on extended learning tasks. Without a source of reinforcement, they do not attend. No child who does not attend and respond can receive the necessary learning experiences. There is no other way to learn. Every child needs a teaching method that guarantees that he or she has the necessary learning experiences. As it is, there is no one in school tracking each child's learning progress, so the problem may continue until the child's clinical deficit becomes clear to the teacher, the successful students, and the child.

All three of those realizations lead to further problems. The nonreading child has the punishing experience of failing, thus learning negative emotional responses to the learning situation, to reading materials, to the teacher, to successful students, and to attending school. Dyslexia involves inappropriate behavior as well as deficits, all due to learning.

My experience and research suggested that attention and participation could be guaranteed if those behaviors were reinforced. That was my motivation for inventing the token-reinforcer method, which consists of having the child select a wanted article that can be obtained by the tokens earned—just like money, but the tokens were much cheaper. The article selected could be a trinket or toy for a young child, or a piece of athletic equipment or a movie ticket for an adolescent. In my first study, which was with adolescents, the child was given tokens (actually poker chips) for attending and participating in the training. When enough tokens had been earned the child was given the real reinforcer, the article the child had selected.

In addition, I worked out a set of reading materials so the child's participation was always rewarded, there were no "errors." When the child did not know something, it was supplied so the child could respond correctly and be reinforced for doing so. Because the child always responded correctly, spontaneously or prompted, and was rewarded, there was no negative experience. Children who have failed need that success. The results of our first investigation (my associates were Judson Finley, Karl Minke, Richard Schutz, and Carolyn Staats) showed that the token system ensured the children attended and worked and that the reading materials used trained them in reading repertoires. Later, William Butterfield and I published a formal study using the methods with a fourteen-year-old Mexican American boy who tested at the second-grade level of reading; he was a rambunctious behavior problem in school; he had never previously passed a course; and he was in a detention home for having vandalized a school. In four and a half months of training using the token-reinforcers, he made over sixty-four thousand reading responses, advanced to the fourth-grade level of reading achievement, passed all his school courses, and stopped misbehaving in class and fighting on the school grounds. Those spectacular results were at the total cost of twenty dollars' worth of tokens (Staats and Butterfield 1965), a trivial investment for the large advancements the boy made and in comparison to the costs of the years of failed education the boy had already experienced. Would the brain of that child have changed? It had to, and that could be easily studied.

Later, Karl Minke and I conducted a series of studies that showed the same success using literate high-school seniors and adults as the trainers and the same methodology (Staats et al. 1967). In a ghetto high school in Milwaukee, the program was supervised by two unemployed women from the same population as the subjects, using black literate high schoolers as the trainers (Staats, Minke, and Butts 1970). The results were equally good. David Ryback and I (1970) established with four white dyslexic kids that parents can successfully use these methods to train their own nonreading children, and Burns and Kondrick (1998) showed definitively not only that this is quite feasible but also that *a child through such instruction can catch up and gain normal reading skills.* It is no wonder that Sylvan Learning Systems, Inc., adopted similar methods in a commercial enterprise. In doing so, they have added a great amount of evidence to the validity of the conception and the methods. That is true also of later behavior analysts who still continue to add evidence that supports the present analysis of reading and dyslexia and other disorders (see Sulzer-Azeroff et al. 1971). Now this framework of methods, principles, and findings is being used even among those who formerly followed a cognitive approach (see Nisbett 2009).

Dyslexic children show a deficit in perceptual ability on psychological tests—such as seeing letters and words backward and upside down. This inappropriate behavior has been interpreted as proof that dyslexic children suffer some brain defect that causes the learning and perceptual ability. Not realized in the traditional brain-oriented view is that perception ability and reading ability are part of the same repertoire. Perceptual ability is learned in the process of learning to read; after all, reading requires learning to discriminate small, similar stimuli. A child who has not learned to read will not have had the learning experiences that produce the perceptual skills measured on perceptual tests. Collette-Harris and Minke (1978) showed this definitively. They gave dyslexic subjects the token-reinforcement training in reading, and *as the children improved in reading, their tests showed their perceptual deficits got better.* My graduate students and I (Staats, Minke, and Butts 1970) also showed that preschool children *first learning to read* make the same "perceptual" errors that dyslexic children do—writing letters upside down and backward. As the preschoolers learn letter-reading and letter-writing skills, those errors just drop out. We all make such perceptual errors until we have learned the required perceptual skills in the process of learning to read and write. There is no intrinsic brain defect involved, just the lack of a perceptual-skill repertoire that is gained through learning.

What has been considered a genetic defect in dyslexics, a learning disability, really constitutes a failure of the learning conditions to which the child has been exposed. Children are dyslexic because their homes and schools have not guaranteed they receive the learning experiences necessary to acquire the reading repertoires.

Shouldn't these findings have by now been used extensively in education? Not so. Instead we have an interesting example of the Great Scientific Error. In its April 19, 2010, issue, *Time* has an article titled "Is Cash the Answer?" (40–47). The article describes a very large study done by Roland Fryer where children in school are rewarded for performance. The study, while making an important step in softening education's resistance to the use of reinforcement, suffers from not using the much more complete knowledge of children's school learning that had already been discovered. Fryer's study still stood out from traditional educational practice despite the fact that what it showed had been proven by many more-exacting studies beginning in the 1960s. Education should actually be restructured greatly based upon that much more extensive use of learning principles and the findings involved, as will be indicated in the last chapter.

Mental Retardation and the Hidden Environment

Each early learning of an important basic repertoire adds acceleration to further child development through learning. When acceleration is stacked on acceleration, the individual ultimately can be advanced a huge amount over the average. A child who talks earlier than usual and walks earlier than usual, who visibly loves his parents, is socially responsive to other adults, and is attracted to and plays well with other children at an early age will not be considered mentally challenged. The child will have the foundation for continued accelerated development. The child will be considered "bright" and will be treated that way, in what amounts to a superenvironment.

On the other side, the child who learns the requisite repertoires late and poorly will be considered mentally challenged. And that child will not have the foundation for making normal progress in development. That will be added to by the special treatment that is given to children considered challenged. *The learning of the basic repertoires (BRs) is the secret of child development whether it is accelerated or retarded.*

This is not to say that there are not types of "mental" retardation that have defective biology as a major cause. Children with Down's syndrome provide a case in point, as do microcephalic children, those with untreated phenolketonuria, Tay-Sachs disease, Williams disease, as well as other conditions. These cases of retardation show brain differences, and it is reasonable to hypothesize that the brain limitations are likely causally involved in the retardation. It is still not justified to attribute all of the child's behavioral limitations to the brain deficit, however, even in these cases. For it is also likely that the learning environments for these children are also not maximal. At one time it was assumed that Down's syndrome children had a much more biologically limited ability to learn than is the case, and such children did not receive the learning opportunities that would have benefited them. With a better understanding, many such children now have learned abilities that formerly were thought impossible.

The problem is that a biological defect (gene or brain) is inferred whenever the child's development runs behind the norm. That inference is made on the basis of the child's behavior. The intelligence test, the bottom-line proof of retardation, can result in placement in a special-education class, and this is not a normal learning environment and will lead to additional falling behind. In the present view, biological defects cannot be assumed. The mentally challenged child's IQ tests only measure deficits in the learned repertoires, not genetic brain deficit. The deficit may be in the child's learning environment. That being the case, any abnormality in the basic repertoires, as measured by intelligence tests, can be due to either defective biological conditions, defective learning conditions, or a combination of the two. This is to say that the causes of retardation can only be established directly. There is no way to conclude anything about cause *simply from observing deficits in the child's performance.*

Behind the usual assumption of biological cause lies the view that the environment couldn't be so deficit as to produce the great deviation that retardation can represent. Especially, how could great deficits be involved when both parents are normal, have produced other normal children, and quite obviously *must* have provided the child a normal environment? The problem lies in that reasoning. For what a child's learning environment has been is never obvious, never known by commonsense, everyday observation. Very generally, parents, and professionals as well, do not have the knowledge of learning needed to analyze the environment in the manner that is necessary.

Let me illustrate a case of the "hidden environment" with a child I dealt with some years ago. A nonassertive three-year-old was not exhibiting normal language and sensory-motor repertoires (including social skill). The concerned parents, both professionally trained, were worried about the child, afraid the child might be mentally challenged. So they consulted a psychologist who, after testing the child, diagnosed him as retarded. These conditions made it likely that a retarded development was to be the child's fate. I observed the family interactions enough to see an abnormal environment at work. The gentle child had a sibling a year younger and one a year and a half older. Both were very active and aggressively competed for the attention of the parents. When a parent would begin an interaction with the "retarded" child, the other two would more actively interpose themselves aggressively in a way that shut out the "retarded" child. The gentle child, rather than competing, was reinforced by surcease for withdrawing and remaining alone. I found by excluding the other two and playing alone with the child that he could join in, be perceptive, and learn. I didn't see any biological defect. My behavior analysis called for the parents to firmly establish measures to ensure the vulnerable child would not be shut out by the two aggressive siblings, so this child too could have quality interactions with his parents. The nanny was instructed to do the same. The loving and caring parents had sought help early. So the child was young and the problem was not yet deep. With the changes we discussed, the child was functioning normally on entering school and went on to develop on the high-level standards of the family.

In this case, a very usual case, the deficit environment that was producing the retardation of the child's development was not visible. An invisible abnormal environment leaves room for the usual conclusion that *the child's problem* comes from some brain defect. And that leads to actions that exacerbate the child's abnormal environment. Ultimately the child's intelligence will be retarded, as would have happened to this child if the psychologist's diagnosis had been followed, indicating the importance of the behavioral analysis.

In the present view, usually when a child falls behind developmentally— with delayed talking, walking, and such—that indicates that learning is not taking place in the usual manner. The learning experience the child is receiving may be the sole or partial cause of the retardation. Despite the cause, the deficit environment may go completely unseen by those who have been intimately involved. The rule is *deficit environments produce deficit behaviors*, like mental retardation. The "hidden environment" plays the role of the cause in various behavior disorders.

Autism

Consider this statement, "Autism is perhaps the most highly heritable behavioral disorder" (Ronald, Happe, and Plomin 2006, 352). Somehow it doesn't come through in the article, since *Science* does not publish speculative articles, but the word "perhaps" in the statement really is straightforward in meaning. When studies present solid proof of something, "perhapses" and "maybes" are replaced by "the evidence shows . . ." It seems, however, that the Great Scientific Error belief adds weight enough to turn "perhaps" into a solid finding. Maybe that is what produces the certainty in the article "When a Child's Mind Is Abducted" (Lowy 2004, C5), which considers the sharp rise in the incidence of childhood autism as of "epidemic" proportions. In describing the enhanced interest in research inspired by the epidemic, the article reflects the same certainty: "While research into potential environmental factors [such as chemicals, foods, toxic conditions like mercury in vaccinations] is growing, the search for an autism gene commands the bulk of the research dollars" (C5). The certainty of belief shown here, despite the uncertainty of evidence, makes this another example of the Great Scientific Error.

Another headline trumpets "DNA Anomalies Linked to Autism." Genome scanning "found in slightly more than 1 percent of people with autism, a chunk of about twenty-five genes had either been duplicated or deleted, mainly in spontaneous mutations not carried by their parents" (Goldberg 2008, A8). Let's face it. Even if the finding is valid, there are various questions that arise. It only pertains to 1 percent of autistic people, so it could explain only a very small percentage of the cases. That is small proof of biological explanation.

Another article (Wickelgren 2005), discusses a theory, proposed on the basis of brain imaging, that areas of the brain are not connected to each other in autistic individuals as they are in normal people. "During this task [one of answering questions about some sentences] the activated brain areas showed far less synchrony in the autistic brains than in the brains of controls. . . . Some of the regions aren't linked up" (1856). The belief is that brain deviation explains autism, that the way to study this is by comparing the brains of autistics and normals. We can see the Great Scientific Error clearly in the conclusion that the finding "points researchers in the right direction" and shows "where the field of autism has to go" (1858).

Study after study makes the same error, overlooking the fact that learning experiences affect the structure and operation of the brain. Differences in the brains of autistics and normals in the present view have at least in great part come from the great differences in the learning experiences of the two groups. In the face of this, which should be clearly understood, the learning experiences of the autistic children should be the focus of scientific study, how those experiences affect the learning of the basic repertoires, how this affects the brain, and how the brain's operation results in the behavior of the child. One thing is very clear: strange as it seems, no conception, no impetus, bursts forth calling for the study of learning in the causation of autism.

The Great Scientific Error beliefs certainly are one reason for this discrepancy. In addition, however, considering the environment in the causation of autism is taken to cast blame upon parents. That has to be corrected, for the possible ways that learning experiences may cause autism must be investigated. If learning is causal, however, avoiding its study would create too much misery for both parents and children in the future. Learning has to be considered assiduously. Parents almost always are well meaning, they do only what they think is best for their child. They do what they have learned. They do what their society has told them they should do. They follow the advice of experts—pediatricians, psychiatrists, psychologists, social workers—they have read and consulted. They do their best. They try to protect their child. It is not their fault that their society and their experts have not been able to indicate to them what they should do. No one is to blame; the state of knowledge is just not generally advanced enough to provide what is needed.

What Is Childhood Autism?

The diagnostic criteria of childhood autistic disorder include disturbances in social interaction, as in play with other children and in making friends. Some autistic children display more serious symptoms of all kinds than others. But generally autistics lack social skills and empathy. Another feature involves deficits in intellective functioning. Severely autistic children typically have little or no language and cannot use language to get their needs met. They cannot understand stories, follow directions, take roles in play, or learn by verbal instruction.

Typically the child also does not display the usual affection for the parent and lacks the interests and activities of the normal child. The absence of affection for the parent seems especially shocking—suggesting some inherent abnormality—because children so generally display that affectionate behavior. This characteristic of autistic children has been shown experimentally: "Psychologist Geraldine Dawson of the University of Washington in Seattle and her co-workers have found that the [autistic] children's brain-wave activity indicates an inability even to distinguish [pictures of] their own mothers' faces from those of strangers" (Bower 2002a, 408). Such findings are believed to indicate that some brain defect causes the autism. That view simply assumes that children naturally have the ability to attend to pictures, to respond to pictures, and to discriminate differences between complex, similar stimuli. My study indicates these are learned skills.

It is the case, however, that autistic children generally do not look at their mothers like other children do, do not seek their mother's attention. Actually, at the very beginning, birth mothers do not have that great attention-getting pull, nor is the mothers' attention that important. These behaviors begin to be learned quickly, however, as mothers feed their infants and do other things that elicit a positive emotion in the infants. We might ask whether infants who later become autistic lose that positive emotional response to the mother. This is unknown, the question never asked.

One must also ask if there is a difference in learning experience for normal and autistic children that affects their behavior in this sphere. Would different learning occur for a mother who provided a bottle propped up as a way of feeding her infant and a mother who held the bottle with her face exposed to the child, talking to the child, smiling and changing expression? Would the face of the second mother elicit a more positive emotional response than the first mother? Wouldn't such a positive emotional response guarantee the mother would elicit the child's attention? A heavy body of evidence tells us the answers are yes. For a child without the positive behaviors that ordinarily bring parental attention and interaction, tantrums may likely become a way of getting attention. In another sphere, the autistic child does not play with toys—like dolls or cars—in the usual ways. Rather, toys are treated as objects having only primitive qualities, like size and weight, and thus they are objects to throw or pound or pull apart. Not realized is that establishing interesting characteristics to toys for a child also depends upon complex learning that produces complex repertoires. A toy car does not have car characteristics until the child learns what a car is.

Unfortunately, society's knowledge of childhood learning does not tell mothers how to conduct that type of learning. Rather it instructs mothers that the genes and brain determine the child's development. And it tells mothers to be alert to the possibility that their child might be autistic, for early diagnosis is important. The parent becomes an observer of the child rather than the child's play companion. I suggest that is a reason for what appears to be an epidemic of autism. I have seen that this instruction makes some mothers uptight, focusing on what the child does that might be abnormal, rather than seeking knowledge of the learning experiences that will make the child normal. It is my belief that the Great Scientific Error acts as an impediment in this way too.

Autistic children also exhibit inappropriate repertoires, stereotypic behaviors, such as rocking back and forth or playing with water for long periods. The child may learn screaming or temper tantrums as ways of getting things. Such behaviors give the autistic child control of the household situation because of the belief that these behaviors are the result of a brain problem of some kind. Head banging can also be learned because it gets attention. The child's movements may be awkward and uncoordinated because of deficit learning conditions. Common sensory-motor skills may be absent or poorly performed. The child ordinarily learns inappropriate repertoires also according to normal learning principles

This is an abbreviated description of the things that lead to a diagnosis of childhood autism. Even in this brevity it can be seen that the autistic child is not typically an intrinsically appealing child to others, children and adults; the behaviors the child acquires are not pleasing to others and they do not prepare the child for a normal life.

Tales of Autism

The project that Karl Minke and I set up in Wisconsin for twelve four-year-old children has already been described. It was like a traditional preschool, but it operated according to my behavioral framework. For example, if a child misbehaved, the consequence was "time-out," which consisted of the child leaving class activities and sitting in a corner until the misbehavior was over. (Our procedure constituted the first introduction of time-out into a group setting.) Each child also left the classroom for three scheduled periods each day to participate in learning reading, writing, and number repertoires. The children received tokens for their participation in the learning, and the tokens were exchangeable for small rewards.

One of the children had been variously diagnosed as emotionally disturbed and autistic. When she displayed inappropriate behavior, such as having tantrums, kicking, and biting, she received overseen time-out until she stopped misbehaving. When she was introduced to the individual learning activities, she first refused to participate and was not urged to do so. But there were two other children at each scheduled training period, and she saw them receiving the tokens and the rewards. So she began fitfully to engage herself in the learning task. In a matter of a couple of weeks, she was completely engaged and learning like the other children, and her results in the seven and a half months of the training were like those of the other children. The same improvement to normality occurred in the classroom. Progressively her misbehaviors became less frequent and then ceased to occur. From then on, she would not have been detected as different from the others. And the effects generalized; her behavior also improved outside of school.

If she had been maintained in this type of program, there is good reason to believe she would have continued to learn normal repertoires, that is, to become normal. However, the research project ended in 1966, when I left for the University of Hawaii. I later received a letter from Wisconsin asking me for a report on this child, for she was being considered for commitment to a public institution for disturbed children. The evidence, thus, showed that the child could function well and learn well when in the special conditions of learning and behavior that we provided. But in the usual conditions of schooling and life, where she was not rewarded for working at learning, and where she probably got what she wanted through abnormal behaviors, she reverted to her former disorder. The case actually had an experimental design, first the presentation of one set of circumstances, then a different set, then a return to the first set. The different learning conditions clearly produced different types of behavior in this mildly autistic child.

At the University of Hawaii, where in 1966 I established a similar classroom-laboratory for preschoolers (Staats, Brewer, and Gross 1970), one little four-year-old boy had also been diagnosed as autistic. He had no functional language. His toilet training was incomplete. He displayed bizarre behaviors, like suddenly screaming for no apparent reason. His social skills were minimal and he had no friends at home or at school. At school, with no language, he was unable to respond appropriately to play opportunities initiated by another child and he became an isolate. The only play interaction he had with the other children occurred when several of them encircled him and he would run at them and they would dodge or push him, baiting him as though he were a little bull.

My treatment program for him focused on teaching language. Much is made of the lack of imaginative play of autistic children and their lack of interacting socially with other children. However, such performances depend on language. A child without language cannot play imaginatively when another child says,

"Make believe you are Cookie Monster and I will be Big Bird." To respond imaginatively demands having learned the necessary language repertoires. This particular child had only minimal ability to imitate some sounds in saying words, so he could not respond to the other children when they gave directions about playing a game of any kind. Barbara Brewer, Michael Gross, and I gave him training in the verbal-labeling of common objects as well as in the verbal-motor repertoire for following simple instructions, using my token-reinforcer system. This training was progressively advanced as he learned and became more competent in language. I instructed his mother on how to conduct this training at home and also told her how to deal with toilet training and how to handle his tantrums at bedtime and in public places (like the supermarket). He was in our program for seven and a half months (when his parents moved to California), during which time he displayed great improvement. He gained rudimentary language ability; consequently his IQ advanced from zero to 50. These new repertoires enabled him to acquire a friend (a year younger) to play with at home. The problem of going to sleep at night dropped out, as did his tantrums, and his inappropriate behavior at school diminished. He had a long way to go, but the path of improvement had begun. Clearly he could learn normal repertoires when he had the necessary learning experiences.

My various studies with children learning behavioral repertoires were convincing, and I began hypothesizing forcefully that autism was a learning-behavior problem that could be treated with token-reinforcer methods (see Staats 1963, Staats and Butterfield 1965). In these studies and in my less formal observations of children, I reached the conclusion generally that children's abnormal behaviors as well as normal behaviors are learned. "[I]t appears from naturalistic observations that many children labeled as mental retardates or autistic children are only victims of poor training conditions. We simply do not know at this time what learning disabilities, if any, are involved with such children" (Staats 1963, 456). My analysis of autism went on to propose that what was involved was the lack of learning of needed behaviors along with the learning of abnormal behaviors. At that time, behavioral studies were limited to laboratory experiments with simple behaviors whose importance was in showing that the usual principles of reinforcement applied to autistic children (Ferster and DeMyer 1961). I suggested that my methods of reinforcement and training should be extended to autistic children and to children with other problems. My graduate research assistant in reading studies, Montrose Wolf, went on to successfully modify tantrum-type behavior in an autistic child using my token-reinforcement and time-out procedures (Wolf, Risely, and Mees 1964). That study had an important impact on the establishment of our new field, called behavior modification—Wolf later used the name "behavioral analysis" (Staats 1963, 459–60) in creating the *Journal of Applied Behavior Analysis*, which became a central mechanism for the development of the contemporary Association for Behavior Analysis International.

Soon Ivar Lovaas also began using our program's methods in training autistic children. No such methods were being used within the radical behaviorism approach. In 1966 he and his coworkers showed that autistic children could be taught imitative speech. He continued to develop his work in creating a full program to treat autistic children with a focus on language training. His extensive research, his systematic program for training autistic children, and the extensions others have made of his work, have become internationally known for their success (see Lovaas 1977). His students have gone on to create additional treatment facilities, and his program continues to advance as an outstanding therapy for the treatment of autism, constituting one of the strongest proofs of the value of behavior analysis and providing foundations for the contemporary move to diagnose autism early so that the young children can be given his type of therapy.

However, there is a general conception that hasn't been drawn from this very valuable program and has not been used; that is, that learning conditions are crucial in the child's development of abnormal behaviors, even those of autism. Lovaas's program trains autistic children; they improve through learning. Framing the findings within the present approach, however, raises a central question. If learning procedures can be used to treat autism after the behavior disorder has developed, then the same types of procedures would have worked preventively, at an earlier time when the child first began to deviate from normal development. If the child had received the necessary learning experiences in the first place, he or she would never have learned the inappropriate behaviors and deficits of behavior that constitute autism. Let me tack onto this the fact that evidence showing that autism is due to the genes or the brain is slim to the point of absence. Evidence that autism is learned is comparatively hearty. We have to extract the central implication; it opens the way to establishing the cause of autism and devising preventive measures. The broad suggestion emerges that even the most extreme behavior disorders may result from learning experiences gone wrong, learning experiences that can be reversed by learning—the earlier the better.

Cumulative Learning and Unlearning Autism

Cumulative learning of behavior disorders has already been described. That process also applies to autism. Once a child has been diagnosed as autistic, for example, the society encourages parents to use the mental-illness conceptual framework. The following letter to the *New York Times* is from a parent who describes an obsession of her child diagnosed with Asperger's disorder, a mild form of autism. The five-year-old boy "desperately needed to see at least one train a day that passed" close by. The parent lauds a *New York Times* article for understanding autism, saying "I hope that teachers today, in knowing so much more about Asperger's, can try to work with the child's obsession" (Skalitza 2011).

Clearly shown is the loving parent's belief, given by society, that the child has a disease—some abnormality of the brain or genes or something—that causes his abnormal obsessive behavior. In society's science view, the child is sick and needs to be treated, specially, as such. The abnormal behaviors are part of the child's nature, wired in. While in a normal child an inappropriate request would be treated as such, the ill child, however, should be treated in accord with his nature because of special vulnerability. Parents and other adults should understand a child's illness and act toward the child accordingly, or so the current societal view holds.

The present conception of human behavior and human nature, and its extension into the understanding and treatment of abnormal behavior, has a quite different view. The child's obsessive behavior is learned behavior. If we had recorded the child's learning experiences in detail, we would see just how the obsessive behavior was learned. As a general rule, the treatment of the child should not encourage behaviors that will be considered abnormal by the world, behaviors that will lead others to consider the child as abnormal and to treat him as abnormal. Rather than the Great Scientific Error conception, parents need a learning conception that tells them to use appropriate methods for ridding the child benignly of aberrant behaviors as soon as they arise. If a child's abnormal obsession or inappropriate fear, or whatever, gets no attention, the behavior will not be learned. The longer such behaviors are encouraged (rewarded), the greater the child's problem will be.

As another illustration that the downward spiral of autism involves cumulative learning, the following case indicates how once the child has acquired deficit and inappropriate basic repertoires, the child's life experiences become more "abnormal," continuing abnormal development via cumulative learning.

No longer does child development continue on in the usual manner. This process is clearly shown in the unusually loving and heroic actions of a family of an autistic child who has been written about widely. His brother reports that when he was three, he became a steward of his two-year-old autistic brother, Noah, and the conditions for both became unusual, with the former a supporting player and Noah the "center of attention. . . . Noah, who can't speak, dress or go to the bathroom completely unassisted, will always be the center of our family. He never earned that role; his needs dictated it" (Greenfield 2002, 54).

We have to esteem the unflagging love of this family for their autistic child and their dedication to the child's welfare. The family lives up to the highest expectations of society with respect to treatment of those who are ill and handicapped. But society's beliefs offer no familial treatment procedures that would remedy the child's grievous problems of basic repertoires. Without that framework, we can see an example of a social environment that has moved away from the normal, in a manner that will curtail the child's development through learning. Children come to do things for themselves if their learning experiences lead them to do so. A child who has not learned those repertoires is set up to learn abnormal repertoires instead. The abnormal repertoires, diagnosed as autism, will get others to accede to the child's wants, giving custodial care because of the biological-medical conception generally held. In this example, the autistic child has come to dominate the household by his abnormal behavior.

With that analysis in hand, let us look at the curative treatment of another less severely autistic child, by his own parents.

> Rowan slept when he was supposed to, walked ahead of schedule, and said his first words before his first birthday. But around the time he turned eighteen months old . . . Rowan wasn't pointing; he didn't respond to his name, his vocabulary wasn't growing. Then the tantrums started, ear-splitting fits that could last for hours. When he was two his demons were given a name: autism. (Isaacson 2009, 87)

The child develops better than average for his first year and a half. Then something changes, his language development stops, he has extreme temper tantrums, he doesn't develop toilet skills, and so on. He is a trial for his parents. Why? The father uses the term "demons" in explaining his child's abnormal behaviors, by which he seems to mean some unknown, abnormal, internal, biological condition.

When a tantrum happened at home . . . there was only one thing I could do: take Rowan into the woods behind our house. Within seconds, the screams would lessen and disappear when he found a patch of sand to run his fingers through, variegated bark to look at. Animals and nature are what motivated him [not to mention the attention and company of his father]. (Isaacson 2009, 87)

The father, demonstrating his love for his son, hits upon something that stops the child's agony of temper tantrums. Actually, in doing so he is unintentionally rewarding the tantrum behavior. But it is clear he is searching for ways to improve things for the child. In doing so he introduces the long-term activity of horseback riding that the child enjoys greatly. That places the father and son into a happy relationship. When engaged in this activity, the child says words and language learning commences again. With this success, the father plans a trip to Mongolia, where they ride and have various types of positive experiences. Language interactions between the child and father multiply. What occurs covertly in the mutually enjoyed activity is the conditioning of positive emotion of each to the other.

By the time we got back to the US my [previously un-toilet-trained] son was starting to take himself to the potty. The tantrums, the hyperactivity and anxiety had left him completely. We had come back with a different child.

At school, his academic status was reassessed. At age five, he was reading at a seven-year-old's level. When a friend brought her stepson, Gavin, to ride with us on our new horse Clue, Rowan and Gavin became fast friends. Soon half the neighborhood kids were turning up for riding play dates. . . . Rowan's social life [a year later] was now like that of any other child. (Isaacson 2009, 89)

Isaacson presents an account that is most informative. Demonstration of the love parents have for their children suffering behavior disorders is touching. Informative also are the things these parents do to improve the child's lot and the effects they attain. Their child had the symptomatic behaviors of autism. The account shows how the father discovered an environmental situation that his son found to be full of interesting and rewarding things and activities, an environment within which father and son could interact with positive emotional affect on both sides. The son, in having such experiences, learns a strong positive emotion toward his father in the context of learning active behaviors that act on the environment, both social and animal. This brings this child into contact with his world in ways he did not have before. The child learns language-cognitive, emotional-motivational, and sensory-motor repertoires. I suggest that Rowan had not been getting these experiences as he developed into autism through learning. Isaacson, however, found the woods setting for his child to begin learning basic repertoires that he had stopped learning. That progressively developed into additional rewarding actions and experiences of riding horses, traveling in Mongolia, and planning and meeting with shamans. Here were rich and numerous experiences.

As the child's behavior changed—where he developed more useful behaviors than temper tantrums—that provided the basis for additional productive parent-child interactions. The child became toilet trained, probably because of the better parent-child relationship and thus the better learning experiences. The child's language advances, and the parents and others are able to introduce experiences by which the child learns to read. They also use their resources to bring their son together with other children in rewarding settings, providing very important social experiences and bonding with children.

The Isaacsons had the wherewithal in resources and knowledge to construct a very rich, broad "therapy" for their son, one that has advantages over that which can be provided in an institution. Another central point stands out: the child's therapy was not by treating some internal biological problem. What was changed was the child's experiences and the learning this produced. That must be considered systematically along with the question, could learning experiences be the cause of the autism? Could Rowan's autism have been prevented? Could the basic principles involved in Rowan's treatment be developed on the basis of systematic knowledge to be used with children generally?

Learning Autism and the Hidden Environment

Autism, although generally believed to have a genetic cause, is not diagnosed by testing for genes. Very revealing. How is autism diagnosed? Diagnosis is made by observing the child's behavior. It is easy to diagnose as autistic a child without language, who does not look at his mother, doesn't play imaginatively with toys, and does not interact with other children, on the one hand, and who, on the other, has severe temper tantrums, bangs his head against the wall, and spends hours rocking back and forth or playing with faucet water. No one would question that. What is not easy for people to see, because society has not developed the needed knowledge, is how necessary behaviors can be absent and undesirable behaviors can be present in families of perfectly normal people. How could the behaviors of the autistic child be due to learning?

Take, for example, the child who does not look at his mother. How could that be due to learning? After all, the mother is in love with her child like other mothers. Let's say, however, she is also a professional with time and interest pressures. As part of her love, she has also made herself well informed about children and child development and about the dangers of the autism epidemic. That has given her society's belief that the child develops via biological maturation, needing only love and good physical care. She loves the child and gives the child excellent physical care. She observes the child closely to pick up any signs of abnormality and questions the nanny she hires about any abnormalities she sees. But she does not spend time with the child playing and pleasurably interacting, not even when the child is eating. She is waiting until maturation brings those interests to her child. Her efforts go into responding to the needs of the child as indicated by crying and other complaints. Her nanny follows her lead. After a few years, the child will not have learned the usual love for the mother. Seeing that, knowing it is a sign of autism, she may then try to get the child to look at her and embrace her, and the child may learn resistance is a way to get attention from the mother. Here is a hidden environment.

There are many ways for hidden environments to occur that produce abnormal learning in children.

Autism Prevention

The American Academy of Pediatrics, at its 2007 annual meeting, pushed for screening children for autism twice before age two, checking on symptoms such as not babbling. What is not included in this excellent advice on early diagnosis is why children do or don't babble. The medical-biological perspective does not include such knowledge, nor does the applied behavioral field that provides therapy for autistic children. *Knowledge about why children don't babble cannot be gained from working with children that have not learned that repertoire. How children learn to babble must be studied.* The kind of learning experience described in chapter 4 is what is needed for children who are delayed in the development of this centrally important repertoire.

Very basic in understanding abnormal behavior is knowledge of normal behavior and how it is learned. Behavior disorders arise when normal repertoires do not. Preventing the learning of inappropriate abnormal behaviors can only be considered when the knowledge of normal repertoires provides the necessary comparison. What is to be taught to the child and what is to be avoided provides the basis for prevention practices.

The Great Scientific Error misleads parents by assuring them that child development occurs via biological maturation, that caring and love are what the child needs. *A central prevention framework for parents is to realize that rearing a child is not just taking care of the child physically, expecting the child to blossom behaviorally like a plant. Parents also must take care of the child's learning, a great need of the child. The types of language, emotional, and motor learning described in chapter 4 will prevent the development of autism.*

Additional Language-Cognitive Disorders

Abnormality in the language-cognitive repertoire constitutes a feature of various behavior disorders. The delusion of the paranoid schizophrenic that he is Napoleon, for example, does not agree with reality and is not adjustive. The conduct-disordered child who lies also exhibits language that does not parallel reality. The manic person who believes in and gives an exaggerated view of herself displays nonveridical language. A language repertoire that deviates too much from reality does not constitute a good mechanism for solving problems, planning, and the cumulative learning of normal repertoires.

Other disorders involve the absence of or paucity of language. That includes the autistic child who displays no language, the child with expressive language disorder whose language is sparse and poor, as well as the dyslexic child. Language is powerful in providing the mechanism for learning, for reasoning, for problem solving, for predicting, for anticipating, and for guiding actions and emotional feelings. When one's language is rich and veridical, it will guide behavior that is appropriate to the life situation. When one's language is deficit or inappropriate, it will guide deficit or inappropriate behavior. That is how the various aspects of the language repertoire constitute a basic, underlying process responsible for abnormal behavior as well as normal behavior and learning.

EMOTIONAL-MOTIVATIONAL DISORDERS

Humans learn a huge emotional-motivational repertoire, that is, a huge number of things in life that (1) will elicit an emotional response, (2) will have rewarding or punishing power, and (3) will bring on either approach or avoidance behaviors. There are so many such things that they are grouped and treated as separate interests, values, desires, needs, attitudes, and many more, including those that are abnormal. That great number means there are tremendous individual differences in what is liked, desired, and loved and what is disliked and hated. Moreover, those differences determine great differences in individuals' sensory-motor and language-cognitive behaviors, for they are determinants of what will be attractive or repulsive. At the heart of some of the behavior disorders lie aberrant learning experiences that have created an aberrant emotional-motivational repertoire. The person does not feel emotionally toward many things as he or she should. That deficit itself may be the problem. Or the disorder may lie in the aberrant behavior that results from a person with inappropriate emotional feelings, like the pederast who loves young boys. The emotional repertoire can be the primary problem. I will describe a few behavior disorders that stem from emotional-motivational problems, although they are not always treated in that way in the customary diagnostic categories. For example, the first behavior disorder, paraphilias, is considered as a sexual disorder, in a way that is different from one of the mood disorders. The present conception indicates that they are both essentially disorders of the emotion-motivation repertoire.

Paraphilias

The paraphilias—defined as having "recurrent, intense sexually arousing fantasies, sexual urges, or behaviors" (DSM-IV 1994, 522) toward unusual and unacceptable things—constitute good examples. In the present view, the problem of fetishism lies in having strong sexual-emotional responses to objects, such as articles of women's clothing or toes or ears. A man, for example, might masturbate "while holding, rubbing, or smelling the fetish object or may ask the sexual partner to wear the object" (526). When a stimulus elicits a positive emotional response—and a sexual-emotional response is very positive—it will further tend to elicit the approach behaviors the individual has learned. The person will approach and strive to have "commerce" with that type of stimulus. If forcing a woman to have sex elicits a strong sexual-emotional response in a man, then he will strive to commit rape. Since children elicit a strong sexual-emotional response in a pedophile, he will have recurrent fantasies about sex with a child, he will have sexual urges when encountering a child in the right type of situation, and he will behave sexually toward the child. He will also be motivated to gain skill in creating such situations. And each time the paraphiliac has a successful experience (reaching orgasm), the behavior that enabled that success will be strengthened and more likely to recur.

> Individuals with a paraphilia may select an occupation or develop a hobby or volunteer work that brings them into contact with the desired stimulus, [for example,] selling women's shoes or lingerie (fetishism), or working with children (pedophilia). . . . They may selectively view, read, purchase, or collect photographs, films, and textual depictions that focus on their preferred type of paraphiliac stimulus (DSM-IV 1994, 523–24).

These usually are employed to achieve sexual arousal and orgasm in masturbation. Humans can learn sexual-emotional responses to anything—dead bodies, feces, urine, animals, pain and humiliation, giving pain and humiliation, dominating, being dominated, pieces of clothing, sucking toes, fires, taking risks, as well as verbal descriptions and films of such things.

Having a paraphilia in and of itself can be maladaptive. The person realizes his or her abnormality and devalues him- or herself for it—a boy driven to steal panties, a man with urges for a dead or dying woman, a woman turned on by beating a naked man, or a priest with a hunger for young boys—but can't control it because of the pleasure it brings. The strength of the emotion may impede the individual from actions important for establishing a traditional life situation. Society may consider the paraphiliac's behavior criminal and levy severe punishment. A psychologist named Raymond conducted the first paraphilia case treated with a learning procedure (see Eysenck 1960). The patient was a man who was arrested and hospitalized numerous times for accosting baby carriages in a sexual way. He had a collection of photographs of baby carriages that he used in masturbation. Raymond used the photographs in the treatment, pairing them with a very aversive stimulus, so they came to elicit a negative emotional response (see Staats 1963, 500–501). For the first time, the man was able to have intercourse with his wife without fantasizing about baby carriages.

Sexual behavior in general is usually considered to be a biologically determined behavior. Thus, the fact that some animals engage in homosexuality is taken as evidence that human homosexuality is an evolved trait. However, human sexual behavior should not be compared to animal sexual behavior. Humans have so many variations in sexual behavior. Those differences have to be learned, including the many paraphilias for objects that did not exist during human evolution. Take Catholic priests. They are actually displaying unusual sex behavior by not having heterosexual intercourse. Would that occur on any other basis than their learning, the strong religious beliefs they have learned, including the learned prohibition of heterosexuality? That can occur through many learning scenarios, and in some circumstances instigate a paraphilia. For example, a boy, because of learning that sex is dirty and sinful, cannot envision a girl when he masturbates and thus uses other images instead. Another boy has a potently pleasurable sex relationship with his mother's boyfriend. I had a childhood friend who had that experience. Another boy masturbates throughout adolescence to images of an eleven-year-old girl he saw naked by mutual consent. Another boy has a first sex experience by forcing a neighbor woman, and he repeats that in fantasy while masturbating and occasionally in actuality. Each case means hundreds of multiple, intensive conditioning experiences. The divergences are numberless for boys and girls, and they all produce the learning that determines later sexual emotionality and sex behavior.

Depression

Paraphilias involve having learned positive emotional responses to things that are inappropriate. Abnormal emotion-motivation can also be the fundamental problem in other behavior disorders. For example, common belief still holds that depression is caused by biological conditions. Elaine Heiby and I constructed a theory of depression within the present framework (1985). Compatible approaches have grown since that time.

> You may have heard people talk about chemical imbalances in the brain, suggesting that depression is a medical illness, without psychological causes. . . . In fact, the chemical imbalances that occur during depression usually disappear when you complete psychotherapy for depression, without taking any medications to correct the imbalance. This suggests that the imbalance is the body's physical response to psychological depression, rather than the other way around. . . . A serious loss, chronic illness, relationship problems, work stress, family problems, financial setback, or any unwelcome life change can trigger a depressive episode. (Psychology Information Online 2010, 1)

Rather than the body's changing in response to depression, those body changes are part of the depression, both being a response to the negative environmental event. But the account is in agreement that depression is environmentally caused, not arising from abnormal biological conditions themselves.

While it is not possible to deal here with the full theory, a few major points can be characterized. Centrally, widely different emotional-motivational repertoires are learned. Of importance here, one person can learn more, and more intense, positive emotional responses and fewer, and less intense, negative emotional responses to life stimuli than will another person. For another person, learning conditions are quite the opposite. That will constitute a type of personality characteristic; in the same life situation, the first person will experience more positive emotional stimuli than the second person and fewer negative emotional stimuli. When a number of emotional stimuli are experienced over a period of time, that will produce an emotional response state that also persists. The person who has learned positive emotion to many life situations will thus have a persisting positive emotional state. The opposite holds for the other person, who has learned negative emotion to many life situations. In common-sense terms, such people are not happy.

A person in a negative emotional state will experience any new negative emotional experience more intensely and experience any new positive emotional stimulus less intensely. A person in a negative emotional state will be more likely to be affected negatively by things not usually found bothersome, to feel more fear of things than usual, and to have disturbing thoughts. On the other hand, a person in a very positive emotional state—such as a person who is tipsy from alcohol—will consider jokes more funny and life more enjoyable, will love more deeply, and will experience anxiety less strongly.

Thus, the person's emotional state depends on the emotional-personality repertoire the individual has learned as well as the person's present life situation. There are life situations that bring on a concentrated host of emotional stimuli. A person who has gone through a natural disaster or who has been sent overseas in the military or who has been imprisoned may have lost family, home, and way of life. Many positive emotional stimuli would have been lost. The person's emotional repertoire may not be positive enough to outweigh the negative happenings, and it may take a long time to change to a positive level.

Within that theory we can see why there are individual differences in susceptibility to depression. Let us take the example of two mothers, each of whom has lost an only child. One mother has a happy marriage, a rich sex life, a fulfilling career, compatible relatives, close friends, and various enjoyed recreations and cultural pursuits. She will weather that loss much better than a mother with an empty marriage, an unhappy sex life, no career, incompatible relatives, no close friends, and no recreations or cultural pursuits. Losing a child will be catastrophic for both mothers, and the behavior of both will be governed by grief. Over time, however, the first mother will have sources of positive emotion that will again begin to affect her behavior and initiate a return to life. The other mother will have no sources that yield positive emotional responses and happiness and strengthen active, striving life behavior. "When positive emotional sources are [unavailable] in the individual's environment, or when the individual has not learned [an appropriate emotion-motivation repertoire] . . . there is not much to counterbalance the scales when negative emotional stimuli are encountered in life" (Staats 1996, 284).

Anxiety

There are various disorders classified as anxiety disorders—phobias, panic disorder, obsessive-compulsive disorder, and general anxiety disorder—all of which in different ways involve learned negative emotional responding or a negative emotional state. They have different ranges of generality and seriousness, a phobia being the most specific disorder and general anxiety the more generally significant. There are too many of these disorders to deal with here. Rather, the general state of anxiety will be described, as it is a component of the various anxiety disorders.

> Let me illustrate, with a mother who has learned a negative emotional response to many things such as dogs, being alone, contagious germs, illness, various failures, being different that others, not keeping up appearances, economic conditions, international tensions, parental censure, sibling competition, loss of sexual attractiveness and performance, loss of husband or his affection, infidelity, loss of friendships, planning trips and social affairs, as well as the uncertainties of children's health and school and social success, and on and on. The mother is concerned that her young child may get chilled, play with the rough boy down the street, eat junk food, get too tired, play too hard, get accosted by strangers, touch contaminated things, not behave properly in the presence of her husband's boss, and on and on.
>
> A mother with such an emotional-motivational repertoire is likely to provide conditioning [via language] to her children that will produce the same type of emotional-motivational learning [that she had]. . . . [R]aised with an "anxious" parent the child will learn that the world contains many "fearful" stimuli. (Staats 1996, 291)

Such a child will have an emotional-motivational repertoire tending toward anxiety, a repertoire that will serve as a foundation of the cumulative learning of negative emotion to life situations. Life circumstances the person encounters can also play a large role in producing an anxiety disorder. Take a career woman with a promotion pending, a new love relationship that could go either way, the responsibility for the company party, a large project that could succeed or fail and add or subtract from her promotion chances, an upcoming city tennis tournament that will determine her state ranking, and the problem of resisting without offense her parents' pressure to move back with them. Worrying about losing important sources of positive emotion, plus the excessive time pressures, could contribute to an anxiety state, a negative emotional state (Staats 1996, 290). This is to say that an anxiety state may occur because of immediate circumstances, or because the individual has learned extensively to have a negative emotional response to many life stimuli.

Additional Emotion-Motivation Disorders

Having a theory composed of principles that apply generally to behavior makes it possible to see the ways in which disorders are related. Take for example the behaviors of the person diagnosed as hypomanic. "The elevated mood . . . is described as euphoric, unusually good, cheerful, or high" (DSM-IV 1994, 336). The individual also behaves differently in various ways, for example, heightened self-esteem, more confident actions and speech, to wit "full of jokes, puns, plays on words, and irrelevances" (336). Here again there are two elements of the behavior. There is the behavior itself, what the individual says and does. And there is the internal emotional state, which is inferred from the individual's manner and actions but which could be measured also by psychological tests or physiological measures, including brain imaging.

Another complex and revealing analysis may be seen in bipolar disorder, where in one phase the person is exaggeratedly euphoric and manic in action. In the down phase, the individual is unrealistically depressed emotionally and in behavior. When a bipolar person goes into the depressed phase, the additive principle of emotion acts to make the condition more extreme. Life experiences that are positive won't be experienced as positively as they ordinarily would. Life experiences that are negative, on the other hand, will be experienced more emotionally than is merited. The individual may break down and cry over something that would not occur with normal people or with the individual himself when not in that phase.

The same process works in the euphoric phase of bipolar disorder, where the individual is in a continuing very positive emotional state. In that state everything is experienced more positively emotionally than is merited. The individual thus comes to feel superpositive about everything, including himself, his abilities, his business deals, his romances and other social interactions, and his future. This is unrealistic, however, and the actions of the individual on the basis of the extreme emotional state do not work out well. Business deals don't go through, the girl he is sure must love him turns him down, he loses social prestige from excessive bragging, his confidence fades, his future becomes clouded. He can then slip from euphoria to depression. Bipolar disorder is generally considered a biological problem. Instead, this analysis sketches the dynamic that takes the individual from overly positive to overly negative.

Pain

Pain is not usually considered in the same breath as depression and the anxiety disorders. Peter Staats, however, has formulated a theory of pain, with contributions from Hamid Hekmat and me (Staats, Hekmat, and Staats 1996), that places it in the same framework. A focal part of his theory is that emotions add together algebraically, positive emotional stimuli add and negative emotional stimuli subtract from the positive level of the individual's background emotional state. The theory treats pain stimuli as strong negative emotional stimuli. So painful stimuli affect the individual's emotional state. A person with a strong chronic pain, thus, has similarities with a depressed person and a person with generalized anxiety. Each suffers a negative emotional state. For the pain patient, a biological condition is central, however.

The theory provides a context for better understanding individuals with pain, as well as for treating the person with chronic pain. Take this example. As a very usual treatment when a person gives pain as the reason for not assuming responsibilities, other family members may use social pressure. But no matter how gently applied, that pressure will arouse a negative emotion in the person, and that will exacerbate the problem. At the very least, the pressure will make more evident to the person that he is not holding up his end as his loved ones would like. The negative emotional response the person in pain then experiences will simply add to the person's negative emotional state and thereby increase the pain as well as worsen the person's life behaviors. The pressure that has been exerted will simply make matters worse. If the pain is physically caused, psychological treatment should aim to reduce the individual's overall negative emotional experience. Peter Staats and Hamid Hekmat conducted a number of studies within the context of this pain theory that verify it.

Unifying the Behavior Disorders

Important here is that these various behavior disorders, from paraphilias to bipolar disorder, and including the negative cases such as depression, are to be explained at least in good part by abnormal learning of the emotion-motivation personality repertoire. The relationship of these disorders, and of the learning conditions that produce them, need systematic study. Differences in behavior disorders can occur on the basis of differing intensities of the emotion, whether the emotion is positive or negative, the nature of the eliciting stimuli, and the nature of the sensory-motor responses the emotion elicits. It is important to analyze the different behavior disorders in the same framework of principles, for that unifies the phenomena and makes them all easier to understand. Such analysis also gives more knowledge concerning the learning causes that are involved.

THE SENSORY-MOTOR REPERTOIRE

Frequently it is the overt behavior that appears to be the disorder even when the behaviors in and of themselves are not unusual—just when they occur and as a response to what circumstances. The more basic problem, however, lies in the language-cognitive or emotion-motivation repertoires. Nevertheless, there are various behavior disorders that center on deficit or inappropriate aspects of the sensory-motor repertoire. That is, sometimes the problem lies in how the individual acts.

We see that clearly in disorders involving the physical violence that is involved in spousal abuse, rape, antisocial personality disorder, and conduct disorder. These behaviors are inappropriate, disadvantageous, and harmful to the perpetrator and to the victim. There are also cases of abnormality that consist of deficits in sensory-motor behaviors. Enuresis, the absence of behaviors of waking up and going to the bathroom at night is an example, as is encopresis, the absence of bowel control, both of which can readily be prevented by the needed learning experiences (see Staats 1963, 1996).

Enuresis and Encopresis

The belief is that children wet the bed at night because the sphincter muscles and nervous system have not matured. In the present view, the problem exists because of learning that should not have occurred, and learning that should have occurred but didn't. What has to be learned is that when the bladder is full, the child must wake up and go to the bathroom. The learning is simple; the full bladder produces internal stimuli, the child must learn a waking response to those stimuli. Enuresis also involves another type of problem, that of having learned to sleep through the discomfort of a wet bed. That occurs because urinating when the bladder is full is immediately rewarded by the decreased pressure, and only later punished by the discomfort of the wet bed. The time difference between reward for a behavior and delayed punishment for the behavior is involved in various abnormal behaviors.

I have already described a simple and easy training method I developed to ensure the child's nighttime toilet training. My daughter would go to bed after urinating. Three to four hours later, just before I retired, I would go to her, wake her up, and get her to stand up. That constitutes an essential part since the child will try to continue sleeping and the training will fail if the parent simply carries the sleeping child to the bathroom. So after she stood up (and at the beginning the child will need support), I would help her to walk to the bathroom and sit on the toilet. There she would urinate, and then I would walk her back to bed, where I would tuck her in. As in all things, beginning learning is a bit awkward, but as it progresses it becomes more and more fluent, with the child waking easily and doing the whole thing without assistance and finally not even needing awakening. Jennifer never developed bedwetting because she had learned to awake herself and didn't learn to sleep in a wet bed. Without some kind of training, a child can continue to wet the bed, for a number of years, calling for the diagnosis of enuresis.

Encopresis consists of the child's lack of toilet-training skills with respect to bowel evacuation. Again, a belief holds that the child's maturation brings the skill and parents can still be advised to wait patiently for that to occur. That is not good advice, and the child who continues to lack toilet skills will face some very negative social experiences that can further deepen the problem—like becoming anxious about defecating, attempting to retain feces, and having explosive, messy, and very negative experiences. On the other hand, straightforward learning-behavioral treatment can produce the necessary training very easily (Staats 1963, 377–79). With my children, I first established the time at which defecation usually occurred. The next step is to place the child on the potty shortly before this time, explaining that he or she is going to learn how to go to the bathroom (like mommy and daddy or an older child). The child should be entertained, as the reinforcement for the response of sitting on the potty. The parent can read to the child, tell a story, have the child color a coloring book on her lap, or whatever. The first time should involve a very brief period on the potty, followed by praise, which reinforces the behavior and conditions a positive emotional response to the potty chair. No forcing or restraint should be employed. When done successfully, the sessions can be gradually lengthened. The experience should always be positive, and the child as a consequence will choose to participate because of being reinforced for the behavior. If this is continued, sooner or later the child will score a "hit," which should be reinforced by the parent's delight and praise. The hits will then increase in frequency. The comparative natural reinforcement for defecating into the potty versus into the diaper favors the former. After a sufficient number of successful potty trials, the child will become toilet trained. The child should also be trained to say when she needs to go potty. This is an essential part of the toilet skills. Azrin and Foxx (1974) further developed this approach.

Attention Deficit Hyperactivity Disorder (ADHD)

Children are diagnosed with ADHD for multiple reasons. They do not listen when spoken to, do not respond well to instructions, have difficulty organizing school tasks, are easily distracted, and are hyperactive. These behavioral characteristics can also lend themselves to classroom disruption. Centrally these characteristics make these children poor learners.

As in all disorders, there is a strong belief in biological causation. Neuroscientists have shown that the brain's cortex thickens from front to back in childhood. The time when the cortex is at its thickest is considered a milestone in brain development. Interestingly, that point in some regions of the cortex—those responsible for attention and planning—is reached by ADHD children several years later than in normal children. "The findings . . . help explain why many children diagnosed with ADHD eventually grow out of it, as their brains slowly become more similar to those of their peers" (Rawe 2007, 49).

The whole thrust of this study is that some abnormal feature of the brain produces ADHD and that medicine treats that brain problem; the Great Scientific Error again, no evidence of cause, and no recognition that learning produces both brain development and behavioral development. The change in the brain probably occurs because the medicine sedated the ADHD children and allowed them to attend to tasks and to learn. Success in drug treatment of a behavior problem does not explicate what goes on to effect the improvement, and success can't be taken as proof that a brain condition is the cause of the problem.

The central point here is that the biologically oriented explanation of the disorder leaves no place for learning to link in as an explanation. Has anyone done a study to see if medicated ADHD children pay attention better and thus have more learning trials in school than nonmedicated children?

The learning explanation of a case of ADHD was clear with an eight-year-old boy of parents with whom I consulted. His teacher, at a good school, wrote the parents that the boy "does not attend well, that he has trouble transitioning to new tasks and finishing old tasks. He frequently does not get the instructions that are given, which frustrates him." The teacher had decided that the child had attention deficit disorder and recommended that steps be taken toward securing drug treatment.

However, my considerations of the various sources of information with respect to this case revealed two things. Concerning the attention deficit, at home the child would not respond when he was calmly asked to do something while occupied by playing, reading, or watching television. After asking in this way a time or two more, the parent then shouted at the child. The child was then startled and upset by the annoyed demand, and in this state he was clumsy in response. The exasperated parent at this point would be unhelpful and critical of the performance. It was clear the child had not learned to respond to normally delivered requests, especially when he was doing something else. Without understanding what was happening, he had learned to feel negative about himself, very unsure about his own ability. The latter was intensified by other interactions. Wanting the child to grow intellectually, the parent concocted thinking problems for the child to answer. The child had come to show anxiety in such a situation, and this was enhanced when the parent (both parents had high-level professional training) then urged him to "think outside the box." But this didn't help the child to understand the problem or what was wanted, leading to the child's bewilderment and poor performance. Doubt was aroused concerning how smart the boy really was. Here was a case of a hidden environment producing deficits in the needed attention repertoire and learning to avoid learning situations.

The treatment revolved around the two problems—that of improving attending to requests and following instructions, and that of converting anxiety about his learning abilities to a positive emotion. It was quite clear from his language and reading ability that he was indeed quite intelligent. The fact that the child could become engrossed in activities clearly showed, as is generally the case, no biological deficit in attention ability.

Resolving the first problem involved instructing the parents how to train the child to respond to normally voiced requests. First, the parent should not interrupt an engrossing activity unless the parent definitely wants a response and is prepared to give the positive training needed to get it, with a good interaction, and without criticism or put-downs. If the parent continues to make a request, and then does not follow up when it is not answered, the child will learn not to attend to requests. If lack of response continues to elicit yelling at the child, then the child will continue to learn to respond only to yelled requests.

The new training I recommended was to begin with requests presented in normal speech. When the child did not respond, the parent was instructed to go to the child and say, with an appropriate, not chastising, loudness, "I need your attention . . . I want you to do [such and such] now." Then the child should be given understandable directions concerning what to do. When the child complies, the child should be told something like "That's good" or "That's the way to listen." To get the attention at the beginning might also necessitate additional parent action—such as physically turning the child's face away from book or television and saying "Listen to what I am saying," in a firm but gentle and kind manner. I also recommended that the problem be thoroughly discussed with the child in benevolent, helpful terms. With a child that age, parents can indicate what the child needs to learn, why, and the consequences of not learning it, such as not doing as well, and being confused, in school. Making evident that the parents' actions are to help the child places the parents on the child's side, rather than in the position of being a watchdog.

The second aspect of the problem involved the child's emotional learning. The child had come to have anxiety when instructed to do something. So all training situations had to be conducted with positive emotionality.

In this particular case, the loving parents were also very adept; they developed ways of dealing with both aspects of the problem, and they incorporated tutoring for the child in arithmetic. The child was already a very good reader, and they used that as a means of showing the child how smart he was. The net result was that the child improved in following instructions and in self-esteem and went on to the next grade, instead of being placed in a less advanced class. Three years later, the child was a top student, far from being classified as having ADHD, definitely with no need of drug treatment. There was no evidence that the child suffered from any biological aberrance. But if the child's ADHD behaviors had led to the wrong treatment, with the consequent loss in learning ability, the child's learning in school would have suffered. His teacher was already recommending he not be promoted because of his learning problem.

ADDITIONAL PRINCIPLES OF ABNORMAL BEHAVIOR

A theory of abnormal behavior has been presented and illustrated. It is based on evidence: a systematically constructed foundation of research findings, applied findings, analyses, and a unified theory of development. This is a theory for extension and elaboration, for further validation. Let me append a few such additions here.

Cumulative Learning of Abnormal Repertoires

Cumulative learning reveals itself in studying complex repertoires. For example, children's acquisition of language occurs through the successive learning of repertoires, where one repertoire has to be learned before another one can be learned. Personality consists of the successive learning of repertoires. Intelligence, thus, involves learning language repertoires that serve as the foundation for learning other repertoires, and on and on.

Lack of knowledge of cumulative learning has held back our understanding of the great force learning has as a cause of abnormal behavior. That applies to various behavior disorders. Abnormal behavior may become extreme because of the long-term cumulative learning that has gone on. There are various inappropriate repertoires and deficit repertoires that provide foundations for learning additional abnormal repertoires. The next section describes one.

Anti-Learning, A Basic Abnormal Repertoire

On the normal side, children generally learn a most important "pro-learning repertoire." A well-developed language repertoire is a strong pro-learning repertoire. So is a positive emotional response to learning situations. Children who have such an emotional response will approach learning situations, be interested in them, because of having found learning situations entertaining and rewarding, worth the effort, in the past. In addition to this emotional-motivational development they will also learn specific sensory-motor skills, including investigative attentiveness to the material being presented and a readiness to make the responses the task calls for. Such behaviors are generally productive when confronting learning situations.

Various repertoires add up to a pro-learning repertoire that increases learning ability greatly. The various aspects of this repertoire are learned to different degrees; those children and adults with the best-developed pro-learning repertoires, other things equal, will be the best learners. A deficit in this repertoire, on the other hand, leaves the child with a less than normal ability to learn.

In dyslexia, for example, the child generally learns a negative rather than a positive emotional response to reading-learning situations because of the negative emotional experience such children have. As a consequence of this negative emotional response, the child is motivated to escape and avoid learning situations. Such children do learn various ways of avoiding learning situations that involve reading, including how to disrupt such situations. These ways of escaping situations, along with the negative emotional responses, constitute a part of the abnormal basic repertoire that I call the anti-learning repertoire, a repertoire that children with learning disabilities have. The ADHD child, as an example, generally avoids learning situations by not attending, by generally flitting around attentionally, by acting impulsively, and so on. Children with well-developed anti-learning repertoires have learned to escape learning situations by such actions as grabbing any materials that are involved and playing with them, by persisting in some other activity, by being disinterested, by responding randomly, by being blank and withdrawn, by not comprehending. Some children develop language-cognitive repertoires that are argumentative, insulting, complaining, and threatening, all employed to divert the training situation from its mission. Another prominent escape from learning situations is to repeatedly throw answers out even before the task has been explained, thereby annoying the parent or teacher and finally disrupting the learning situation. Sensory-motor elements typically are present in the anti-learning repertoire. A child can learn to escape unpleasant learning situations by being hyperactive, by playing dumb, by crying, by having a tantrum, by grabbing the materials (pencils, letter flashcards, books), thus taking control of the situation so the task is not presented and confronted. Acting out in class, baiting other students, and getting in fights are among the many disruptive behaviors whose effect is to get the child out of learning situations.

Children with a learning disability will have learned the anti-learning repertoire. Much of that learning experience will occur in the school or be school-connected—for example, when a child is called upon in class to display a skill the child does not have. He may be teased because he cannot read, be admonished by parents for poor school performance, and experience various other signs of his failure. Thus, along with the other elements of the anti-learning repertoire, the child becomes sensitive concerning his or her ability.

The anti-learning repertoire can continue despite its long-term disadvantages because, in the short-term, it removes unhappy experiences. I knew of an adult who, although of very high-level accomplishment in many areas, was anxious about and had an anti-learning repertoire about acquiring new athletic sensory-motor skills. The repertoire included posing challenging questions for the instructor concerning why making some move was necessary, that it was not helping. The instructor would be told that this was a bad day for going into what the instructor wanted to do, repeated to the effect that this just was not the time. The behavior simply meant that various aspects of the game could not be dealt with, and much time was wasted in the confrontations the person with the anti-learning repertoire produced.

As the anti-learning repertoire develops, the child will come to appear as though she or he has a diminished learning capacity. And, indeed, she or he has. For what the child has learned is "how not to learn." We need systematic study of the anti-learning repertoire in its different forms, in how it is learned, and in the way it is involved in the cumulative learning of various behavior disorders and behavior problems. The anti-learning repertoire is one of the chief basic repertoires that leads individuals in what I have called the downward spiral of learning abnormality (Staats 1971a, ch. 15). Future studies should address the prevention of learning the anti-learning repertoire, including study of the pro-learning repertoire and how to produce it, for that is preventive.

Normal and Abnormal Competition: A Reciprocal Relationship

The general concept here is that normal and abnormal repertoires generally are in a competitive relationship in the sense that the learning of one interferes with the learning of the other.

> [Let us say] when the mother sees the child reaching for some object and grunting or whining the mother will obtain the object for the child and give it to him. . . . [If this experience] is general in the child's history . . . the child will learn the grunting-whining-gesticulating repertoire. . . .
>
> The important point here, however, is that such a repertoire becomes competitive with the normal language repertoire. The child learns the grunting-gesticulating repertoire more skillfully in place of learning the language repertoire and insists the parent respond to what he has learned. It is language that will be needed, however, to adjust to society outside the home. (Staats 1971a, 305–306)

The principle of repertoire competition has implications for treating various cases of abnormal behavior. Usually it is the abnormal behavior that is the focus of attention. Thus, the therapist may be concerned with reducing behaviors like temper tantrums, whereas the more fundamental treatment would be to train the child to a language repertoire so the child could make his wants known more quickly and more easily, as well as open the door to other adjustive learning. The adult individual may be imprisoned for burglary in order to reduce this behavior, when a more effective treatment would be job training and job placement. This is to say that when an abnormal repertoire interferes with learning a normal repertoire, the focus should be on teaching that normal repertoire. Treating the normal repertoire will replace the abnormal repertoire; it will not need to be dealt with.

BIOLOGICAL FACTORS

Some readers may ask whether I am saying that abnormal behavior is created entirely by the environment and learning. The answer is no. For one thing, the learning theory that I have set forth states that the individual learns basic repertoires that constitute personality. Learning, however, including the learning of personality repertoires, takes place only via the individual's biological mechanism. Anything that makes the mechanism malfunction can result in a malfunctioning of learning. A child born with a lesser brain structure—such as microcephalic or Down's syndrome children or those born of an alcoholic or drug-addicted mother—can suffer learning limitations that decrease their acquisition of the basic repertoires. A central point here, however, is *limitations in learning because of the biological mechanism must be definitively established.* Biological limitations cannot be inferred on the basis of abnormalities in the person's behavior, because that behavior will also be a function of the learning experienced.

The examples given of the biological limitations on learning have been of a permanent nature. However, biological abnormalities may also be ephemeral. Let's say that a young man has learned a normal sensory-motor repertoire of driving a car. But one night the individual goes to a celebration party and gets drunk. On leaving the party, he drives faster than usual, past legality. He drives erratically, tending to go right, and then has to pull back in his lane. When cars slow down in front of him, there is an atypical lag before he reduces his speed. Finally, he bumps into another car because he has not braked promptly enough. Nevertheless, the next day his brain is cleared of alcohol and his basic driving repertoire will have returned. In this same way, temporary biological conditions can also affect basic personality repertoires and can also lead to learning behaviors in an inappropriate way.

Biology in another way can help determine the learning of the basic personality repertoires. There are many studies in social psychology that show that the attractiveness of the individual is an important factor in how others behave toward that individual. Experiments have shown that subjects sit closer to attractive people than to those who are less attractive. They also talk more to and are more likely to ask for dates with those who are more attractive. Attractive children receive better grades in school. Members of our society behave differently to little girls than to little boys. It can also shake us to realize that even mothers interact more closely with their child when the child is attractive than when the child is not. Tall men are more likely to be CEOs than short men. The very word *attractiveness* tells the tale; an attractive person elicits positive emotion and approach, supportive behaviors in others.

What does that mean in the present context? On the average, being attractive, a biological characteristic, means that the person has a different learning environment than others not so attractive. That special treatment begins early and lasts through life. That pertains also to abnormal behavior. A child with strange or unappealing looks is more likely to have nonauspicious social experiences than a handsome and appealing child. One reason the twin studies of intelligence and mental disturbance cannot be taken at face value is because there is no control for the effect of physical features. If physical features do in part determine how a person is responded to, and the physical features of identical twins are the same, then they will have learning experiences that are more alike than will two people who are not physically similar. That will have the effect of making identical twins' personalities more similar than usual. If one twin gets better grades than others because of being attractive, then so will the other twin. If one twin is not as warmly mothered as is usual because of being unattractive, then the other twin will receive the same treatment. I suggest that there is a large, extensive, long-continuing, and cumulative biological/learning effect here that affects abnormal behavior in ways not well considered.

CONCLUSION

In concluding this theory of abnormal behavior, different behavior disorders have been categorized as though they involve either the language-cognitive, emotional-motivational, or sensory-motor repertoire. This simplicity of analysis is never the case, of course, but one repertoire may play a more prominent role than the others. The human environment is exceedingly complex and varying and human learning is hugely complex, as are the repertoires produced, and both abnormal and normal human behaviors occur in response to complex situations.

A general conception of human behavior must deal with behavior that is considered abnormal as well as the behavior that is considered normal. The present conception does that, in both basic and applied study, empirically and methodologically and theoretically. Moreover, the same principles explain both normal and abnormal behavior, an important feature. The new theory opens new heuristic avenues to pursue. For example, the present theory of abnormal behavior calls for longitudinal learning research, simple, straightforward studies where trained researchers are in homes where they can observe what learning experiences are occurring that produce a deficit or inappropriate behavior in a child. We need records in such cases of the interactions of parent and child, the behaviors of each. That needs to be done very early. For example, observers could be placed in a number of homes in which children of a year and a half have not yet begun to speak, to conduct a stipulated program of observation of the behavior of the child and the parents. It would be possible to ascertain in detail what is occurring with respect to the child's language learning. The same could be done for young children exhibiting early signs of autistism, attention deficit hyperactivity disorder, conduct disorder, and so on. It would also be possible to make such studies experimental by giving half of the parents training in how to train their child to remedy the behavioral defect the child is beginning to develop. And, after a suitable period of observation, the other half of the parents could also be given the same training, and all of the parents could be offered professional therapeutic services. It would be possible to establish how the various behavior disorders are learned in these and other studies. The fact is that such a study has never been conducted; we really have no evidence of the learning that takes place to produce abnormal or normal repertoires in childhood.

Isn't it strange that the importance of learning for the development of behavior is well recognized but we have not even begun studies of how human behavior, both normal and abnormal, is actually learned? If it is learning that produces the behavior the individual displays, normal or abnormal, and if it is learning that drives the development of the brain, then if we wish to understand human behavior, if we want to deal effectively with problems of human behavior, there is no other way to do so besides studying the learning involved. Not to make heavy investment in that study is the height of folly, surely a *Great Scientific Error*.

Part 5

HUMAN EVOLUTION AND MARVELOUS LEARNING

Chapter 8

ON THE ORIGIN OF THE HUMAN SPECIES

C harles Darwin made a great contribution to the beginnings of biological science, to understanding life itself, and to understanding the nature of all animals, including humans. His theory changed belief from divine creation of all creatures to a naturalistic, scientific account. The conception of evolution would have been basic in Ivan Pavlov's and Edward Thorndike's discovery of the fundamental principles of learning, and in the works of men like John Watson, B. F. Skinner, and Clark Hull in systematizing and extending those principles.

With an inconsistency not recognized, however, Darwin's evolution also contributed greatly to the Great Scientific Error that shuts out developing the study of the great importance of learning for understanding human behavior and human nature. For his theory included the belief that human behavior was to be explained by evolutionary principles and processes. Intelligence, for example, was considered a trait of human nature that evolved. This fundamental error turns the study of human behavior toward biology.

This fault in Darwin's theory was shown clearly in that no role for learning was carved into the treatment of human evolution. Despite the fact that humans have the greatest learning ability of all animals—so dramatically different, as shown by human achievements, and so different from any other animal—human evolution was considered in the same terms as the evolution of other animals.

A theory of human evolution is tremendously important. It is central in defining humanness. Christian theology holds dearly that humans were divinely created. That belief is a central feature of religion, giving humans special qualities, a special place in the universe. Humans are unique, not like the other animals. Humanity is to be known through religion, through religious belief. Darwin's theory said instead that humans are not special, humans were created in the same way that other animals were. Humans are an animal and are to be known by scientific study, not through theological knowledge. Much hinges on this difference, for example, whether knowledge of human behavior and human nature can be found by studying animals, and whether human actions are to be judged by and guided by religious or science study.

Human origin plays a great role in the conception of who humans are and what explains their behavior and their nature. A paradigm that treats humanness, as Darwin showed, has to address human origin. That is true in the present case. Does the paradigm of humanness that has been developed here have something to say about human origin? This chapter answers that question.

EVIDENCE OF HUMAN EVOLUTION

Paleontology and paleoanthropology, in studying fossil species that trace human evolution, have special problems because that development occurred long ago. Because of the time involved, and the manner in which various conditions break down animal remains, the evidence has great limitations. The specific skeletons are largely destroyed; frequently only deteriorated parts remain. The finds also are limited to rare fossils, only a tiny portion of the prehumans that must have existed. That goes also for the possessions of those fossil beings, such as their tools; only stone implements and such survive. So the record cannot be complete or known with detail. What must have been a rich and continuously progressive development comes down to us now only through a partial, damaged, and interrupted sampling. This makes it difficult to piece together the specifics of human evolution, what actually took place. For example, what may have been large variation between two individuals within one species may appear to represent two different species. What may have been a process of continuous changing of the prehumans may thus appear as discrete species.

The course of human evolution must be pasted together via fragmentary findings. However, the systematic quality of science and its enormous capacity for devising observational methods and constructing analyses—using and relating products from various fields—display themselves fully in this quest. Remarkable scientific ingenuity has been employed in chasing the most elusive quarry, changing what began as occasional peeks into a more continuing picture. When taken with the sciences of genetics and molecular biology and the rest of evolutionary biology, paleoanthropology has traced the course of human evolution clearly in general process, if roughly in detail. A sketch of its fascinating story helps fill out our understanding of human origins, and with it our understanding of human behavior and human nature.

Fossils

By forty million years ago, animals existed that had evolved important human features: hands with a thumb that could grasp, and eyes in the front of the head that yielded both depth and color vision. There is evidence that apes lived twenty million years ago, and it is thought that orangutans evolved from that stock fifteen million years ago, gorillas about nine million years ago, and chimpanzees about seven million years ago. Seven million years ago was also the time when the human (hominid) line began.

The fossil record between eight million and four million years ago remained barren until recently. What the first step was in the human journey may never be firmly fixed, for it will always be possible that an older species than those found might have existed. But, remarkably, good and accumulating evidence picks up the descent of primate species that walked upright at an early time. A find in Kenya, named *Orronin tugenensis* (*orronin* meaning "original man"), has been given a date of six million years ago. Said by its discoverers to have walked upright and to be closer to the origins of humans than any other fossils, *Orronin tugenensis* as a species in this context remains controversial (Balter 2001). But findings that take us back in time and give us a richer picture of what occurred then are coming more rapidly (see Gibbons 2010). An even more recent finding, *Ardipithecus ramidus*, changes a long-held view that hominid development began with a move from the forest to the savanna, where walking upright was an advantage. *Ardipithecus ramidus* findings show an early species that lived in the forest and walked upright.

A richer record can be traced, beginning with the discovery of the species *Australopithecus*, which lived several million years ago. The first and most abundant fossil remains of such creatures—including the several *Australopithecus* species *anamensis, afarensis, africanus, aethiopicus, robustus,* and *boisei*—with dates between about two and four million years ago, have been found in eastern Africa. A new find dating about 3.5 million years ago has also raised the possibility that another distinct species, *Kenyanthropus*, may have lived in the same time and the same East African space as the australopithecines. Whether this is a distinct species or a variation within a species is still being determined.

Some groups of *Australopithecus* were robust in build and some slimmer, called *robustus* and *gracile*, respectively. The robust australopithecines appear to have been vegetarians, the slender types were omnivorous. That difference could have produced further separation. The brains of these species were not much larger than contemporary primates, but newer species arose (or perhaps the same species evolved), and brain size later increased, a significant matter. (Brain size will be used here as the determinant of learning ability, although more features than size determine that ability.) More generally, the advancement of different organ systems in evolving from ape to human did not occur together. Bipedal structure, for example, might advance more quickly than brain capacity. Thus, *A. afarensis* still used powerful arms to climb trees while the arms of *A. anamensis* were further along toward humanlike development.

While the steps of transition during the period—beginning four to six million years ago—remain unknown, in the present view these hominid creatures had already begun the journey from the primate stock species on the way to the distant *Homo sapiens*. (This is not the general view; only species whose name includes *Homo* are recognized as being in the human lineage.) That lineage then continued to develop through various species. It is also clear that hominid species overlapped in time. The australopithecines, not yet advanced enough to be included in the *Homo* class, continued living between 2.1 and 1.1 million years ago. However, the remains of the new species, *H. habilis*, date from 2.4 to 1.5 million years ago. So the *H. habilis* and *Australopithecus* species coexisted even though *Homo habilis*, as the name denoted, was much more advanced in the human direction. The face of *habilis* was less prognathous (protruding); the teeth were smaller (but not yet human); and the brain had advanced from the *Australopithecus* size of from about 375 to 500 cc to an average of 650 to 800 cc. The brain shape had also become more humanoid, *with an area involved in speech present in one specimen.* Locomotion was bipedal, and the pelvis, narrower than for humans, probably made for an even more efficient gait. There were probably variations among *habilis* creatures in the extent of these developments, which we cannot know from the limited number of fossils. Also, the evidence of tools associated with *habilis* specimens is more substantial than for the australopithecines, and the tools are more advanced.

Another step forward in human characteristics came with the *Homo erectus* species that existed between about 1.8 million years until 300,000 years ago. This means *H. erectus* overlapped with *habilis* and even with remaining australopithecines. So the three species, with their subtypes (and variations), appear to have coexisted contemporarily in Africa. In comparison to modern humans, the jaws of *erectus* still protruded, the teeth were large, the thick brow ridges jutted, the chin was lacking, and the skull was low and long. But the brain size had increased, ranging from 750 to 1,225 cc. Toward the later part of the one and a half million years involved, the average brain of *habilis* increased to about 1,100 cc. Generally, as in this case, when fossils of a hominid species date over a long period of time, there is a progressive increase in brain size, which is an important fact. Something was going on that produced a continuously enlarging brain.

Because of the relatively small brain (head) size, the pelvis of this species could be narrower than for modern humans, which must have aided upright locomotion. While *habilis* fossils have only been found in Africa, *erectus* specimens have been found also in Europe and Asia (as in China and Java), indicating that behaviorally a new ability to adapt in new and different environmental circumstances had arisen. This probably indicates that population pressure to move to new pastures also increased over time.

I will suggest, in an anticipatory way, that increasingly widespread geographic penetration was made possible by the development of more advanced skills is such things as tool making and language. Another species, *H. heidelbergensis*, appears to be transitional, derived from *erectus* but possibly ancestral to *H. neanderthalensis* or archaic *H. sapiens*. Again, we see in *H. heidelbergensis* the increase in brain size, 1,100 to 1,400 cc, that was occurring. There are *heidelbergensis* fossils all over Europe and in Africa spanning a period from about 600,000 to probably 200,000 years ago. How rich the evidence has become.

In any event, further evolution of anatomy did take place in the various hominins. Sometime around 500,000 years ago, archaic *Homo sapiens* appeared that had features like a mixture of *H. erectus* and modern humans. Thus, an adult male cranium deposited in China about 200,000 years ago had a capacity of about 1,120 cc but with brow ridges, a flat forehead, and thick cranium walls. There are fossils from around 130,000 years ago that more evenly have modern features—brain average of 1,450 cc; high brow; chin; and diminished brow ridges—suggesting that modern *Homo sapiens* probably dates from 100,000 to 200,000 years ago.

Even this summary depicts a dramatic tale, from an apelike animal to a human being in four to seven million years, with a large increase in brain size. My own interpretation is that rather than advancement by distinct species, what was involved from the beginning of upright apes on the savanna were many groups living in their localities and evolving separately, spreading their advancing genes with other local groups with whom they had some interaction. Would it not be fascinating to be able to actually see what those beings were like and how they behaved and lived—what their language development was, their social behaviors, their technology? Unfortunately, we cannot go back in time and must satisfy ourselves with the bits and pieces paleontologists and paleoan-thropologists provide us. Still, their rich particles weave an intriguing tale that tells us of a creature with a head perhaps a little larger or better formed than chimpanzees of that period and also with a body in transition from a primary tree dweller to one that had bipedal locomotion. It arouses our wonder to think that this species arose and went through continuous change.

The descendants of those animals evolved into a fully two-footed upright creature, with changes in hands and arms toward modern human characteristics. More slowly, but very progressively and very consistently, the creature's head underwent evolutionary change that produced expansion of the brain in size and structure. That brain development brought other anatomical changes to the head: the muzzle of the apelike being changed in the direction of the face of a human, becoming flatter, located beneath the frontal lobes of the brain—versus an ear-lier location of the eyes and muzzle in front of the frontal lobes. The large teeth grew smaller, losing their canine ferocity; the ridges over the eyes disappeared; and bones to support the bridge of a nose appeared. A chin was developed instead of a receding front jaw; the cranium's walls thinned; a crest on the top of the head disappeared, as did a ridge on the back of the head to support heavy muscles. The larynx and other features changed to permit full speech. In these changes there was much variation among the fossils of the various species that appeared, variations in the extent to which apelike characteristics were retained.

The new finds suggest that evolution included a number of species, or vari-ations of a species, that lived at the same time and had a mixed bag of modern and primitive features. It is important to note that despite variation in the com-bination of features over the multiple fossils, a consistent progression in the size of the brain occurred. Something maintained that developing feature on a track of advancement through the many other variations, at least from the australo-pithecines to modern humans.

Finally, think of the huge behavioral journey those upright apes had to tra-verse from no or practically no language, from a little developed emotional system, and a paucity of motor skills to what is customary in modern humans. That journey was not just one of physical structure. The account developed in the preceding chapters tells us that, comparatively, humans today have learned huge repertoires of behavior. Those repertoires are not inborn, they had to be learned over the millions of years probably between *Ardipithecus ramidus* or *Australopithecus* and our species today.

Culture

Those ancestors not only left bones that enable us to know them in some bare form. They also left some of the things they used in living. Closely related to the search for fossils by which to trace the evolutionary history of our species has been the search for the products of our progenitors in the form of tools, decorations, artwork, indicators of behavior (such as burial customs), and so on. Those products are the primary means by which to consider the behavior and nature of our progenitors. The same problems exist here, however, as they did for biological remains. The great time involved provides for destruction of evidence; only those things remain that are resistant to deterioration. Wooden, leather, and fiber implements have little chance to survive the long time periods involved in going back to the beginnings. The time dimension thus makes the evidence fragmentary and incomplete. Moreover, behaviors of those ancient creatures, such as language, leave little or no mark. Even how they gained their subsistence is largely inferred from the bones and seed remains. How foods were gained through hunting and gathering must be inferred from other artifacts that were preserved and from knowledge of hunting and gathering peoples that live today.

But again, and surprisingly, as the various remains have been systematically subjected to scientific analysis in giving an outline of the physical hominid advancement, so too evidence of the products of those hominid groups have been progressively discovered and analyzed to gain a general picture.

Tools

The production and use of tools have sometimes been taken as a unique mark of humanness. But accumulating evidence has revealed that chimpanzees use twigs to fish for termites and rocks to break nuts; dolphins use sponges on their noses as protection in bottom-feeding; and sea otters crack shellfish with rocks, among other tool uses. But no animal shows progressive increases in using, finding, or making tools.

Tool *working*, a more advanced skill, has not been associated definitely with the human line prior to *H. habilis*. However, the australopithecines appear to have systematically *used* tools, extending a good deal the very rudimentary use of tools by lower animals. In the present view, that should be considered as a large behavioral advance. Later finds definitively show progress in a tool-making industry. It is divided into Olduwan, Acheulian, Mousterian, and Paleolithic (Lower and Upper) stages. Olduwan tools have been found that date to as early as 2.4 million years ago, associated with *H. habilis* fossils, and as late as about 1.5 million years ago. This creature had already carried the advance in brain size clearly past the apes. Their tools, unlike with the australopithecines, were not simply sticks or horns or rocks that came easy to hand and were used in their unaltered state to dig, smash, break, or cut something. Olduwan tools included specimens where flakes had been struck from a core, on one or both sides, with a hammer stone in order to sharpen the core for scraping or chopping. Sometimes, it has been concluded, the tool product was the sharp chip struck from the core that was used for cutting.

The early tool development represents a huge leap forward from a chimpanzee opportunely picking up an available stone to crack a nut. The extent to which the tools were worked indicates the complexity of the effort, and thus gives information about the nature of the creature involved. The skill clearly must have been developed by learning that spanned many generations. That can also be true of the site where the tools were worked. When the source of the raw material for the tool is located at a distance from where the creature lived and worked the tool, then long sequences of behavior are demanded. That gives clues about the behavior of the creatures. The investment in the tool activity would have been considerable, and it called for an emotional-motivational investment to sustain the activity. The extensive nature of the task would have called for some behavioral process bridging the time from discovery of the raw material to its working and then its use. That strongly suggests the use of language as the mechanism for maintaining the extended sequences of behavior intervening between searching for stone materials that would only much later be used, just as language serves us in that role today. Language with that function would have meant considerable learning had to have been involved. Modern apes' behavior is pretty much bound to immediate stimulation, using implements that are close at hand. So the tool making of a hominid two and a half million years ago was a tremendous advancement behaviorally and learning-wise.

About 1.4 million years ago, the Olduwan tool making gave way to the Acheulian. This industry produced new types of tools—the hand ax and the cleaver—that, over time, involved progressively fine working. The first find in Ethiopia was associated with a *Homo ergaster* jaw. This species was between *H. habilis* and *H. erectus*. Definitely humanoid, the apelike qualities of early hominids were disappearing and the human qualities were increasing. The greater working and variety of the tools suggests, again, that even with just a 900 cc brain size, this hominid had psychological features that enabled it to select materials that would not be put to use until a good deal of sustained effort had been applied and a considerable amount of time had elapsed. This represents a great advance behaviorally over picking up a handy rock to pound something or tearing a limb off a tree to plumb the depth of a stream. The tool making even at the *H. ergaster* level must have involved individual differences in skill level that go unrecognized because of the paucity of fossils. To me this again suggests some rudimentary language had to be functioning.

It was not, interestingly, until 500,000 years ago that this tool industry appeared in Europe, which supports the contention that the human species arose in Africa and migrated to other parts of the world. Again, there is a large step involved in the advancement from the Olduwan to the Acheulian tool industry in Europe as well as in Africa.

About 300,000 years ago, tools appeared that could be classified as a higher level of tool production. Called the Mousterian industry, this industry lasted until forty thousand years ago. It centered on preparing a core stone for removal of a large flake with a sharp edge that then could be further worked to produce the implement desired. The other way was to work the core to mold it to its desired shape. The work showed much more knowledge of the materials and was more finely wrought. Spearheads with hafts were produced, as were double-sided scrapers and cutting tools. Creating spearheads attached to a shaft with some type of binding, such as a leather strip, involves a combination of materials. The use of such combinations has been interpreted to mean that hominid manufacturers had a higher conceptual ability than those who came before. I will suggest more specifically that the progression from the Olduwan to the Mousterian industry pictures, in gross form, a cumulative learning process of continuous growth. The missing-link conception lends itself to this theory of development. Great advances in learning ability must have been involved as a consequence of the basic repertoires that already had been acquired.

That was by no means all. The Paleolithic industry, which began in Africa and the Asia territory north of it, went from forty thousand years ago to the Upper Paleolithic period that carried to ten thousand years ago in Europe. The humans in this period continued becoming more and more skillful along various dimensions. Stone tools were made to work bone and wood. Characteristic was the manufacture of long, thin blades, skillfully struck from carefully selected cores of certain kinds of stone. Heating was a method for producing fine work that could not be done with cold materials. There were also arrows, bone/stone harpoons with several barbs, spear-throwers of wood and bone, and finely worked pieces for aesthetic as well as utilitarian purposes. What a great advancement in what had been learned and in the behavioral achievements the advancement made possible. The missing-link conception pictures the tool industry in general form as a cumulative learning process of groups of prehuman creatures that took place over millions of years.

The traditional conception considers the advancement in the tool industry in terms of biological evolution of the brain through mutation. Thus, for example, it is noted that worked tools were not general during the reign of the australopithecines. Rather, the working of tools occurred only after the advent of a *Homo* species with a larger brain. Then, after several million years of slow advancement in tool industry, about forty-five thousand years ago, the core technology arose. Again, the conception is that the hominid brain had developed through mutation and this caused the behavioral advancement shown by the tool refinement. In light of the present learning framework, we must ask if there is another way of looking at these facts, in the way I have begun to suggest.

Artistic and Symbolic Artifacts

In the period of about forty thousand years ago, hominids began to do things and produce works that smack of modernity. Burial sites, for example, have been found in which the deceased have been arranged in particular postures, and objects (such as flowers and tools) have been interred along with the body. This has suggested that those who were doing these things were performing a ritual based on a spiritual conception of some kind, for example, that the deceased had an afterlife in which they would use and enjoy the things with which they were buried. That type of belief could have taken place, it may be added, only with people who had a well-developed language.

Art objects have also been found—like ivory carvings of animals and humans—some of which have symbolic features, such as a figurine of a woman that accentuates fertility features. Jewelry and decorations have also been discovered that show aesthetic interests. There are also decorative carvings on tools and weapons. The new finds in the southernmost parts of South Africa have shown geometric designs that have been interpreted as having symbolic characteristics.

And marvelous drawings in caves have generally suggested not only artistic skill but also symbolism and perhaps religious worship. Further development of art forms is shown by instruments like a flute made of bone, indication that music had become a part of life for those prehistoric beings. During the same time interval, tools became much more sophisticated, including bows and arrows, spear-throwers, harpoons, and needles made of a variety of materials. This advancement in technology and art is conventionally explained by genetic changes in the brain.

In concluding this section, let me indicate this is just a quick look. In the search for fossils, in their discovery, in the study of the materials found with them—including food particles and their digestion and elimination products, climate analysis, analysis of the fossils of other animals of the time, and other science findings and analyses besides—there is vast science prowess. Those working within evolutionary theory have made their point. In science fields there is no doubt that humans have evolved. Evidence exists of primate species some four or more million years ago that had begun standing and moving on two feet, but still had tree-climbing arms, hands, and feet—a real "ape-man" combination. This species had brains only slightly larger than those of a chimpanzee. The fossil evidence shows that by two million years later, species had evolved that were firmly bipedal, had larger brains, and used tools chosen for shape or roughly shaped. This shows the pace of human evolution; in comparison to today's rate of advancement it was tortuously slow. It took another million years to evolve a hominin species with a larger brain and a more fully human structure, along with more refined tools. Archaic *H. sapiens* took another million years to appear. The behavioral achievements were a gradual but continuous development.

HUMAN ORIGIN?

The wealth of the fossil evidence shows beautifully what Darwin proposed, that the principles of natural selection applied to human origin. Ten million years ago there was a primate ancestor that was common to humans and chimpanzees. Humans did not arrive in one fell divine swoop a few thousand years ago. It took millions of years of progressive biological evolution of erect apelike creatures to modern humans. That evolution involved intermediate species that gradually developed the distinctive human features, first in body and more slowly in head and brain. That evidence is undeniably strong.

Less strong is the explanation of the dynamics of that lengthy progressive development toward humanness. The strongest view, of course, has been Darwinian, that the explanation of human evolution is the same as the explanation of the evolution of other animals. The goal from this approach is to prove in various ways that human evolution occurred by natural selection, that we too are an animal species, not unique.

> There is absolutely no reason to believe that the rules of the evolutionary game had been even slightly bent in paving the way for our [*H. sapiens*] arrival. . . . [H]umans have sought to deal with their own past . . . as if the process of becoming human has in some way been different from the processes of becoming one of the ten million other species inhabiting the world today. . . . [H]owever, there is absolutely no reason to believe that the rules of the evolutionary game had been even slightly bent in paving the way for our arrival. (Tattersall and Schwartz 2000, 242)

> If we look back to the time of the australopithecines, some 4 million to 1 million years ago, it is obvious that . . . [they] were constrained and directed by the same evolutionary pressures as the other organisms with which they shared their ecosystems. (Johanson and Edgar 1996, 21)

> It is not mere hubris to argue that *Homo sapiens* is special in some sense—for each species is unique in its own way; shall we judge among the dance of the bees, the song of the humpback whale, and human intelligence? (Gould 1996, 354)

> We may never know exactly what happened in the evolution of human brains and language, but this does not indicate a cosmic mystery about their origins. There is every reason to assume that the same processes that produced the grace of flying swallows, the marvelously skilled and sensitive trunks of elephants, the stupendous size of sequoia trees, and deadly strains of gut bacterium *Escherichia coli* also produced the underpinnings of our diverse natures: our amazing brains and our equally amazing linguistic abilities. (Ehrlich 2000, 163)

Not everyone has agreed with that view. Some have had the view that despite having evolved, humans nevertheless had done so in ways that made them unique. One important way conceived was that unlike other animals, humans use tools. That has been answered by evidence that primates and other animals also use tools.

> In a swampy forest clearing, a female gorilla yanked a roughly three-foot-long branch from a dead tree and waded into a deep pool of water. Keeping the stick

in front of her and her upper body above water, the gorilla slowly advanced about thirty feet into the pool as she tested the water's depth. . . . The scientists, led by Thomas Breuer of the Max Planck Institute of Evolutionary Anthropology . . . had for the first time witnessed and photographed tool use by a wild gorilla. (Bower 2005, 253)

Chimpanzees living in western Africa's Tai forest, and only there, "stockpile stones at places with broad tree roots or stumps that serve as anvils for cracking nuts" (Bower 2002b, 195). The rocks employed are sometimes transported as far as several hundred yards. Chimpanzees can also be trained how to make chopping tools, albeit crudely, although they have not themselves invented that skill.

These findings are all interesting and provide information regarding our primate brethren. But the humanlike behaviors involved are very primitive, actually not like the human behaviors they are considered to be like. Chimpanzees stockpile stones crudely; they only make chopping tools crudely and only after being taught by a human. Tearing a branch off a tree to advance in a pool is similarly crude. Although these findings are interpreted to show that these animals are indeed like humans, they do not do that—for, as is fundamental, none of them show the human-defining cumulative learning.

In the present view, such studies only show the chasm that exists between the behavior of other primates and that of humans. The ephemeral, occasional, fortuitous primate use of tools and humans' long-term, continued, growing development of tools from hand axes to spacecraft is as different as night and day. The difference resides in the presence or absence of cumulative learning.

Language Evolution

Another characteristic of special interest in the consideration of human evolution has been language. Are humans the only species that has language? Within evolution considerations there has been a prime impetus to prove that animals too have language, that human language is the same, only further evolved. Elephants and whales have been studied for that purpose, with conclusions that they do have language. Chimpanzees have been observed in the wild in order to establish their language ability.

Chimpanzees communicate with a wide range of calls, postures and gestures. The food calls—a mixture of food grunts, barks, and pant hoots—[alert] other chimpanzees to the whereabouts of food sources. A special intensity of excited calls of this type indicates that there has been a successful kill after a hunt. Each individual has his or her own distinctive pant-hoot, so that the caller can be identified with precision. A loud, long, savage-sounding Wraaaa call is made when a chimpanzee comes across something unusual or dangerous. (Jane Goodall Institute)

Chimpanzees and other primates have been trained to small languages also, as have dolphins. It has been found that not only can these animals learn a language, but they can then put words together creatively in small ways.

This has meant to human evolutionists that animals have language ability. The language trait is there, but just in a lesser nature than for humans. Evolution, indeed, remains the king explanation; such findings are used to support belief that nothing new or unique occurred in human evolution. The same principles that apply to animals explain human behavior and language: they are natural selection and genetic evolution. "So language . . . has been built into our biology as the most efficient and effective means to communicate our thoughts" (Johanson and Edgar 1996, 107). Werker and Vouloumanos, though, have another version of this view (2000). The various versions, interestingly, do exhibit great variation in the time of the language mutation, from a half million years ago with the first appearance of archaic *Homo sapiens* to thirty-five thousand years ago.

We see in Tattersall and Schwartz (2000) a more complex explanation of human evolution by mutation. They consider that unknown natural selection forces, over several million years, brought hominins' brains to the point where about thirty-five thousand years ago, some early *H. sapiens* person was born with an unknown mutated neural development that was advantageous to the species in some unknown, nonlanguage way. But that mutated gene, completely by happenstance, also carried with it language ability. Because the neural development was advantageous, so the theory goes, that development became general to the species through natural selection, thus also giving all *Homo sapiens* the potential for language. At this point another fortuitous event occurred. Some *H. sapiens* person happened "to invent" language. The other members of the species, with a brain already prepared for language by natural selection, then just took up language since it also was valuable for survival and procreation (Tattersall and Schwartz 2000).

Again, language herein has been considered as learned. As will be additionally treated a bit further on, that means that such explanations are quite wrong. Human language is very different from the "language" that these theorists see in other primates. Again, the difference lies in whether or not cumulative learning is involved.

A New Awakening

These beliefs fall within traditional human-evolution theory. We can see, however, a new opening that is actually an inconsistency of belief—sometimes in the
same person. It is closer to the present view, which has considered learning in
some form within the context of human evolution, beginning in 1971. That conceptual framework is employed in considering central types of human-behavior
traits.

> To inherit specific behaviors as a member of the human species would have
> been maladaptive. A Stone Age man who had a repertoire of mathematics or
> chemistry, courtly manners, perfect pitch, ethical behaviors, a pacificist con-
> ception of human interaction, or what have you, would have had a useless set
> of skills. . . . [T]raining that would have produced these skills was absent.
> Rather such men *learned* to shape rocks, fight savagely, throw spears, club
> prey, make fire, plan group hunts, stitch furs, carve fishhooks, find and eat
> insects, discriminate subtle cues in tracking prey, communicate, and so on. . . .
> *[T]he marvelous adaptive powers of man are due to his nonspecialization—to
> his generalized adjustmental (learning) capabilities.* (Staats 1971a, 48)

More generally in the human-evolution field, the idea has begun to arise
that experience, culture, or learning needs to be considered in explaining human
evolution.

> New Guineans whose fathers lived in the Stone Age now pilot airplanes,
> operate computers, and govern a modern state. If we could carry ourselves back
> forty thousand years in a time machine, I suspect we could find Cro-Magnons
> to be equally modern people, capable of learning to fly a jet plane. They made
> stone and bone tools only because no other tools had yet been invented; that's
> all they had the opportunity to learn. (Diamond 1992, 51)

> The brains of our hunter-gatherer ancestors probably were growing because of the
> positive feedbacks of an increasingly complex social system in which those most
> socially adept were outbreeding the less skillful. Plotting, formation of alliances,
> planning, and thus abstract thinking and clear communication all add up to social
> adeptness. A developing protolanguage—one with minimal syntax—would be
> most effective if long lists of words were available. Advantages would accrue to
> those who could remember more words and use them more effectively in com-
> munication, and thus, quite likely, to those with bigger brains. (Ehrlich 2000, 160)

[M]arkedly increased brain size in human evolution may have ... added enough neural connections to convert an inflexible and rather rigidly pro- grammed device into a labile organ, endowed with sufficient logic and memory to substitute nonprogrammed learning for direct specification as the ground of social behavior. Flexibility may well be the most important determinant of human consciousness. . . .

Why imagine that specific genes for aggression, dominance, or spite have any importance when we know that the brain's enormous flexibility permits us to be aggressive or peaceful, dominant or submissive, spiteful or generous? (Gould 1977, 257)

Our basic claim is that biological thinking about heredity and evolution is under- going a revolutionary change. What is emerging is a new synthesis, which chal- lenges the gene-centered version of neo-Darwinism that has dominated biolog- ical thought for fifty years. . . . It is a view that may relieve the frustration that many people feel with the prevalent gene-centered approach, because it is no longer necessary to attribute the adaptive evolution of every biological structure and activity, including human behavior, to the selection of every chance genetic variation. . . . When all types of hereditary variation are considered, it becomes clear that induced and acquired changes also play a role in evolution. (Jablonka and Lamb 2005, 2)

The notion of a cumulative cultural process is in agreement with the cumula- tive learning conception developed in the present approach (see Staats 1968, 1975). It has begun to be employed (see Sterelny 2003, Tomasello 1999). Tomasello, a psychologist, calls his cumulative-learning principle the "ratchet effect." That ratchet effect, however, is set in the primatology framework of his co-worker Christophe Boesch. In this context, McGrew (1992) also studies chim- panzees in order to gain knowledge of humans. Unfortunately, human cumulative learning does not occur in any other animal. So these approaches, despite making an important contribution, cannot conceptualize the way human cumulative learning was the central determinant of human evolution. Actually, humans are unique. Human cumulative learning is unique. And human evolution was unique.

That learning uniqueness needs to be considered in understanding human evo- lution. Perhaps we have to open the traditional evolution-theory box, so that we can consider specifically how learning was an important part of human evolution, a unique part. The human evolution story should be considered in terms of the con- ception that has been developed here, behavioral development that was occurring, and the new type of learning that made that behavioral development possible.

HUMAN SELECTION THEORY

I will begin the new theory with consideration of the body changes that occurred that changed the behavior of the species as well as the species environment.

Chimpanzees' bodies enable them to use their hind limbs to move on the ground, with some aid from their front limbs. They are far from having humans' bipedal locomotion, however. Instead, their ability to survive and reproduce has rested on a tree-centered structure, on learning highly skilled motor behaviors in that environment, and on learning other behaviors relevant to that environment. Some time, perhaps about six or seven million years ago, primate species began to appear that had a more upright posture that increased the ability to move on the ground.

The Move to the Savanna, Behaviorally Speaking

These species that walked more upright perhaps still lived in the forest. But that body structure had to be a watershed type of development, for it opened new horizons. Since these animals moved on the ground more rapidly than the tree-centered primates, new types of behavior must have been developed, such as running after small game or fleeing from predators. Although the arms and hands of members of the species were still made for tree climbing, their increased bipedal ability must have suited them for new things. Their bodies didn't suit them to replace other primates in the forest, but they did have the potential for opening a new niche for themselves. Even if they still resided in the forest, they could venture on the savanna for various purposes, such as digging roots or trapping small game, with greater range than a slower-moving chimpanzee.

Such species would thus have been in an environment where further development toward humanlike bodies in terms of posture-locomotion and arm-hand dexterity would have been advantageous. Bipedal structure that further increased running capability would have been an advantage, for example, and genetic changes that brought this would have been naturally selected. The addition of leg skills and the savanna environment probably also made advancement of the brain/nervous system advantageous. Six million years ago, however, the species would still have been a "glorified chimpanzee" with a body somewhat more human in terms of structure and function and a brain a bit bigger than that of the primate cousin.

What this fractional jump toward humanness had done, however, was open a significant new evolutionary environment that would, according to natural selection, lead to further evolution. For one thing, the move to the savanna would have made group behavior important. A single *Ardipithecus ramidus* creature moving out on the plain would have been in greater danger than a group of one hundred. Moreover, a single *Ardipithecus ramidus* creature whose shoulder, arm, wrist, and hand were better structured for throwing stones would have been more valuable to himself and to the group than would another whose structure remained better for climbing in trees.

We might assume that the very early upright *Ardipithecus ramidus* had at least the chimpanzee level of sensory-motor ability, and probably a bit more. Let us consider in behavioral terms *the environment that* faced that early hominid creature. Its increased walking skill made it possible to cover greater distances when on the savanna. However, doing so would have left the creatures open to attack, for the refuge of tree climbing would have been less accessible. So traveling in groups and *carrying* weapons, like rocks to throw, torn-off tree limbs to use as clubs, and tools like sticks for digging up roots, would have enhanced survival. Tools would have become more advantageous. And that would have made the ability to learn tool use more advantageous in terms of survival and reproduction. Much of the time involved in transition from the forest to the savanna must have been for that learning to occur.

These new practices, moreover, would also have brought new experiences. Tool use in other primates was a sometime occurrence, not much developed. Probably the first development of the early species that ventured on to the savanna involved a more constant carrying of tools and weapons. Constant use would have brought new experiences that would have increased skill in the manipulation of tools and weapons. Anyone who has seriously played golf, tennis, or baseball knows how awkward one is with the "tool" at the beginning and how comfortable and skilled later use becomes. Certainly the young in an early primate group that lived on the savanna and constantly used even unworked tools would have extensive new experiences that a young chimpanzee, bonobo, or gorilla—with their infrequent use of tools—would not.

Moreover, consistent use of stones by generations of early bipedal apes would have yielded instances where an accidental chip would have made the stone a better tool. And such experiences combined with other behaviors, such as skill in retaining tools and frequent use of tools in obtaining food, would have resulted in learning a positive emotional response to them. That would have made possessing them and working them positive occurrences. Those early beings would have become emotionally motivated to go further to find and select—and much later to begin to improve—their tools. A better-working tool belonging to a fellow creature would be taken as a model. Attempts to make tool improvements would have led to skill development, for example, in producing enhanced eye-hand sensory-motor skills. What we know of animal and human learning presupposes such developments.

Observation of chimpanzee working of tools reveals their imprecise actions. And their skill in this is not self-achieved, it occurs only through training administered by a person. When advancement depended on original learning, as was the case with the evolving prehistoric upright primates, progress had to be exceedingly slow. It took several million years to advance from what must have been the chimpanzee-plus level of *Ardipithecus ramidus* to the tool use displayed by the *Australopithecus* species of Lucy fame. One must realize the great advancement that this comparison indicates in a learning sense. No contemporary primate can approach that level of ability. And there was no human trainer to teach what the australopithecines had learned to do. The chimpanzee mother can teach her progeny to fish for termites with a stick. *Australopithecus* parents had to provide their progeny with learning experiences that yielded a hugely larger repertoire of motor skills. This represents a great advancement; no amount of training of a chimpanzee could match that of the usual australopithecine, a species not evolutionarily advanced enough to be considered in the human, *Homo*, lineage. How did that huge jump come about?

Before addressing that, let's first consider another major type of learning that had to occur in the evolution of humans.

Language in Human Evolution

Four million years ago there were upright apes like *Ardipithecus ramidus* walking the woods and savannas of Africa. With a brain not much bigger than a chimpanzee's, its vocal development must not have been much greater. But this species, or another like it, lived on the savanna. And that brought new circumstances that would have made an addition to the vocal calling repertoire more valuable in terms of natural selection. Doing most everything in a group of one hundred individuals, for example, would have been very different from the experience of forest primates, where social interaction involved smaller groups. Vocal communication, even of a very primitive sort, would have improved coordination of action of a larger group. Mutual dependence also would have had the effect of producing bonding.

In any event, at some place in the evolution process, the upright hominid or prehominid groups began adding vocalizations, words, to their repertoire. The fossil record shows that the australopithecines used tools and had repeatedly used "camps," perhaps with some domicile arrangement efforts. They also went appreciable distances to find tool material. Those are complex behaviors. They had to be learned; they wouldn't have come out of the air, or by mutation, to the individuals who performed them.

Behaving in such complex ways at some point must have involved at least a small repertoire of learned vocalizations, words made to name things (such as plants, tools, and threatening predators). Those words could be uttered to affect the behavior of others. Such a simple language, when it began, would have made each vocalization that was added to the australopithecine group's language repertoire an important enrichment of that repertoire, making possible increased communication and coordination of behavior in the group.

It should also be noted that the australopithecines existed for over a million years, and their behaviors advanced during that time. The experiences they had during this period must have resulted in new vocalizations, that is, new words in the verbal repertoire. Each added element would be valuable in a traditional evolutionary sense, in terms of aiding survival and reproduction. To realize how valuable adding to language is to a hunting-and-gathering group, the following field research by Diamond is very informative.

> New Guineans with whom I work typically have separate names for about a thousand different species of plants and animals living in the vicinity. For each of these species they know something about its distribution and life history, how to recognize it, whether it is edible or otherwise useful, and how best to capture or harvest it. All this information takes years to acquire. (Diamond 1992, 69)

That describes a complex language and some of the ways it would function to aid a hunting-gathering group's adjustment. Diamond also describes the length of time necessary to acquire such "information" (such a repertoire). A thousand-word repertoire is large, and to that number of nouns would be added the other classes of words involved in the above description, plus a complex grammar. In the context of the present conception, it is apparent that a language repertoire such as that of the New Guineans must have been originally acquired only gradually, piece by piece, through learning, not through mutation and invention. After all, in our society biologists are still discovering and naming new plants and animals and thus adding incrementally to our language, as are many other people. Mutation or reproductive genetic change is not involved, as neither is capable of implanting words in the brain.

The language repertoire Diamond describes would have been light-years beyond the repertoire of the earliest hominids and prehominids. The New Guineans' distant forebears must have added new word units excruciatingly slowly. But over many generations the language repertoire would have grown, word by word, each one introduced by an individual and then learned by others. There is no other way than through learning. Each new word would be an addition to the group's basic language repertoire.

And that leads us to the crucial understanding. *That increasing repertoire, which becomes language, acted back on the biological development of the species. At some point in human evolution, what was learned took the determining role in the biological development toward humanity, for the group's language repertoire would have been part of what had to be learned by the next generation of the group. That would make the language repertoire an instrument of evolution.*

Missing-Link Learning and Human Evolution

That last sentence brings us to a crucial concept in this theory of human evolution. It requires specification.

First, words added to the group's language repertoire would make the repertoire more difficult for members of the next generation to learn. Thus, there would be a continuing interplay between these developments, that is, an increasing number of words in the group's language repertoire, an increasing value of the language, and an increasing difficulty in learning the language repertoire. As the language grew, it demanded greater learning ability to acquire, and that meant demanding an increasingly large brain for those who were to learn the language.

That tells us what had to occur. The increasing difficulty of learning the group's language repertoire *would have constituted a selection device, a selection device for human evolution, over and over and over again.* We have to recognize that within the group of the evolving hominin species there would be individual differences in the size of the brain and thus in the ability to learn. And those differences were functionally important at that time, for the repertoire involved, language, had not reached a peak level of development. That extent of learning ability would determine the ease and amount of language learning of which the individuals were capable. Some of the children born in the group would have the capacity to learn readily the group's language repertoire because their brains were larger and they had better learning ability. But other children would not have brains as well developed. Some of the children would be more like the upright ape *Ardipithecus ramidus*, some more like an advanced *Australopithecus*. Those less advanced would not be able to learn the language sufficiently well. Remember, their parents would not have been expert trainers; a child would have to pick up the language of the group without special instruction, as is mostly the case today.

Language thus became a selection device, not natural selection by some environmental entity. The language repertoire, even in its early simple form, would constitute a *group basic repertoire*, a required repertoire to be learned by successive generations of the group. The language repertoire, as it expanded by learning, would become a more generally useful repertoire. Speech would occur regularly in various group activities such as gathering, hunting, tool preparation, food preparation, "home" finding, and caring for offspring. A child with poor language or no language would have been handicapped behaviorally in describing happenings, in responding to instructions, in making plans, and thus in behaving in coordination with others. Such children would constitute problem children. The weakness in linguistic ability would make the individual undesirable, handicapped in terms of surviving and reproducing. *The result of that differential ability would be the weeding out of those without sufficient language-learning capacity, that is, without sufficiently advanced brains.* Through this process of continued cumulative learning over the innumerable generations of the group, successive weeding out of poor learners would occur, and those with larger and more effective brains would be selected and would reproduce. The members of the group, over generations, would progressively come to carry the genes for larger brain size and greater learning capacity.

Here was the operation of a new evolution process. The environment no longer selected for the larger brain, in the traditional natural-selection process of evolution. It was the learning of the creatures themselves that constituted the criterion for selection. The selective mechanism was not nature; it was the fellow creatures in the group. Human evolution was then learning-based and the mechanism was human selection, not natural selection. Natural selection continued to

play its role, but as a basic process. This new process of human selection, oper-
ating in a succession of generational steps, would have been central in pro-
ducing a succession of fossils judged as new species. Generational cumulative
learning and human selection fundamentally changed the process of evolution
at some point in the hominin line that led to Homo sapiens.

I said earlier that cumulative learning does not occur with lower animals, certainly not in any continuing way. But this type of learning is characteristic of humans; the unique human learning principles constituted a missing link in understanding human behavior. Let me propose now that generational cumulative learning, especially in the language area, was also a *missing link* in the explanation of human evolution, an explanation that makes human evolution unique among animals. With limited action at first, group cumulative learning played an increasing role until the advent of *H. sapiens*. But it wasn't only language that was learned cumulatively and over generations, motor- and emotional-repertoire learning was also fundamentally involved.

What I am describing is the beginning of a cumulative-learning development that took place over thousands of generations. When we speak of learning, ordinarily we are considering the progress of an individual. The human-selection process for producing human evolution, however, involved the learning of innumerable beings. Actually, it would have taken many generations to learn even the beginnings of basic repertoires of sensory-motor behaviors. As simple as is the tool repertoire of the australopithecines, it nevertheless would have represented great complexity and achievement for the early creatures involved, complexity much beyond that of the learned repertoires of any of the forest-dwelling primates. The same would have been true of the increase in the language repertoire from the *Ardipithecus ramidus* species to the *Australopithecus* species.

IN SUM

An early erect-ape species began venturing onto the savannah, perhaps a species like *Ardipithecus ramidus*. This species had combined features; a more upright posture that would enable the animal to move on the ground more rapidly and for longer periods than chimpanzees, gorillas, and orangutans as well as a brain that was a bit larger. Over a period of several millions of years there would have been physical evolution more and more suited to savanna living, as occurred with the *Australopithecus* species. Savanna living also would have provided experiences by which new repertoires could be learned—new tool use, opportunities for digging and preparing roots for eating, perhaps stalking and hunting opportunities, rock-throwing skills, probably group skills for defending against attack, new skills of care for the young, and new opportunities in these various activities for using new words—many of them emotional. Learning such repertoires would have constituted an integral part of the creature's ability to survive and reproduce, according to natural selection, on the savanna.

The savanna represented a new environment, and it favored development from *Ardipithecus ramidus* to *Australopithecus*. The shoulder, arm, wrist, and hand would have undergone changes suitable for life on the savanna, for example, for throwing rocks, digging roots, grasping tools, and lugging things and babies, rather than for swinging in the trees. The spine, hips, legs, ankles, and feet would have undergone changes that enhanced walking, running, throwing, lifting, and pulling.

These changes would have improved the ability to hunt small prey, to drive other predators away from prey they had killed, and to escape from larger predators. The environment would have rewarded new behaviors, like group actions in hunting, gathering, and defending against attack. These life conditions and group activities also would have evolutionally "rewarded" emotional, motor, and language learning. Over many generations, millions of years, learned repertoires would have been built, bit by bit. Slowly, too, the physical growth and interaction with other predators and prey would have enlarged the brain so that it was a little larger than the erect ape that first made the move from the forest to the savanna.

At first these results could have occurred according to natural selection. The physical and behavioral advances would have been selected because they yielded additional efficacy in the environment. However, at some point the larger brain, along with the new behavioral repertoires learned, would have produced a new dynamic for evolution. Being able to *learn* those repertoires, which depended on brain development, at some point would have become a cut point for successfully contributing one's genes to the developing line of hominin species. This opened up a whole new ball game.

For in such an evolving species there would be individual differences in brain size and consequent learning ability. Some individuals would have greater learning ability than others. They would more easily learn the repertoires acquired by previous generations of the group, and that learning would make them more able to learn new things themselves—new words, a broader learning of emotions and thus broader motivation, better skills for hunting, tool making, and childcare. Their advanced behavioral repertoires would favor these individuals' success in survival and in procreation. That is saying that at some point the group's learned repertoires gained causal status themselves in the further evolution of the species.

That was the new ball game. Over the generations, the bar was raised for selecting hominin specimens according to their learning ability and thus brain size. That raised bar would not have been a result of gene change by mutation or by sexual reproduction combinations. The change would have resulted from the generational cumulative-learning process that made those capable of learning, those with the larger brains, more reproductively successful. That causation would have taken over as the cause of evolution toward *Homo sapiens*. At this point, each step of the selection process for brain size would have become the ability to learn the basic repertoire that the group had learned. *Then and thereafter it was the group's learned nature that the new generations had to measure up to for successfully perpetuating themselves.* Each group of those beings had a basic repertoire. Learning that group behavioral repertoire, then, became the determinant of evolution.

The "judge" regarding survival and gene reproduction was no longer the natural environment, no longer natural selection, it was the *social group itself that selected.* Those doing the selecting had no idea they were conducting a program that would lead to the evolution of *Homo sapiens*. But they inadvertently produced such a program. A child who did not learn language well was perceived and treated much like mentally challenged children are in our society. They were less successful in the group, less likely to survive and reproduce as a consequence. Moreover, the human selective process—reducing the reproduction of such children—did not have to be perfect to have had its selective outcome. Since the selection process operated over so many generations and so many individuals, a small statistical advantage in reproduction for good learners (larger brains) would have produced the human-selection effect. That imperfection of selection helped make the process of evolution exquisitely slow.

For those who might be skeptical about the reality of this explanation for the millions of years of human evolution, let me point out that *this process is exactly that which the social Darwinism eugenicists proposed, that is, limiting reproduction opportunities to those with ability in order to weed out those of lower ability to, in this way, raise the level of the human "race."* The same occurs, without planning, to developmentally disabled individuals in our society. They are taken care of, but they are not given normal reproduction access. The principles of the cumulative-learning and human-selection theory of human evolution apply in human actions; they are not imaginary.

There is no other case in nature, with other animals, whereby the group makes the selection for evolutionary success based on the individual's ability to learn what has been *learned* previously by the group in a long-term, cumulative manner, in a process involving millions of years of progression. As the learning accumulated, the bar for successful survival and reproduction was progressively raised higher, demanding higher learning capacity, a bigger brain. And that higher learning capacity in turn made it possible for members of the species, then, to have new experiences, learn new behaviors, and contribute new elements to the cumulating behavioral repertoire. The groups that learned best, which had better-developed repertoires, and thus selected members with larger brains, survived and reproduced better than groups with fewer of these features. So human selection of this type actually was working for groups, not just for individuals.

Two million years after *Ardipithecus ramidus*, the fossil remains referred to as Lucy showed important advancement over "Ardi." Still not human enough to be considered conventionally in the *Homo* species, the fossils of *Australopithecus* specimens had gained a larger brain size and a more thoroughly bipedal body. Importantly, there was also evidence that these anatomical advances—based on learning advances—were indeed associated with advances in behavior.

> Mary Leakey's "living floors" sites [at Olduvai] were replete with bones and artifacts, and, famously, in one instance, with a circle of stone walling . . . [suggesting] that some of these sites were "home bases" . . . that by . . . 1.8 millions years ago . . . a set of new adaptive strategies, including toolmaking, transport of food and materials, eating of meat, and sharing of food, had been attained. The older Gona stream artifacts in Ethiopia led [to suggestion] that such behavior patterns may have been acquired much earlier there than at Olduvai. (Tobias 2003, 1193–94)

This is a description of a species of beings whose learned behaviors are vastly advanced in the direction of *Homo sapiens*, beyond the behaviors of a chimpanzee or *Ardipithecus ramidus*. Tool making, transporting food and materials, eating meat and sharing food, constructing a stone wall—these are the behaviors of a group of individuals working together in complex, coordinated behaviors. These involve not just single behaviors but complex learned repertoires. The activities are social and would have required at least a primitive talking language, perhaps like that employed by three-year-old children interacting with each other. The progression of learned group-behavioral repertoires was already advanced, and that would indicate that the cumulative learning process of human selection was already at work. *Human selection, it is suggested, even then may have been determining evolution on the way to* Homo sapiens.

CHANGING PERSPECTIVES

The origin of humans calls out many questions and interests, for matters of origin concern characteristics of human nature. Let's consider a few of them, in the context of human-selection theory.

Why Did Other Non-Upright Primates' Brains Cease Expanding?

Why did those not in the upright-ape lineage stop evolving larger brains before reaching the size of the *Homo sapiens* brain? Chimpanzees, bonobos, gorillas, and orangutans essentially stopped evolving in brain size. The same millions of years passed by without measurable increase, just as they also passed by for the upright apes like *Ardipithecus ramidus*, *Australopithecus*, and *Homo habilis*. But the other primates were already successful with the brains they had. So if some chance mutation or some male-female reproduction combination produced an offspring with a larger brain, that would not have made that animal more successful in surviving and reproducing. Nothing in natural selection would have favored chimpanzees with larger brains. Just the opposite. The larger-brained chimpanzee would have had to spend much more time hunting for food, wasted time with respect to seeking sexual partners. For larger brains demand more food.

Australopithecus, *Homo habilis*, and *Homo erectus* were also successful in their environments. Why didn't they stop evolving in brain size? Why did they continue to evolve larger and larger brains? Larger brains meant they had to invest more time in getting food and more time in getting food for their offspring. That larger brain and head also added jeopardy to women in childbirth. The obvious answer is that continued evolution of brain size depended on something going on in the group itself, not in the environment.

Traditional evolution theory lacks the fundamentals for explaining questions such as the present one concerning the evolution of the brains of humans and other primates.

The Human Brain Not Only Grew, It Grew Fast

In addition to the question of why only primates in the human lineage grew larger brains, there is also a question of the time it took for the human brain and nervous system to evolve from its original chimpanzee size and state. *That evolutionary development took place at a staggering speed.*

After having taken some six hundred million years of animal evolution to reach a volume on the order of 450 cc in our simian ancestors, the size of the hominid brain went through an astonishingly rapid rate of expansion, virtually jumping almost three times its original size in little more than two million years (De Duve 2002).

The time in reaching that 450 cc volume, let me suggest, reflected the speed of ordinary natural-selection evolution. That ordinary evolution through the mutation and reproductive variation that powers natural selection is slow. The unusual speed from that size to the huge human brain provides evidence that something other than natural selection was at work. That something else was cumulative learning and human selection. The principles of natural selection still functioned in the process, of course. But the criterion for selection had become learning ability, and the mechanism was human selection. The random process of mutation is slow. That slow process of natural selection was bypassed. As the accrual of learning accelerated, so did the growth of brain size and learning ability.

The Cultural Explosion

Well, how about the advance of human culture? Was that explosive too? One of the mainstay beliefs in the traditional view of human evolution is that modern human traits appeared abruptly. The cultural advances, such as the drawings in the caves at Lascaux, appear to spring from nowhere, a sudden jump. That seems to hold for permanent settlements, jewelry, and musical instruments. These clearly were the marks of human beings. What explains their appearance?

> [T]he modern human sensibility—what has been called "the human capacity" —was acquired as a package, rather than bit by bit over the millennia (Tattersall and Schwartz 2000, 241).

> Is it possible that the brains of early *Homo sapiens* were simply not yet wired for sophisticated culture? The modern capacity for culture seems to have emerged around fifty thousand years ago. (Johanson and Edgar 1996, 43)

This view then infers that some thirty-five to fifty thousand years ago, a mutational change gave the *Homo sapiens* species real human abilities; new genetic structures produced the rapid development of human culture.

The theory actually is in answer to a traditional evolutionist's dilemma, that is, how to explain the fact that the human species existed one or two hundred thousand years without much change in human culture. If the genes drive human behavior advancement, then why didn't recognizably human culture begin to emerge then, when the *Homo sapiens* species was established? Why was there such a long wait, until only some thirty-five to fifty thousand years ago? The answer in the traditional framework is that a genetic change did occur, despite the lack of evidence for it.

The Cultural Explosion and Instant Hollywood Celebrity

There is a standard Hollywood joke about seemingly instant jumps to star status. Actually the appearance of the rapid rise to stardom occurs because the years of struggle to get to that celebrity role have gone on in anonymity. I have been surprised a number of times when, while watching old movies, I see a famous star is in the movie, much before I would have thought, but in a distant supporting role. I may even have seen the movie before without recognizing that the star was in it. That is why stars may chuckle when asked about their instant success. They worked hard and long at their craft in relative obscurity before getting to that "instant success."

We have very well shown the effect in our research. Dyslexic children are considered to have a perceptual brain problem because they see letters upside down and backward and normal readers do not. The reason for that erroneous belief occurs because the behavior of dyslexic children is observed, it's "abnormal." The behavior of children first beginning to learn to read is not abnormal, so it isn't observed. The long learning involved in becoming an expert reader just isn't generally known. Actually, all children beginning to learn to read see letters upside down and backward, and their beginning learning is slow.

A phenomenon like that has been involved in conventional wisdom's view of the human origin. Interested in finding genetic explanation, scientists haven't noted or understood what was involved in the laborious journey to the point where human cultural progress took off. There was no instantaneous advancement to being human, no magic time thirty-five thousand to fifty thousand years ago, only the lack of consideration of the long development toward humanness, what was involved, and what was driving the development.

Cumulative Learning: Slow and Fast

Let me suggest, rather, that the culture explosion of thirty-five thousand years ago was just a usual feature of cumulative learning. Cumulative learning generally follows a particular pattern. It takes place at a very slow pace in its early stages and it accelerates as it goes on. I have described that with the way a child learns language. It takes a year before a first word is said, even though the child has a fully human brain and even though the child is exposed to fully developed language experience. To show the acceleration that occurs in cumulative learning, twenty years later, that child in college learns huge numbers of words in a school year. In human evolution not only did the creature involved have far less than a human brain for learning, those with language had a very small and incomplete language to impart. The language itself had to be learned as the process advanced.

So two processes were involved. There was the biological change of the size of the brain. That had to occur. And there was the learning of the repertoires themselves. Both took time. Learning and genetic advancement would have been very slow.

But advancement would have also slowly accelerated. The more that is learned in a repertoire, the more valuable the repertoire becomes in learning expanding repertoires and in learning new ones. That would have been working. What human evolutionists see as the takeoff point for humanness is actually the point in this great learning voyage when the repertoires produced learning rapid enough to be noticeably familiar to modern progress.

In the quiet years following attainment of the full biological level of *Homo sapiens*, but before the "cultural explosion," learning was going on. If it were possible to see the difference in language development from theirs to the language of contemporary Yonomamo Indians in the Amazon, there would undoubtedly be much growth, just as there would be in comparing the language of those Indians to our own. Our language expands every year, still. The expansion in the language of the early *Homo sapiens* to that of thirty-five thousand years ago would have suited the modern beings to do things much more effectively. Moreover, repertoires other than language were also expanding in that same period, which would make them more effective in advancing through learning.

Not only were they creating names of objects in their lives, they were advancing the grammars of their languages, making them better instruments for reasoning, planning, and thinking. At some point they learned how to tell stories both true and imagined. They must have concocted myths of their own origin as a central part of world origin. They also learned to clothe themselves, use fire, refine their tools, create instruments for making pleasurable sounds, construct shelters, hunt and fish effectively, trade with other groups, and live in larger groups. With this infrastructure—especially their language—learning the many things that compose a culture began to come more rapidly. Once they began living in settled groups and developed some specialization of skills, their advances came more and more rapidly. Our advancements now are lightning fast in comparison to then, just as their advancements were lightning fast in comparison to the advancements of the first modern *Homo sapiens* of a hundred thousand years ago, and their advancements were lightening fast compared to earlier hominid species. That's the way cumulative learning operated in the generational human-selection process.

The accelerating path of cumulative learning can be seen in the number of patented inventions produced yearly in the United States between 1850 and 1903. In that period of time, the annual rate of inventions patented increased by about thirty-four times. The inventiveness grew at ten times the population increase. The acceleration in inventiveness can be predicted from knowledge of cumulative learning; it starts slowly and speeds up as the basic repertoires are acquired. The cultural advancement from species like *Ardipithecus ramidus* to *Australopithecus* was actually great, but millions of years slow. The learning from *Australopithecus* to *Homo erectus* was still slow but faster. By fifty thousand years ago we had gotten to a more rapidly accelerating part of human learning, but it was not accelerating nearly as fast as it has in the last fifty years, and even that will be slow compared to what occurs in the next fifty years, barring a calamity like global warming.

No Cultural Evolution

The concept of cultural evolution is misleading notwithstanding the fact that it is used so widely. The concept says that culture develops according to the principles of natural selection, that genetic changes lead to cultural changes. That belief was criticized in chapter 1 using the example of science advancements being considered as cultural evolution. There is no cultural evolution. Cultural advancements do not take place by the natural selection of mutated genes.

The human-selection theory presents a very different explanation. The amorphous term *culture* is defined as the behaviors of humans in some particular grouping along with what those behaviors produce. Cultural advances are advances in behavior and products. The cultural advances of all the humanoid species, probably from *Ardipithecus ramidus* or some such group to modern *Homo sapiens*, occurred through learning. The learning was cumulative learning. Cultural learning explains cultural change. That says that questions about the explanation of a particular aspect of culture are better answered by the study of the cumulative learning that led to the development rather than by studying possible genetic causes.

There is no evidence of a genetically produced change in the human brain for a hundred or so thousands of years, during which time there have been fantastic advances in culture. This holds that the cultural advances in technology, entertainment, politics, language, economics, education, inventions, and other social features are due to learning. There is no cultural evolution.

New Perspectives

The new conceptual view yields new views of things not yet studied as well as of things that have already been considered. As an example of the latter, take the cave paintings at Lascaux in France and Altamira in Spain, which were done some thirty thousand years ago and are now onsidered marvels. Were those paintings simply the result of a mutation that abruptly produced a marvelous artistic trait? Rather, such cultural developments need to be analyzed in learning terms. Did cave paintings involve the same kind of basis as the cumulative learning of artists today, whose skills are built upon the learning of centuries of prior artists? Was the prehistoric cave artists' skill built up from thousands of years' experience of predecessors? Had many prior hominins drawn pictures in the dirt like kids frequently do? Had prior hominins drawn pictures on rocks with chalk that rain later washed away? Was it the case that those prehistoric cave artists could display their learned skills because some prior individuals had learned to manufacture more durable paint? Or was it because only paintings done in caves survived and that makes that type of painting seem to have emerged abruptly?

Could a history be done to trace the cumulative learning of artistic skills that occurred subsequent to the work of the cave painters? Over the many subsequent years, was there an increase in the various aspects of art skill that made it possible to paint more and more realistic portraits? Could the development of various human behavior achievements also be shown by historical analysis to be cases of cumulative learning?

Take religion as another example of cumulative learning. About thirty-five thousand years ago, Neanderthal and other peoples buried some of their dead with flowers and valued objects. This action has been taken to indicate that Neanderthals had a belief in an afterlife.

Isolated groups of people found around the earth who live in Stone Age cultures have systems of beliefs about the world, its creation, their own creation, and spirits and gods. These beliefs constitute an important part of their language-cognitive repertoires. Moreover, other evidence shows that religious accounts advanced in a cumulative way. A book titled *Man and His Gods* by Homer W. Smith (1952) provides a historical description of how aspects of the religious beliefs of a people are drawn upon by a later people in creating their religion. Smith showed that developments in the Old Testament (Hebrew Torah) have prior existence in Egyptian and Babylonian religious beliefs. Moreover, those developments do not end there. In later generations this development continues, with the Christian Bible and the Muslim Koran being built on top of the Old Testament.

Farther down the line, taking us into contemporary times, we can pick up these developments as they move into science. A. D. White, in his book *A History of the Warfare of Science with Theology in Christendom* (1955), traces how the discovery of findings in the various science areas disagreed with the biblical accounts. In each case the difference in what had been learned forced the formation of a science that became opposed by theology. For example, the biblical belief holds that Earth was divinely created only a few thousand years ago. But science study revealed that Earth had existed far longer and that it had been in great and perpetual change. The point is that the religious conception of the world appeared first and would have acted conceptually as a foundation, a steppingstone, for later interest in scientific discoveries.

Science development may be seen as a later, cumulative learning activity that built upon past religious knowledge of the world, continuing on a tremendous learning voyage in many fields. The point here is that, while there was and is a paradigm conflict, a division of views, the growth of knowledge is continuous. It is the result of cumulative learning that began with ancestors perhaps a million or more years ago learning language-cognitive repertoires that we today consider primitive myths.

We can trace the same type of development in every human area of achievement, some whose later histories lie in our own times. Look at automobiles, airplanes, and iPads® and iPhones®, at government, law, warfare, and languages, at schools, medicine, civil rights, economic and welfare systems, and so on, all with histories of continued development through cumulative learning.

That kind of historical tracing should be done for the various important cultural-learning developments of humans. I believe, for example, a book that traced the development of painting as a cumulatively learned, very complex skill over generations of humans would tell us a great deal about artistic ability, a topic of perennial interest. Such works should be produced in various important human areas; they would give us a new view of humanity.

I hope this chapter has shown that analyses of human origin cannot take place only in a traditional framework. Analyses also need as a basis a conception of human behavior and human nature and how human learning was central in human evolution. To illustrate, there are questions and controversies concerning when and with which species the human line of evolution began. Species such as *Ardipithecus ramidus* and *Australopithecus* are not granted that hallowed place. *Homo habilis*, as the name indicates, plays that honored role. The decision is made mostly on the basis of the physical fossil, especially the size and shape of the skull. Was the species' head large and developed enough to be judged as humanoid? The context for the decision is biological.

There is no question that brain size and upright posture are centrally important. But the point at which the human evolutionary line should be considered to begin is when some species had learned enough that this became a criterion for successful selection in the group. Human evolution began when the ability to learn became significant in biological evolution of the species. Paleoanthropologists should begin studying the evidence of species behavior in terms of the level of learning it indicates. When does the complexity of the behavior shown advance past that of the chimpanzee to such a degree that significant learning ability had to be involved in passing it on generation to generation?

I believe, just in terms of the *behaviors* described by the Leakey study of australopithecines, that for them learning was already significant. That learning had accumulated for a couple million years. Not only was that learning going to continue to cumulate, perhaps even at this point it was the selective device that drove the increases in brain size that occurred in the late australopithecines and in *Homo habilis*, and on and on. I suggest, thus, that paleoanthropologists must examine the behavioral evidence in human-evolution findings.

The point is that the new conception of human evolution has implications for future developments. Paleoanthropologists should expand the conceptual framework they use to include a full conception of human behavior and human nature as learned. For the principles involved operated in human evolution.

DIVINE CREATION, ANIMAL EVOLUTION, AND HUMAN CREATION

Tales of human creation have a long history. God created humans in the biblical explanation. For many, placing humans at the apex of God's creations gives meaning to existence and the comfort of being special to the Lord, supported by the Lord. But there is no evidence that God created humans. Divine creation has little to defend it as soon as one's criterion slips from substantiation by faith to substantiation by observable proof. Science has brought such tremendous knowledge of the world, as well as many ways to better deal with the world.

Natural selection, in contrast to religious belief, has great empirical confirmation, and it has made great contributions to human life. But it does not distinguish human evolution from the evolution of other animals and plants, so it thereby takes away a person's special place. Many people derive special meaning to their lives from the belief that they have been created by a divinity and that their actions derive from that divine inspiration. The theory of evolution takes away that belief and that meaning to life. "[T]here is in fact no way in which natural selection can single out one particular trait to favor or condemn" (Tattersall and Schwartz 2000, 242).

This approach assumes that mutation produces new behavioral traits and that natural selection then takes over. Evolution theory makes human origin a chance event, dependent on mutation and chance gene assortment in sexual reproduction. Conceiving of human evolution as being the same as the evolution of the other animals has an unfortunate side-effect: it loses much public support for evolutionary science by making humans just another animal, without fundamental uniqueness.

Belief in human uniqueness provides motivation for clinging to the divine theory of human creation. Vigorous attempts have been and are being made to reject, weaken, or isolate Darwin's natural selection. The contemporary evolution opponent has the name of "intelligent design," which holds that humans are too complex to have come about through the chance operations of natural selection. Intelligent design, of course, also paves the way for a belief in an intelligent designer. Although it may not be stated, that designer would be God.

Let me suggest that straight natural selection, by making humans just another animal, inaccurately deflates the wondrousness of human creation, thereby drawing more opposition than should occur. Humans are not just another animal. Humans really did evolve in a unique process. *Human ancestors themselves created the human species by their learning and by their selection for learning ability.* That is true of no other species. The human creation standard was progressively raised through the millions of years of evolution involved. It was not intelligent design and God that created humans. It was not natural selection alone that created humans. The present theory could be called *the human selection* or *the human-creation theory of human evolution.* Whatever it is called, it involves a theory that recognizes that human evolution was different than the evolution of other animals.

Moreover, the evolution process was not chance. In a sense, human evolution had a "goal." Or, at least, human evolution had a path to be followed, not preordained but nevertheless obligatory according to natural law. Tattersall and Schwartz say that "any new structure must arise [by mutation] before it can assume a function." But that has the causation backward. Rather, some generations of a hominid group learned a significant repertoire and thereby created a criterion of learning, a criterion that later hominids had to meet. The learned function came first. Later, that learned function *drove* the biological change in the species. While the very distant end product (the large human brain) did not reach back and cause earlier-occurring evolutionary changes—things yet to happen cannot determine earlier happenings—the process was also not a case of chance mutations that took those *australopithecines* on a multimillion-year journey to *Homo sapiens.* Once human selection of learning capacity became a criterion of human evolution, the "goal" was set. The goal that functioned for humans was great learning ability, ultimately maximal learning ability. Modern humans' learning ability would be the ultimate product of the human evolutionary process.

How can that be said? Isn't it outrageous to suggest in a scientific account that the six-or-so-million-year evolutionary journey was a fixed race? Could such a long and tortuous path have been inevitable? That sounds like some form of predetermination or *intelligent design.* But that is not what I suggest. Let me explain. It is relatively straightforward to recognize that once the to-be-human line descended from the trees and entered the savanna, its bipedal locomotion, barring some interrupting occurrence, was going to evolve into a fully upright creature. That was going to happen because upright locomotion was advantageous for the species in that environmental niche. So that end state of physical structure was inevitable. If the species could function successfully and reproduce when its body was not yet fully adapted to upright locomotion, then the individual differences in the species in the direction of strengthening, lengthening, and straightening the legs and bringing the arms and hands into congruence would have made the species even more evolutionarily successful. The conditions were thus set in the time of the *australopithecines*, or maybe in the time of *Ardipithecus ramidus*, where any developments that would make for more effective bipedal locomotion and improved arm-hand functioning would be evolutionarily rewarded. All that was needed was that there be variation of these physical characteristics within the species. Sexual reproduction guaranteed these variations at the genetic level. There were thus straightforward conditions that made it inevitable that full bipedal locomotion would evolve. In that sense, natural selection had a "goal."

The same is true for the human selection that yielded *Homo sapiens*. When the *ability to learn the group repertoire* became a criterion for reproductive success, evolution toward a species with a huge learning capacity became inevitable. That occurred millions of years ago. From then on, increased levels of learning became the evolution criterion, thus increasing the size of the brain, enlarging the human head, as well as increasing the helplessness of the human infant and the length of childhood. Other behavioral developments were also guaranteed. Human selection played the determining role in human evolution until the members of *Homo sapiens* generally came forward with full learning capacity. Once underway, millions of years ago, that process had to go on until humans became capable of the learning level that presently exists.

It will be important for many who resist the explanation of human origin through evolution to realize that *human evolution was not the same as evolution for other animals*. For the long line of hominids themselves, as they learned new repertoires over many generations, were creating the human species and human nature. The process in that sense was human creation of humanness.

The acceptance of intelligent design demands rejecting the weight of fossil and other evidence of human evolution without any justification for doing so. Evolution theory, joined with genetics, widely and greatly underlies important scientific fields involving great development of methods and apparatuses of study and analysis, huge funds of evidence, and great products of application. An intelligent-design approach that discards such enormous scientific developments, totally and with no methodology to add to that knowledge, with nothing to replace that scientific knowledge, constitutes a great, nonsensical, disadvantageous error.

Human-creation theory does not reject evolution theory. It glorifies that theory, recognizing its great contribution to science, to culture, and to the practice of life. Human origin took place in great part according to the laws of natural selection, bringing about a body structure made for behavioral versatility in some early hominoid creature. That included development of the upright posture that separated the functions of the fore limbs and hind limbs. Further evolution of the body, as with an opposable thumb, led to additional behavioral generality so important in human achievement. These changes in structure also had to involve an increase in the development of the nervous system, including the brain, and an increase in learning ability.

The beginnings of human evolution must have taken place according to usual natural-selection principles. However, a point was reached when what had been learned became so important that it—rather than natural selection—began to determine the evolution of brain size and learning ability. That would have been the point when the species entered the *Homo sapiens* track of evolution. At that point in human evolution the laws of human selection became central, dependent as they are upon learning.

The human-creation theory lies fully in science, and it aligns with natural selection. But it avows that human evolution was special, unique. Humans, indeed, are special, unique.

Chapter 9

WHO WE ARE

Like Darwin's theory of evolution says, we humans are animals, a species separate from other animal species. We have special qualities of structure that give us unique variability in motor behavior. Humans can thread a needle, do brain surgery, play a violin, and put together wristwatches. But humans can also block three-hundred-pound defensive linemen, run one hundred meters in less than ten seconds, and lift weights of more than four hundred pounds. In terms of motor skills, humans can coordinate a number of organs in emitting a wide range of vocal responses. All of the motor responses humans can make can be learned in complex sequences and skills. Humans, like other animals, can respond emotionally, positively or negatively, in varying intensities. And emotional responses are learned vastly, to thousands and thousands of things, actions, words, songs, people, values, and so on.

It took billions of years for animals to evolve physical structures for motor, emotional, and vocal responding. By the time evolution had progressed to mammals, much behavior had already become learned, with only some still wired in. Jumping ahead to primates, probably almost all behaviors are learned. Humans no longer have wired-in complex behaviors, only simple reflexes. Humans learn incredibly complex repertoires, like language, and complex sequences of emotion, like those involved in deciding to enroll in a university program to study high-energy physics. Humans can do such things because of cumulative learning that involves learning a complex repertoire that provides the basis for learning another repertoire that then enables the learning of a third repertoire, and on and on. Humans are unique in the world in being able to learn cumulatively in this way, which is the reason why human behavior is so complex and so variable within the individual and across individuals.

When a child is born, she or he is a unique creature because of the learning potential. The child does not have any wired-in or instinctual processes that by growth and development will give the child human characteristics. The human infant is not born with a personality, a temperament, or any such proclivity. The infant does not possess inborn cognitive characteristics that will blossom as the child matures.

The infant is a highly developed learning animal with a huge brain that is ready to learn. The child must learn everything. The child has fantastic learning ability because of the human body, nervous system, and brain. Because of this, the child will learn from what occurs around him or her. The child, for example, "picks up" language from parents, even without formal training. However, what the child learns will depend on the child's experiences, and that continues through life.

We are all the things that have been described in the previous chapters. We are human beings because of our bodies and the great variability of the actions our bodies enable us to make. We are human beings because of our evolved monumental learning ability. We also have unimaginable amplification of that ability through our unique "missing-link" human learning ability. That ability comes from the hugely complex repertoires that we begin to learn from infancy, repertoires that have been learned over the period of human evolution. We are human beings because each of us learns a huge emotional-motivational repertoire, a huge sensory-motor repertoire, and especially a huge language-cognitive repertoire. These repertoires constitute our personality, unique from everyone else's. As human beings, we can learn repertoires that are disadvantages for ourselves and others, sometimes to an extent that is considered abnormal. We are human beings because our repertoires over the group add up to our culture—and that culture impinges on us from our beginnings as infants. We are who we are as a species not only by evolution through natural selection but also by evolution through human selection.

These qualities, which no other species has, make us a unique species in very special ways. Humans have other characteristics, of various kinds, that cannot be considered here, but a few others of common interest will be.

WHAT ABOUT FREEDOM AND WILL AND MIND: WE AREN'T JUST PUPPETS

In expounding on humanity in terms of learning, there are purely mental operations that are ingrained in our language and in our beliefs that haven't been mentioned. Before moving on, thus, it should be indicated that saying our brains are just the mechanism by which learning experiences produce our behavior doesn't mean we are just computers—type in some stuff and the program inside will arrange it according to some rules, and out will come a response.

Our own experience with ourselves is not like that at all. When we face a significant life situation, we don't just respond mechanically. We consider the various aspects of the situation, we think of various ways we can respond. That arouses us to consider how other people would react to our different responses; what the legal, ethical, and moral implications might be; and what might arise. We encounter all kinds of situations that present various alternative responses, and we have to choose a response according to our considerations. We do not just react mechanically. We experience having freedom of action in such cases, having choice, and deciding what we will do. We have innumerable experiences of exerting our will, to do or not to do.

So a mechanical explanation of human nature meets lots of resistance. How are "freedom of choice" and "will" to be explained? Science, after all, is concerned with lawful causation, where one event causes another. Free will implies spontaneity. In answer, it is true that the basic repertoires and cumulative learning operate according to law. A human's behavior is a function of her or his lawful experiences. At the beginning, that behavior very much depends upon the learning experiences the parents control. However, progressively, increasingly, our basic repertoires become us; they direct what we like, what we want, and what we strive for. The basic repertoires also determine what we can do, what we can learn, and the further repertoires we learn. Our memories of past performances and choices and our present repertoires compose us. More and more as we go on, our choices determine our experiences and what we become. At any point, then, what we have done is responsible for what we are, what we do, and how we affect others and the world. We make decisions, take actions, affect others, raise children, assume beliefs, all based on our experiences that we have been responsible for. Our repertoires are very much dependent on what we have already chosen innumerable times. If our actions are considered wonderful, then we become to ourselves and others wonderful as persons. We are indeed products of what we have experienced, but much of that depends on what we have done, on the choices we have made. Yes, in that way we have freedom and will and are responsible for what we do and who we are.

And that is not all. We know we have a mind. We dream about things, with pictures in our heads. We do that when we are awake, too, experiencing again in pictures and thoughts in word form occurrences from the past and projections for the future. We have worries and anticipated joyful experiences. How can we not feel we have a mind?

However, although all those experiences inside do occur, it is not because we have a mind composed by our brain. We have thoughts and feelings and pictures in our heads because of what we have learned. Some of those experiences are of a language kind and function when we are planning something. The plans may be for things yet to happen. But we couldn't make those plans if we had never learned our language repertoire. An autistic person with no language could not make those plans. A child of three who has learned only a limited language repertoire could not make those plans. Our planning, our reasoning, our worrying, our beliefs, our feelings, depend upon our learned repertoires.

That also pertains to our thinking that is composed largely of pictures, of what went before or of what might happen. Although I did not have room to treat the matter in this book, when we sense something in life, those sensations are responses. And as responses they can be and are learned. We can see something "mentally" that has happened to us because those sensory responses were conditioned to some stimulus. When that stimulus occurs, we experience the conditioned sensations—we see a picture of an old girlfriend or boyfriend when a song that the two of us used to listen to many years ago is played today. We may even purposefully revivify something from the past by bringing up a stimulus associated with that experience. We get a picture of our bygone friend by thinking of the song. We have all kinds of experiences, even seeing the future, that convince us that we have a mind.

Humans evolved a marvelous learning mechanism. The most complex of experiences produces in our brain a mind comprised of learned repertoires of the most complex nature. And that gives us the mental activity we know. The experiences that formed that mind are too complex to know about; the choices we have made and the experiences that have resulted are our responsibility. What we are seems free to us and everyone else.

WE'RE EQUAL. REALLY. IT'S EVOLVED.

Part of my disaffection for the Great Scientific Error stems from the fact that the evidence on which it rests is the same as that which fosters a horrific error: racism. Racism consists of the view that different peoples of the world have genetically caused behavioral traits that vary in quality. This constitutes a social problem of long standing, for various groups are and have been considered inferior and undesirable. Nazi Germany's treatment of Jews, Gypsies, and Slavic peoples dramatized the social evil of the belief. Racism, also, was an intimate part of slavery, and it has been causal in various wars. Today, as a consequence, most everyone condemns racism and would deny holding racist beliefs.

Nevertheless, as Stephen Jay Gould's *The Mismeasure of Man* (1981) and Allan Chase's *The Legacy of Malthus* (1977) both describe, there is "scientific racism," for (in ways not labeled as racist) scientists have given support to the belief in a racial hierarchy of fitness deriving from evolution-genetics. Gould (1981, 291–302) performed an important service in writing about how some of the prominent early figures in psychology's intelligence-measurement field contributed to erroneous racist conclusions. He referred to how psychologist H. H. Goddard and statistician-psychologist Karl Pearson conducted research and wrote things that are considered racist in nature, namely that whites have higher intelligence than blacks, on a hereditary basis.

Let me add that right up to present times there have been others who have done twin studies and other types of research, composed theories, and written articles and books claiming genetically determined differences in human behavior across "racial" groups (see, for example, Herrnstein and Murray 1994; Jensen 1969, 2006; Rushton 1999). (I put the word "race" in quotation marks because designations of race are vague—the label "African American" includes people of all kinds of genetic mixtures. Tiger Woods, for example, is Caucasian, Asian, and African.) Some of these scientific statements consist of scales of superiority (see Rushton 1999). Rushton's tract is built on the evolutionary theory that Africa, because it was warm and "the living was easy," did not select for intelligence in its people. He says the lands where Caucasians lived were colder and less hospitable, in contrast, and thus provided a natural-selection process that created a more intelligent race. The lands where Asians lived were even colder and selected most strongly for intelligence, he says. In Rushton's view, thus, Asians are the most evolutionally advanced race, Caucasians are close behind, and Africans are much farther back. Rushton, as is the case with such conceptions in general, considers that the less advanced race generally has an inferior nature, being less moral, less judicious, more impulsive, and more primitive—generally lesser humans. These are actually old ideas. Rushton merely puts on a scientific patina by using biology "evidence" and "scientific" arguments.

It is instructive to realize that *the same evidence that is a fundamental part of the biologically oriented conception of human behavior and human nature also lends itself to the racism conception.* Most of that evidence has been gathered in studies that do not refer to race. They are studies interested in biological-genetic-brain factors as determinants of human behavior. But those studies can be used as proof of racism. It should be understood that if genes are taken to explain differences in the behavior of individuals, by the same token they can be taken to explain differences in the *behavior of groups*, for groups may also share genes.

As Rushton's screed indicates, the tie of racism to evolutionary and genetic views involves other types of study that I have described: brain studies, pharmacology studies, genetic studies, behavioral-genetic studies, developmental-psychology studies, and brain-damage studies. To illustrate, take the study where brain imaging shows different brain activity in normal readers versus children with a reading problem. A prominent view is that there is a characteristic of the brain that determines reading ability. Dyslexic children thus are believed to share a characteristic of the brain that causes their problem. The same thinking can be used to "explain" differences for racial groups as well; blacks as a group show less achievement in reading than whites. So it could be argued that some defect in blacks' brains makes it more difficult to learn to read. That interpretation, in turn, could be used to suggest that blacks should receive a lesser level of education (see Herrnstein and Murray 1994, Jensen 1969). By the same token, blacks are represented in US prison populations disproportionately to their numbers in the general US population. The traditional biological orientation would make it reasonable to believe that blacks genetically, racially, have personality traits that predispose them to criminality. That such views exist in powerful form is not just speculation. Federal research monies have been spent in an attempt to isolate those genetic, read *race*, "causes" of violence.

This is to say that racist interpretations trace themselves back to the same bedrock evolution-genetic conception. That makes racism an ingrained part of the generally accepted conception of human nature and human behavior. There are many individuals who are vehemently against racism but who nevertheless believe in the concepts, principles, and findings that provide the foundation for the belief in racism. Recognition of the Great Scientific Error, in contrast, derails the justification for racism. I am suggesting that racism rests on bad science and that it thus leads to bad conclusions and social projections, one result being an obstacle to doing good science of the kind that is needed. Racism will not be eliminated from our society until a conception of human behavior is accepted that does not support the unproven intellectual belief in the biological causation of human behavior traits.

One of the fundamental differences between the present conception of human nature and the traditional conception concerns the question of human equality or inequality.

Inequality: Racism, Genderism, Poorism

Nesting among the broad and hugely productive implications of Darwin's evolution lurk some erroneous principles and unproductive implications. One of the worst constitutes the fundamental basis for racism, genderism, and other individual superiority-inferiority conceptions. Generalizing natural-selection principles as the explanation of behavioral characteristics led to belief that such traits as intelligence vary widely and are genetically determined. Many studies suggestive of that type of causation—including brain-imaging studies of abnormal groups and twin studies of intelligence—have been taken as proof.

How else can performance differences among groups be explained? Conventional wisdom holds that groups that perform better in life or on tests are genetically superior to those that perform less well. The fact that Asian Americans score more highly on intellectual tests and in the classroom than white Americans leads to the conclusion that Asian Americans have better brains. The fact that black Americans and Hispanic Americans perform less well in school and on tests than white Americans leads to the belief that white Americans have better brains than the other two groups. (Ignored is the inconsistent fact that Native Americans are also Asian Americans, having migrated from Siberia via Alaska about fifteen thousand years ago.)

This conception is not limited to "racial" or ethnic groups. Women as a group have not achieved as much as men in music, art, science, math, and politics. So the "genderism" conclusion has been drawn to explain those differences by inferring a genetic liability. Women are widely considered to be inferior to men in certain brain-mind characteristics, the most recent infamous example coming from casual remarks of a president of Harvard. (The belief in genderism, of course, is now taking a demolishing hit by the fact that females have been exceeding males at every level of education.)

Also, "poorism" can be used to name the belief that lack of economic success comes from personal, genetic limitations. The widespread belief that different peoples of the earth by virtue of their genetic quality have achieved different levels of development can define "people-ism." Since the general belief holds that the quality of human behavior shows the quality of the brain-mind of groups as well as individuals, various types of performance become the sign of personal, biological superiority-inferiority. The wealthy, the powerful, the educated, the leaders, the talented, and the athletically or artistically or scientifically gifted, on the other hand, are considered to be blessed with genetic superiority.

A centrally important premise of the present conception is that natural-selection principles apply to physical characteristics. Genetic principles also apply to differences in learning ability among species lower than humans on the animal scale. But they do not apply to the behavioral characteristics of humans; there are no genetically determined superior or inferior groups. Whether of different gender, "race," color, ethnicity, religion, nationality, education, whatever, we are all the product of millions of years of recent evolution and billions of years of total evolution. That human evolution has made us so that we all can learn a complete human nature—except for a very small percentage identified as damaged (with direct physical proof, not "proof" inferred from behavior).

An Evolutionary Conception of Equality

It is important to understand why humans are equal, despite evident achievement inequalities among individuals and groups. The explanation is that hominids with lesser learning ability were winnowed out of the human lineage over a period of millions of years of human-selection evolution. They just did not have the cumulative learning ability necessary for acquiring the complex repertoires that had been learned by previous generations. When a hominin child was born with a brain that could not learn the basic repertoires for the group, the child's genes over evolutionary time were eliminated. *That happened innumerable times until all individuals with lesser brains were eliminated from the species.* That elimination process continued for millions of years, countless generations. It continued for the hundreds of thousands of years of archaic and modern *H. sapiens*.

In the present view, all humans born without damage have the learning capacity with which to learn language, and that means a maximal learning capacity. Being born into the human species now, and for thousands of earlier years, means having a human brain. The millions of years of *human selection* for learning capacity has made us all equal—we all meet the learning-ability standard. *Whatever individual differences separate us lie in what we learn.* If performances of individuals differ, the cause lies in their learning circumstances. For example, for many children in the Hawaiian Islands, pidgin is their first language. It is a full language, showing they have full learning ability. But that means those children will not do as well as usual on intelligence tests or learn as well as usual in classes taught in English. That says nothing about their natural ability, however.

Not only is language learned, so are morals, values, kindliness, honesty, consideration, cooperation, ambition, responsibility, learning ability, sociability, truthfulness, industriousness, egalitarianism, humor, art appreciation, music appreciation, and creativity. We could all be expert mathematicians, artists, musicians, psychologists, physicists, surgeons, authors, or schizophrenics, criminals, welfare dependents, attention deficit hyperactive people, obsessive-compulsive patients, drug dealers, and prostitutes. What differentiates our characteristics are our learning experiences.

That makes humans equal in competence and character. Does this statement deny that there may be *individual* differences in brains in the number of neurons and the capacity for learned connections? The answer is no. But since all have the learning capacity to learn a full language, without special systematic training, just by the informal interactions of usual family life, having more neurons than others makes little or no difference. That is because the high criterion of learning capacity that all had to have gave us the capacity to learn any of the types of knowledge that humans have gathered. Whether there are individual differences in learning capacity that make a difference has yet to be shown. What we can conclude now is that all members of the human species (born without damage) are equal in that we are all capable of learning language and are thus capable of cumulative learning with which to learn everything humans learn. The present conception declares human equality, *not just as a politic and humane thing to say and do but, rather, as an actuality.*

Am I saying that there are no *group* differences in behavioral traits? No. Let us take intelligence, and my introduction of a concept of group intelligence. Recognizing that learning creates repertoires that intelligence tests measure, the conception is that a group whose members are in interaction with one another, whose technology and culture and other similarities of experience are shared, will learn to some extent common basic repertoires that characterize the nature of the group. Different groups learn different basic repertoires. Consequently, they display different human natures, different characteristic behavioral traits. And they will test differently. A group of nomadic peoples from the Middle East who are white would no doubt have lower *intelligence-test* scores than a college-educated African American group. The intelligence of those nomadic people would be similar to other groups with similar life experiences, like similar groups in Southeast Asia and Africa and the Philippines, regardless of being of different "races." Groups of college-educated African Americans, with parents and friends of equal backgrounds, would be more similar to white Americans and Asian Americans with the same education, on the other hand, than they would be to Africans considered to be of the same "race" who belonged to a society having a primitive technology. Characteristics of a group, with respect to intelligence-test measures of basic repertoires, depend upon the experience of the group's members. The group's language is such a repertoire, as are basic motor skills, such as driving a car, using drawing utensils, playing sports or a musical instrument. A child in a hunter-gatherer group may learn a complex repertoire involving responding to complex and subtle visual cues in tracking animals; a child in our culture in contrast may learn a complex reading repertoire involving learning responses to complex and subtle visual cues. Those different repertoires will give the two groups different behavioral characteristics. They will learn differently, will solve problems differently, will think differently. When the members of a group share certain basic repertoires, a child born into the group will be exposed to learning that produces such repertoires. When groups are in isolation from one another, what the child experiences is limited to the learning that the repertoires of the one group provide. Even though all human groups are of the same species and have the same native abilities, there can be great differences in the repertoires they acquire through learning, and thus in what they are prepared to experience and learn and how they will behave. *Each such group, thus, will have its own human nature.*

The same principles hold true for any group. In a complex society, there are groups within the larger group. Those groups may in part have different experiences and will thus learn different basic repertoires. Black Americans, Hispanic Americans, Asian Americans, and white Americans have sufficiently different learning that their performances on IQ tests differ. Moreover, their school performances differ because of those same learning differences. The black American experience, for example, beginning with slavery, was generally restricted with respect to learning the repertoires that are measured on intelligence tests. There were laws and practices that systematically excluded blacks from education, skilled occupations, and professions, not to mention access to museums, art galleries, and many other intellectual and cultural experiences. Such circumstances, in addition to limiting scholarly repertoires, constituted severe obstacles to learning the working and studying skills and motivations that are needed for academic achievement. Moreover, the restrictions and limitations on black Americans' experiences and learning did not end with the Civil War. Many black Americans are still restricted in a variety of ways from having the full experience open to white members of the American society. A frequent view is that slaves were freed almost one hundred and fifty years ago, so the restrictions of slavery cannot explain poor life performance now. But deficits in learning persist across generations. A parent who has lacked the experience that would have produced the basic repertoires that are displayed by the white group, who lives in a stunted social environment with inadequate income, cannot be expected to produce for her child a learning history that will give the child the set of basic repertoires of a middle-class white child. The disadvantages once produced, without some intervention, will continue generation after generation. Many black Americans have overcome the disadvantages society has imposed on them since slavery. However, the black-white difference on intellectual tests and academic performance that remains will only end when conditions are changed and the general learning experience for all black American children fully equals that for white American children. *Then all differences in intelligence and in performance will disappear.*

Interestingly, what eludes the notice of those interested in "race" differences in performance as an indication of genetic superiority-inferiority (see Jensen 1969, 2006; Herrnstein and Murray 1994; Rushton 1999), are superior performances of black Americans in certain areas such as football, basketball, baseball, and other sports; music; and other entertainment varieties. That superiority has been dramatized in the movie *White Boys Can't Jump*. When black Americans were allowed into football, as duplicated in other sports they have entered, they became highly successful, in far greater numbers than their representation in the overall population would predict. That was generally explained by the assumption of genetically given physical advantages. Quarterbacking in football was still held out as beyond the capacity of others besides white men because, it was said, quarterbacking demanded intelligence and leadership. Now, however, black Americans have "invaded" this territory with notable success. Coaching, another "holdout" because of the belief that only white Americans can meet the intellectual-leadership demands, is also being set aside as more black coaches are allowed into the field. (The two teams meeting a couple years ago in the Super Bowl had black American coaches.) For those who sniff that football is only a physical sport, let it be noted that success on the professional level generally demands a high level of knowledge, work ethic, learning ability, and emotional-motivational learning. Sure, many professional athletes have undeveloped traditional intellectual skills. But that is true in many occupations that demand great dedication of time and effort beginning early in life. After all, many intellectual professionals are klutzes on any playing field, and some are social klutzes, nerds, as well.

Progressive people, at least, disavow racism and other isms for their projection of terrible actions and social policies. But, while their hearts are in the right place, those same progressive people believe in a fundamentally wrong conception, a conception of human nature that assigns the cause of individuals' behavioral characteristics largely to biology. That makes performance the measure of innate ability, the fundamental and erroneous conception that underlies racism. For what applies to individuals will also apply to groups of individuals and their behavior. It is that conception that must be changed to eradicate racism, genderism, poorism, peopleism, and all the other pernicious isms. The "missing link" of human learning principles, the learning theories of child development and personality and mind, along with the human-creation theory of human evolution, constitute a fundamental conception that humans are equal, not unequal. Equality is what human evolution guaranteed, and understanding that gets rid of the isms of individual and group superiority-inferiority. If we want equality in performance across all groups, as we should, then we have to look for ways to equalize learning experience.

Who are we? Well, for one important thing, we are equal in terms of learning ability.

CULTURE

Everyone knows what culture is. Yes. But it is squiggly knowledge. It jumps around. Sometimes it has one meaning, sometimes another. One meaning gives culture an ethereal quality, in the sense that the term *soul* has—a cause with some mystical properties. Some emphasize the things that a people make. Others emphasize the arts and customs of a people. Religion and other beliefs may be considered. Less likely are government, economic, and military systems to be included, although being warlike, capitalist, or democratic could be. For the most part, what *culture* means is taken for granted; with the addition of the concept of cultural evolution, a biological meaning is added in.

> What are the roots of human culture? No animal comes close to having humans' ability to build on previous discoveries and pass the improvements on. What determines those differences could help us understand how human culture evolved. (Kennedy and Norman 2005, 99)

An example of the type of mixture that prevails, the authors recognize that humans build on past advances and then pass on what they have gained. But they then consider culture to evolve, adding a dollop of the Great Scientific Error. The new paradigm, in contrast, introduces a conception of culture that is systematic, not subject to definitional vagaries.

Personality has been defined as the repertoires the individual has learned. Those repertoires derive from learning experience, but those repertoires also provide the basis for later learning. Culture is very similar, except it applies to groups, not individuals. Culture consists of the repertoires the people in a group have learned as well as the products of those behaviors—an artist has skilled behaviors that produce a work. Not all the repertoires of everyone in the group are part of culture, of course, only those that at least partly are general to a group. In this sense, a language constitutes a behavior common to most of the people in a group. The behaviors of smaller groups within the larger body may also be part of a culture. The medical-pharmacological-biological "industry" is part of the culture of the United States of America. This group does not constitute anywhere near a majority of people in the country, but it is a large group. Those in many occupations have repertoires and make products that most others in the culture do not, repertoires and products that are important parts of the culture.

Not all repertoires that occur within a people are part of their culture. Some are idiosyncratic. A couple may work out some personal things they do in sex, an accountant may have particular ways of handling accounts, and an individual may work out prayers and religious concepts that are not shared, none of which are conveyed to anyone else. Some behaviors are enacted only by the individual or a small number of individuals. Those repertoires are not part of the culture of the group.

Culture Is Learned

I must state emphatically, culture does not evolve; the concept of cultural evolution dispenses error. The principles that operate in cultural development are not principles of evolution. There is no evidence that cultural changes, any at all, have emerged because of genetic changes. Repertoires that compose a culture are learned. *Culture consists of the learned basic repertoires of a group and the products that are produced from the resulting behaviors.* That conception, when joined with the human-selection theory of evolution, yields a new view of culture that carries with it various explanations, such as the time involved in developing a culture and how culture determines the nature of humans.

The genetic view of evolution proposes quick cultural development as a result of new genetic mutations. That view thus holds that language appeared only one hundred thousand years ago or less, in a fully formed character, due to mutation within modern *Homo sapiens*. Earlier hominid ancestors are not considered to have had language.

In the new conception, culture was gained over a period of millions of years. Lucy and her people built walls and used crude stone tools and were learning additional repertoires that expanded the culture of later australopithecines and *Homo erectus* creatures. Somewhere along the line, a rudimentary language repertoire began to grow through learning. It took millions of years for the features of a full language to be added, bit by bit, to form modern language in its many variations. That progression also involved the human-selection process that produced bigger brains. But no language gene produced language. It was learned over thousands of generations of hominoid creatures. That goes for all human culture, all learned over an extended period that becomes progressively more brief because as culture is added, human learning ability increases.

This is central to understand. Hominins from their beginning were learning and cumulating that learning, storing their behavioral achievements that became parts of future generations' cultures and the foundation for more group learning.

NO CULTURE, NO HUMAN BEING

Let us imagine an adult human, with the marvelous structure of the human: fine and varied senses, upright locomotion, adept hands, a structure capable of many extraordinary sensory-motor coordinated actions, plus a huge brain and an extensively developed nervous system. This specimen possesses the unique learning potentiality of humans. This specimen is a member of the human species.

However, let us say this marvelous animal has been raised like an isolated animal, with the constituents for good physical health, with opportunity to move freely and explore a varied habitat, but without contact with another human being. Let us say this being, from a child *Homo sapiens*, has been taken care of by mechanical-looking robots. This being has been cut off from all the learning that had taken place during human existence, even learning how to make a stone hand ax or to say a few words, or to respond in an understanding way to the words of others, or to have a positive emotional response to other humans.

Placed into a world situation and lacking the products of human learning, this member of the human species would be like a grown-up baby. He or she would have no hunting-and-gathering skills and would be dependent on the robots for food. This marvelous animal would lack the many skills common to any and every culture. He or she would have no language, would not respond to verbal or gestural directions, would not hold or follow any values, ethics, laws, mores, manners, aesthetics, or human expectations. This being would display no reasoning, no planning, no projection of the future, no past learning of other hominins, no principles of justice and morals, no tool-making skills, no altruism, no aesthetics, no music, no number concepts, no creativity, no art, no ambition, no responsibility, and no interests, standards, self-concept, ego, intelligence, aptitudes, or creativity—none of the behavioral characteristics that generally provide the definition of humanness.

Moreover, this creature would not display the usual human's potentiality for learning. That would be the crowning deficit of a central species characteristic. The loss would come from the deficit of the necessary basic repertoires central to human learning. It would be impossible to teach a mature creature of this kind, as we do with our own children, to drive a car, to use a telephone, to perform a dance step or play football, to sing a song, to display good manners, to use a toilet, to put together the parts of a bicycle, to write a few letters of the alphabet, to read, to count, to do the dishes, to appreciate art, to want to do well in school, to compose a letter, or to like music. Our usual instructional means would be entirely useless.

The creature would look like a human, would walk upright, would be able to grasp things, would display sexual interest and behavior of a solitary sort, and would eat a wide variety of foods. But we do not even know whether a female animal of this species would display such a basic thing as maternal behavior if it had been raised in isolation and were impregnated artificially. Harry Harlow (see Harlow and Harlow 1966) raised monkeys in such a situation, and the females, even after having experience with other monkeys, never functioned well as mothers or as sexual mates. Probably the human creature with no human experience would not even display fundamental sanitary behaviors.

Let us say that this human creature was introduced to a group of like creatures, each raised in the same manner. They would be alien to each other. They would not interact. Harlow also found that monkeys raised in isolation later do not learn customary emotional and social responses to other monkeys. That suggests that a human raised in isolation would lack even the social and emotional behaviors we consider part of human nature. Although a marvelous human animal, a *Homo sapiens*, he or she would not be a human being.

A popular adage attributed to Hillary Clinton is that "it takes a village to raise a child." For the reasons developed herein, let me amend that to say "it takes the learning from our human ancestors, from the prehistoric past to the present, to raise a child." The marvelous animal, via evolution—natural and human selection—has inherited the marvelous human biological structure. That structure gives the human infant the *potentiality* for becoming a human being.

But that is what the marvelous learning animal *Homo sapiens* represents, only a being with extraordinary potentiality.

We are what our genes give us. But we also are what we learn and what our forebears learned. If anything happened that removed that vast past learning from all of us, the human species would be thrown back to the beginning. Our human nature would be gone. We would be bodies with no humanness, no human learning. It would take even a fully evolved group of *Homo sapiens* thousands or millions of years to develop the basic repertoires all human beings display today. We, each of us, from native tribes living in the Amazon to those with doctoral degrees, have inherited the learning that began with our prehistoric ancestors. We haven't realized the learning gifts we have received from our distant lines of hominin ancestry that make us the marvelous learning animal.

This view has wide implications. For example, some states in the United States of America are legislating that human life begins at conception. Underlying that position is a belief that at conception the God-given soul is imparted to the embryo. Abortion thus murders a young human. The present view is that a beginning embryo, rather, is a simple two-celled organism. Its abortion is not murder. Only after some months of gestation will the fetus become capable of life outside the mother and capable of learning to be a human being. Science rather than religion should be the basis for abortion's legality.

Like Personality, Culture Can Be Abnormal Too

Mental health sciences search for abnormal behavioral repertoires and for methods of diagnosis, treatment, and prevention of behavior disorders. There are also abnormal cultural repertoires, in the sense that they are not good for the people that practice them. Such cultural repertoires should be described and diagnosed so that they too can be treated, changed.

Let us say one cultural group holds fundamental religious beliefs very strongly and another culture emphasizes science more. As a consequence of this difference, important life situations—like global warming—will be responded to differently by the two cultures. As another example, one culture has beliefs about gender differences that result in restricting women to the home, without education and career. Another culture views the sexes as equal in cognitive ability and women feature prominently in education and careers. The productivity of the two cultures should differ in any sphere that demands educated and skilled people, one culture having many more of those people than the other.

Being warlike, also—as in valuing military exploits, spending excessively on the military, having military opinion in political and economic matters—may be an abnormal feature of a culture, as it was in Nazi Germany.

There has been a belief in cultural relativity, studying cultural differences without considering cultural characteristics good or bad. The present conception calls for study of different cultures in terms of advantageous and disadvantageous characteristics, theirs and other peoples', as a means of gaining knowledge of how to change and improve. "According to this older worldview . . . [the individual's] job is to struggle daily to strengthen the good and resist the evil [within]" (Brooks 2012, A23). Brooks also describes a psychology view that humans gained violent tendencies from human evolution. In the present view, for cultures as well as individuals, there is no good or evil that is inherited, no genes or brain developments of that sort, only what has been learned.

INFINITE LEARNING AND THE END OF HUMAN EVOLUTION

A popular belief is that *H. sapiens* will continue to evolve, to gain greater intelligence and higher moral fiber, to be less violent, to become finer beings in various ways—all hardwired into humans by the evolution yet to come. Evolution is seen as the way that humans will advance past the selfish, violent, stupid ways that have been displayed throughout history. Many also believe that humans have never progressed, that humans have the same weaknesses and errors in living that have always been present. That belief comes from the notion that humans form a one-time creation of some kind.

Rather, I am going to present a view of human progress. But first let me state that progress will not occur through evolution. Humans have reached the end of their brain evolution; they will never through evolution become more peaceful, more socially just, more kindly to others, better parents and spouses, more honest, more intelligent, and so on. Humans will never gain a finer nature through changing biologically. What is the basis for concluding that? Let us see.

Evolution first produced a creature that made crude stone tools and weapons and hunted game and gathered plant food. After that came five million years or so of evolution to produce humans, by human selection, about one or two hundred thousand years ago. Although the products of those early humans were vastly superior to those of hominids a couple million years before, their living conditions and the level of advancement were very primitive by our standards. What does that say about the genetic causes of human behavior?

> The conditions which, in an evolutionary sense, selected for man's biological superiority in brain structure occurred long ago [the paleoanthropology estimate was then fifty thousand years and now ranges up to four times that many]. . . . This means that man's biological structure has remained unchanged during a period when his intelligence—his [learned] skills of various kinds—has changed fantastically. (Staats 1971a, 50)

> We have no evidence that the modal form of human bodies or brains has changed at all in the past one hundred thousand years. . . . Everything we have accomplished . . . from the origin of agriculture to the Sears Building [now called the Willis Tower] in Chicago, the entire panoply of human civilization . . . has been built upon the capacities of an unaltered brain. (Gould 1996, 220)

That indeed removes genetics as an explanation of human culture. So we must ask, how could the brain that produced the Stone Age learning and behavior of one hundred thousand years ago be the same one that accomplishes the wondrous and wondrously complex thinking, creativity, and behavior that takes place today? This must be emphasized as fundamental. The difference between the behavior of the human of one hundred thousand years ago and the behavior, culture, of humans today is enormous. *That advancement, all of it, has taken place through learning and not biological evolution.*

So great advancement in learned cultures occurred without any addition of neuron capacity of the brain and consequent increase of head size. What, however, does this have to do with the termination of human evolution of learning and other abilities? The answer is very straightforward: the great cultural advances took place without any genetic change; *human brain evolution by human selection is over. Human evolution has stopped.* The great change in human behavior through learning, with no biological change, suggests that *Homo sapiens* as a species is capable of infinite learning. How could that be possible?

Across Space and Infinite Learning

The underlying answer is yes, learning over the human group has become limitless, in a manner that takes away the scenario for additional evolution by human selection. That can be seen by the utterly fantastic cultural advancement since the advent of the *Homo sapiens* species. Several prominent reasons make clear why human learning has become infinite.

Remember, lower animals influence others of their kind only through personal contact, as chimpanzees do in grooming and sex activity and when a mother shows the young chimp how to fish for ants. So there is some learning induced by another's actions. Animals that hunt in packs, such as lions and wolves, also may be influenced by seeing where their cohorts are and what they are doing. Dolphins hunting in packs may communicate among each other much more complexly and extensively via clicks and other sounds in a manner that allows them to adjust their behaviors. And apes and other animals may also communicate via vocal calls. But the ability of animals to affect one another across space is very limited; this is entirely different than for humans.

With the progressive development of language and language capacity, humans acquired an instrument for complex communication across space. At first the range was limited by the carrying power of the human voice. At some point, signal systems were devised, such as the drums used in Africa or the smoke signals used by Native Americans. Later systems include writing, telegraph, telephone, radio, television, newspaper, movies, e-mail, Internet, and iPhones® and iPads® and such. Each development of an instrument for increasing the ability to communicate across space has increased human learning. Experiences can be transmitted via the various systems by which individuals far removed from the source can learn. So learning from direct experience has been multiplied greatly in the ways that the learning of myriad separate individuals is brought together. This constitutes a marvel of human learning completely unknown in the rest of the animal kingdom.

Across Time and Infinite Learning

We can see the limitations on lower animals' learning that the time barrier imposes. Each elk, raccoon, pet dog, rabbit, or eagle has about the same skills as the generation that went before. Even if by some unusual chance an animal learns some skill that is uncommon, the progeny of this animal may not display the skill. For the "learned" animal may be unable to teach the skill. If I teach a pet dog to sit on command and that dog has pups, the pups will not display that behavior. There are a few learned skills that are passed from generation to generation—such as the chimpanzee using a twig to fish for termites or banging one rock on another with a nut in between. We have no indication, despite the great interest in teaching language and other cognitive skills to apes and dolphins, that there is transmission across generations that will sustain the skill and advance it. Lower animal learning appears to be finite, pretty much static.

Human learning, on the other hand, presents a picture that is so different as to be beyond compare. Our history reveals in just several thousand years how human knowledge has grown generation by generation in an astounding manner. That has happened in thousands of specialty areas. And it has happened for the common person. Ordinary educated people today know more about the world than the great Greek philosophers two thousand years ago. Human language and means for recording language enable humans to leap time barriers. "Great Books" programs can serve a historic role, but for intellectual advancement, modern great works are more valuable.

> Man develops *intelligence* as he learns in an historical or phylogenetic sense. That is, we may look upon the history of man as a species as the acquisition of a great variety of behavioral skills through the principles of learning. The original acquisition of any skill may be seen as a laborious, time-consuming, imprecise process—like trial-and-error learning. . . . One can see, historically, this slow and uncertain development in any area of complex behavioral development—music, athletics, religion, business rules, government, language, science, and so on. (Staats 1983, 51)

> As one example, in the author's lifetime the athletic skill of high-jumping has gone through distinct developments. At one time the skill consisted of the "scissors" jump. Later the jump was performed by a side roll. From there the variations in the behaviors tried produced a more effective jump, one where the bar was crossed as the individual was face down. The winner of the 1968 Olympic Games, on the other hand, trotted up to the bar, swung around and catapulted himself over backwards, going over directly with his back to the bar. . . . (Staats 1971a, 291)

> The major point here is that man develops his skills phylogenetically—as a species or group, over the generations. The principles of learning are involved for the species development as for the individual. . . . At any point, however, what man has so laboriously learned he passes on to the individual child. The child then receives in his learning the learning of the human race—as provided by its representatives in the form of parents, family, peers, teachers, and so on. The child learns in his lifetime . . . [what] has taken his people . . . from the beginning [to learn].

> Of course, the child does not have to go through the trial-and-error discovery process. He can be directly trained to the skill. It may have taken centuries before a germ theory of disease was developed by man, as [an] example. But it

composes part of the general language skills of the present generation of our children—imparted with relative ease in the language learning of the child. Thus, the intelligence of the individual advances as does the intelligence of the "phylum." (Staats 1971a, 294–95)

This conception, presented some forty years ago, stated clearly that human learning accrues across generations, thus establishing another reason why human learning is without limit. Human learning did not begin one or two hundred thousand years ago. The learning had to be accruing for five or so million years. Those early primitive creatures were, laboriously slowly, learning language repertoires, especially, and complex emotional and motor repertoires too. The learning of modern humans' abilities did not begin with the advent of the *H. sapiens* species. When *H. sapiens* emerged, the species already had a large part of the learning task accomplished. Actually, if we could establish a dimension of difficulty of learning, we might find that learning a full language from nothing was a greater task than that involved between the Wright brothers' airplane flight and landing an astronaut on the moon. On that basis, the *Homo sapiens* people of a couple hundred thousand years ago already had traversed much of the learning that rocketing to the moon rested on. There is a cute commercial on TV these days depicting a "cave man" living in contemporary society where people think of him as simpleminded, and he says, "Oh yeah, didn't we invent fire and the wheel and other things basic to all of you?" Constructed for humor, the ad is actually prescient.

Infinite Learning by Specialization

A powerful feature of human learning that also makes it limitless takes place via a division of labor. One could cite specialization also in lower animals. For example, ants are adaptive in part because there are groups that do different things. Reproduction is allocated to the queen, so not all ants have to shoulder that burden. Other ants are workers and are responsible for the collection of food. And there are soldier ants that guard the nest.

Despite such examples, there is very little development of specialization in the animal kingdom, at least in comparison to humans. We have thousands of specialists, ranging from actors to zoologists, who have knowledge and skills in great elaboration that are not possessed by the ordinary person. Mathematics and science are just such developments, with precursors in common behavior. Most usually, a special knowledge will begin with someone learning beyond what is commonly known. As the knowledge or skill in the area grows, it reaches a point where special study is necessary if it is to be encompassed. There was a time after science began when it was still possible for one person to learn the knowledge existing in multiple areas of science. But the knowledge in the various areas of science has grown so that only relatively few members in the specialty area can learn it. Commonly, such specialized knowledge grows so large that study through the doctoral level (a work of twenty-plus years) provides only an entrée to the subject matter.

Unlike with ants, human specialization is *learned*. That means it can develop in any direction and is available for every demand. Human knowledge is infinite in extension through having different individuals responsible for studying particular areas in great detail. The specialized learning that results can then be abstracted in part and transmitted to the multitudes of nonspecialists; the result is general cultural elevation of the level of general knowledge. Or it can be transmitted in full in teaching others to be specialists. Human learning is unique in having no limits in part because not everyone in the group needs to learn everything that the group has acquired through generational cumulative learning. Yet the entire group may profit from that specialized knowledge. As already indicated, we long ago passed beyond human selection. No longer can the ability to learn what past generations have learned be the criterion for survival and reproduction. No child in today's advanced societies can learn all that is known, which would demand a brain so large it is unimaginable, surely biologically impossible.

Infinite Learning by Compacting and Archiving Knowledge

Human knowledge has been said to have doubled every fifteen years or so since 1910. That means the amount of knowledge today is at least thirty-two times more than it was early in the twentieth century. In recorded history, the gain in knowledge has been astronomical. I have suggested that this has been possible via various means. An additional mechanism can be considered, that of "compacting" elements of knowledge after they have been discovered in less efficient form. Customarily the discovery and beginning formulation of knowledge lacks the precision that will be added later.

One way of simplifying knowledge is through the creation of a theory. For example, Galileo's laws of falling bodies, Kepler's laws of planetary bodies, and the laws of the tides once existed as isolated pieces of knowledge, each stated in its own special language and each requiring a separate learning task. Newton later formulated the principle of gravitation that analyzed the three elements of knowledge in terms of the same principles, markedly simplifying the task of learning not only these three types of phenomena but many more as well.

Theory represents one way of compacting knowledge and making it simpler and easier to learn. Recording and archiving of knowledge also plays a great role. As the body of knowledge grows too great for even the specialist in an area of study, most of the knowledge can be placed in books, recordings, and computers where it can be accessed as needed. The compacting of knowledge results in an increase in the amount of human learning capacity.

EVOLUTION OF THE BRAIN IS OVER

In conclusion, many speculate about what humans will eventually become through evolution. What powers of the mind will emerge? How big will the human brain and head become? Such popular questions abound; there are pictures of more evolved creatures, and there are movies about them. I hate to be a wet blanket. But let us consider human evolution. What drove it after the physical body was complete were the continued, successive advances largely in language. However, when humans' brains generally attained the capacity for learning a full language, the conditions for further evolution were withdrawn. For that capacity made human learning infinite in nature. From that point on, there could be no evolutionary reward for a larger brain. Such a brain would demand additional upkeep, and it would make birth more difficult, both of which militate against survival and reproduction, in addition to being a waste. Further advances in the human "mind" will not depend on evolution. Those advances will come from learning and from finding ways of increasing the communication and storing of that learning. *Human evolution for gaining human-nature development is over.*

Who we are as a biological species is set.

PROGRESS

If human evolution of mind is over, does that mean there will be no human progress? There has been a continuing difference of opinion about progress. Those who believe humans were created by God believe that humans will be the same forever; what appear to be changes are superficial, dictated by changing external circumstances. Those who hold that human nature is laid down in evolved genes believe human nature is fixed until further mutation occurs. The fact that a nuclear weapon could kill a million people and that Stone Age humans killed only one person at a time with a club demonstrates only a difference in weaponry, according to these views. It shows only that humans basically have always been violent and brutal and always will be.

The question of progress also pertains to the evolution of species. The expression "lower animals" has been rejected for suggesting that *H. sapiens* represents a higher species. As Gould (1996) points out, Darwin himself was inconsistent on this point, which probably engendered the continuing controversy. Gould's position is that Darwin was convinced that evolution, natural selection, entailed only local adaptations that were retained by survival and reproduction advantage. As local adaptations they could move in a random direction, forward or backward. There are animals, for example, that have lost organs their predecessor species had, such as cave-dwelling species that have lost organs for vision. Rather than consider that as degeneration in evolution, Gould takes the position that progress does not really occur in evolution, that Darwin's inconsistency on the matter represented an attempt to placate his society, which believed in progress.

Is evolution a progressive process? I agree with Gould in the short-term and specific case, as in the loss of senses in some troglodytic animals, but not overall, not generally, and not in the long term. Let us look at the principles of learning in this regard. Simple one-celled animals exhibit the most minimal of behavioral flexibility as the result of experience. From this example of zero there has clearly been a huge increase as the learning ability of animal species grew, terminating with *H. sapiens*. If we can label that change, unquestionably an increase, as progressive, then evolution certainly has been progressive with respect to learning. Perhaps there were setbacks of a local nature, where a particular species lost some learning ability through a shrinking nervous system. But long-term, and considering all species, learning ability has advanced from nothing to marvelous learning ability.

This consideration involves the physical mechanisms by which learning occurs. But how about what is learned? How about human societies? Utopians have claimed that humans will ultimately develop a wondrous society in which all live in happy conditions. Karl Marx viewed society as determined by economic systems, with progression from feudalism, through capitalism, to communism. He and later communists believed also that in that process human nature shows advancement. The historian Arnold Toynbee (1947) posited that societies moved through *stages* of human development, as do individual humans in the life span. Western civilization in the womb began with the Greeks; the Roman empire was pregnancy; the "Dark Ages" constituted birth; the Middle Ages constituted childhood; and since then have been the prime years. Talcott Parsons and Kenneth Clark saw sociocultural evolution as biological evolution that proceeds "by variation and differentiation, from simple to progressively more complex forms" (1966, 2). Such conceptions at least implicitly posit change or progress in societies over time. Steven Pinker (2011) gives another partial view of progress in considering that violence has been in long-term decrease (which actually represents a sharp turnaround from the nature views of his 1994 and 1999 books). But none of the positions presents a conception that explains why progress should or should not occur.

In my view, there is no question that humans have progressed in being able to live together in increasingly large groups, from bands of early hominins to villages, towns, cities, and countries. Within those increasingly large groups, intragroup violence has lessened and there has been growth of benign relationships. Practices such as human sacrifice, torture, and public hangings are no longer sanctioned, and brutality has lessened. Some years ago, Sweden passed a law barring corporal punishment of children by parents, and this has been followed in other societies. Laws have progressively lessened the workweek, and children have been protected from exploitation. Employers must protect their workers more from the dangers of the workplace. Many advanced countries guarantee medical care and retirement pay plans, unemployment insurance, parental leaves, and the like. There are civic laws to protect individuals within a society, as well as traffic laws, domestic laws, laws to protect workers, and gun laws. Nations still exploit other nations, but not so openly as in earlier times. Those are all very modern progressions, if not common to all societies.

No question there has been progression in all the major societal institutions, from education to science to government to healthcare—at least in advanced societies. The reason that we no longer have serfs and slaves is because these social systems, although yielding reinforcement for some, did so by disadvantaging others. Those who profit from a social system will support it and struggle against change. Those who are not well rewarded will not struggle to protect the social system. In fact, they will be motivated to bring it down and to create a social system that offers benign features toward more people, including themselves. Even in my own history I have seen this type of revolutionary effort occur in various countries. That dynamic has been the force that has produced social change in the past, and it will continue to operate. Such change does not occur rapidly because societies are very complex; it is not easy to detect the drawbacks of a society, and it is not easy to see what progressive changes will remove those drawbacks.

Cumulative learning when starting from scratch occurs slowly. And the early modern humans of one hundred thousand years ago, while clearly possessing much prior learning, were still early in the cumulative-learning process. Beginning as animals, it is reasonable that the earlier hominids had no more social organization and no better ways of social interaction than cousin primates. But as they learned language over many generations, they could profit from circumstances that occurred in the past as well as those occurring in the present. They could progressively learn how to function effectively as a group. When their technology had advanced sufficiently, they were able to live in larger groups and in permanent settlements. With the advent of agriculture, larger populations formed in certain advantaged locations. Larger interacting populations made it possible for the learning of more people to add together. Further advances led to larger groups until cities and states emerged. Such developments opened new possibilities for learning of all kinds. They also created problems concerning the best ways for people to live together in large groups.

Advancement characterizes learning. But in complex developments the path may wander, taking wrong turns as well as right ones. That characteristic gave rise to the term *trial-and-error learning* to describe performance in laboratory study of individual subjects. But group human behavior is light-years more complex than such laboratory study. In complex social situations it typically is not clear how best to deal with the distribution of the wanted things of life, in a manner that satisfies all, and what kind of social organization will facilitate that distribution in the best way. So human history is one of learning how to live together with mutual benefit, always in situations of great complexity. We have to assume that *H. sapiens* had no knowledge of the ways by which to maximize social living in the large groupings that formed. Humans have been and are on a long journey of learning how best to behave in large groups and how to arrange the many complexities involved, including the allocation of the many important lands, mineral resources, commodities, objects, education, health resources, and many other things to the peoples of the world.

We have to remember that the basic principles of animal learning are those of classical conditioning and operant conditioning. That means that individuals are governed by what will elicit positive and negative emotions and be reinforcing for them. Those principles by themselves set people into competitive situations with behaviors to match. With no other effective principles than "dog eat dog," the strong impose their wants on the weak. But even then there are learning-behavioral forces working to mitigate simple selfishness, for the weak also possess personal qualities that are desirable to the strong. The weaker woman has strong sources as a potential good sex partner and mate. The weaker male as a slave performs important functions for the owner that can vary in effectiveness depending on the slaves' feelings. People who are mistreated and exploited experience hate and can respond in ways that bring negative happenings to the strong. Emotion and reward/punishment conditions for the members of a society can be met variously by different ways of social living. But how those conditions can be maximized for living in groups has to be learned as those groups became larger and more complex.

Some of the past steps in the political, social, and economic laws and actions that have progressed through human history have been noted. There is no reason to believe that all the learning important to good social living has been attained already. It is too evident that enormous problems remain. Many people suffer wars, illness, malnutrition, ignorance, and misery while the favored minority live in relative luxury. It is very clear that benign and efficient ways of living that include all the peoples of the world have not yet been learned. At this late date in *Homo sapiens'* journey, individual and group struggles to obtain the good things of life, without consideration of the needs of others, still take precedence over an organization to share. Wars and preparations for war involve unimaginable waste, when ways of living would be possible that would yield abundance for everyone without strife, destruction, and killing. If just the money spent upon competition for the good things of life—military personnel, weapons, repair for war destruction, secret services, propaganda, political organizations, police, and other governmental agencies—were spent instead on raising the material and social conditions for all, then struggles to obtain such things selfishly would not be needed. Benign conditions would reduce the motivations for rebellion, insurgency, terrorism, and passive resistance.

Humans will move inexorably in the direction of complex social behavior that maximizes reinforcement for all. But a social system composed of many individuals is very complex, and there will be contending individuals and groups with respect to supporting or changing a system. Force may be used to maintain a system and force may be used to change it. It is very difficult to establish large and complex groups with complex social-political-economic systems in a way that maximizes reinforcement. A social system that is reinforcing for the feudal lord is not for the serf. It could have been predicted from this that feudalism would ultimately be replaced, since there are more serfs than lords. But there are various influences on individuals' behavior besides material reinforcers. There are religious experiences, communication media, and cultural customs, as well as conceptual beliefs. People with personal advantage to be gained will to the extent of their resources attempt to persuade others to continue the ways that yield that advantage. A newspaper owner who gains advantages from low corporate taxes may not choose news reporters and columnists likely to call for raising those taxes. That is one example of a mechanism that resists change. Social experiences are very complex for individuals and groups. So the movement forward of human social behavior takes place along complex and inconsistent paths whose progress can only be seen in historical perspective.

My conception of human evolution indicated the complexity involved in human selection for learning capacity, which is why it took millions of years for that evolution to occur. That is true also of the progress of humans in terms of social behavior. The lines of progress may be obscure. It may take a great deal of time for changes to take place in complex societies. But, like the laws of human evolution, the learning-behavioral laws are invariant and always in force. In this sense, Mother Nature will not be denied. Humans will progress.

Chance, Foreordination, Progress, and Utopia

"We tackle the greatest of all evolutionary questions about human existence— how, when, and why did we emerge on the tree of life and were we meant to arise, or are we only lucky to be here . . . ?" (Gould 1996, 8). Gould is addressing the question of whether humans were foreordained, as in divine creation, or whether humans arose through natural selection of chance variations from mutations and sexual reproduction. In that vein let me raise another thought-provoking question: Did the advent of life in its simplest form, given the abundant and varied environment of Earth, guarantee that one day the *H. sapiens* species would emerge? Was it guaranteed once single-cell life emerged that multicelled organisms would evolve that had an inchoate flexibility of response to environmental conditions that can be counted as primitive learning? And once learning had been established in organisms as advanced as mammals, was it inevitable that progress would occur in learning capacity until reaching its ultimate in the human being? And when that learning capacity became general to the human species, did that guarantee that humans would learn fantastically complex abilities and performances? Likewise, is it a foregone conclusion that there will be a time when there is no race or other prejudice or mistreatment, no war, no exploitation of people by other people, no brutality, equitable abundance for all, and general kindness and compassion? Utopia?

I believe these things. That belief derives from the principles and concepts that have been presented in this conception of human nature. My theory of evolution indicated that once the hominin line was established, it was inevitable that the end point of human evolution would be *H. sapiens*. For once learned repertoires became criteria for evolutionary selection, the process had to proceed to the point where learning was maximal in balance with the costs involved. The inevitability of the genesis of full human beings was guaranteed billions of years before that, when learning in its crude forms first appeared in simple animals. For learning enhances survival and reproduction, and that provided the guarantee that learning ability was going to continue to increase to its maximum. This reasoning also says that if there is another planet in the universe with the physical conditions of Earth, then that planet will sooner or later develop creatures like *H. sapiens*.

The same inevitability holds for the development of human nature. We consider a culture, like that of the Aztecs, and wonder about their practice of human sacrifice. How could they have done such horrible things to other human beings? Or we look at cannibals and think they must have been a qualitatively different species with a different, degenerate nature. Or we see cultures in which there is gender inequality to such an extent that it seems criminal to us. But if we go back into our own history, we find practices that would now be eschewed as evil. In the history of our group there is human sacrifice, killing for spectacle, mistreatment of the mentally ill, great gender inequality, enslavement and exploitation of conquered peoples, child labor, racism, caste systems, and many other deplorable social practices. Nor have such practices been limited to one particular culture.

What our history tells us is that social practices change, progress. What was considered usual and moral and accepted at one time may be shunned as brutal at a later time of development. The fact that a group practices something we consider reprehensible does not mean that the practice will exist forever, due to the immutability of the biological nature of the group. The Spanish *conquistadores* met the Aztecs about five hundred years ago. By the present time, had the Aztecs been left to their own devices, their practices of human sacrifice most surely would have changed. After all, in a matter of fifteen thousand years or so they learned how to construct a remarkably advanced technological, artistic, and organized society. And certainly European social practices have progressed greatly in the past half millennium. We would have to expect the same for the Aztecs or any other group with sufficient resources to support large populations. One thing we know for sure is that positive reinforcement strengthens behavior and negative reinforcement weakens it. That principle applies to all people even though there may be a very complicated mixture of reinforcements acting on different individuals, so they retain practices negative for others. Nevertheless, over the long haul, we can expect that in a society the learning-behavior principles will inevitably have their way; developments will occur such that more people will be positively reinforced and fewer people negatively reinforced. Such developments result in better conditions for all; the opposite developments result in worse conditions for all. We can see that in gender equities. Even as late as Roman times, society gave men roles as rulers of the household, holding control over wives' and children's lives. Our society has advanced far past that in equity because doing so brought greater general happiness, reinforcing changed behavior.

There are certain variables that hasten group learning. For example, when environmental circumstances enable living in large groups, having members of the group with spare time, and having access to resources—including trading with other large groups with whom knowledge is exchanged—those conditions will hasten group learning. The opposite conditions will foster much less learning in the group. Much human learning remains to be gained before any general utopian culture can emerge. But the many people in the world now are capable of intimate communication. And the other factors needed for acceleration in learning exist. The industrial nations of the world already are utopian in comparison to the grossly underdeveloped countries and in comparison to the past societies in the same geography. Utopia beyond that is to be expected, and for everyone, not just for the dominant countries or the dominant people in a country.

The long term of human progress is very bullish. But the short term appears pessimistic. Because of ignorance about more effective social living—*prominently including ignorance about human behavior and human nature*—solvable problems remain in contention. The planet continues to be befouled at an advancing rate with industrial and political leaders unable to learn from the scientists' call for changes in our way of living, such that the inevitable global warming will ultimately yield catastrophic conditions. The problems involved cannot be resolved yet because large-scale changes of an economic nature are demanded, changes that would lessen the advantages of powerful individuals and groups. The needed changes thus are effectively resisted. It appears the world population will first have to suffer catastrophic consequences before it will clearly become informed that changes must be made for the common good.

Unfortunately, the impending catastrophe may interrupt for a very long time the benign advances that cumulative learning would otherwise produce. Barring the ultimate catastrophe, however, humans will continue the process of learning repertoires that serve as the foundation for learning additional repertoires such that conditions are improved for larger and larger groupings of humans until this extends worldwide to all. There will be never-ending progress; the time line is infinite because human learning is infinite. The learning capacity of our species guarantees that the species will continue to make progress in its behavior of the broadest social kind, just as for individual behavior. *Progress is a central part of who we are.*

Part 6

A HUMAN PARADIGM

Chapter 10

THE TIME HAS COME, THE WALRUS SAID, TO TALK OF MANY THINGS

T his new paradigm of humanness—unlike the biologically centered Great Scientific Error paradigm—places the center of explanation in learning. This change carries with it a very different view and calls for general change— of basic science, of human science, and of applied science and practice.

Biological science gives us a model of science development. Darwin's theory and Mendel's genetics melded to provide an important part of the foundation for a group of sciences ranging from basics like cell biology to paleoanthropology. This science grouping has a shared, consensual basis. That coherence is not just verbal chicanery, the widespread science areas considered, and their phenomena, are really interlinked, unified in their endeavor and in their contributions to each other. They constitute in appearance and in fact a connected mass of knowledge power. This unification is based in science development and it projects science advances.

From the field of child development to political science, and all that's in between, there is the attempt to draw explanatory concepts and principles from the biological sciences. Despite the unimpeachable demonstration that all kinds of human behavior—whose study is broken into various fields of science— heavily involve learning, a unified group of learning sciences does not exist. Moreover, there is no impetus, no interest, in working toward such a development. There is no realization that the great facts of basic and human learning are not being used to understand and deal with human behavior. The lack of science development is as clear as it is inappropriate, given the great effect that learning has on human behavior. In concluding this work, we should address that.

Like the Walrus says in *Through the Looking-Glass*, the new paradigm that has been depicted has many things to talk about—not of ships and sails and sealing wax, but of science and other things.

OF BASIC SCIENCES

Let's talk, first, about basic sciences needed to provide the foundation for understanding and dealing with human behavior and human nature.

Biology and Learning

The nature-versus-nurture opposition continues despite its obvious lack of sense. As many have stated, both biology and environment are involved in determining human behavior. The question that remains, however, is in what way are they both involved? The new paradigm says this has to be confronted, not just swept under the rug. To really achieve unity in action, how biology and learning are interrelated in the explanation of behavior must be set forth. That is necessary for a science unification to proceed in this area.

Evolution and Biology of Learning

Learning is biological. It involves the stimulation of sensory organs, the conduction of nerve impulses to the brain, the action of brain neurons, and the conduction of nerve impulses to response organs, both emotional and motor organs. The process in terms of learning must be recorded in the brain. Learning ability determined by the extensiveness of the learning mechanisms has evolved biologically, along a dimension from none to the complexity of the human species. What are the biological mechanisms that progressively make learning possible?

The way that process took place has been studied already. Jelly fish, for example, have a simple nervous system called a *nerve net* that enables the animal to make general, overall, movements. Nervous systems advance in various animals, like insects, enabling the behaviors of ants and cockroaches. Vertebrate animals have a more complex nervous system that makes possible more complex behaviors. The neural development of mammals provides the basis for the two types of conditioning. Advancement also in the means of sensing the environment, of responding, and of connecting the two, which is basic in learning advancement, has also been of science concern, as is the biological mechanism for recording the learning

In general, the learning-biology relationship needs to be systematically studied in intimate detail as the basic end of *the learning sciences*. What occurs that produces learning at the cellular level needs to be known. Learning as it occurs in simple organisms up through and to mammals must be systematically specified. Much of this has been done, for the biology of learning has been a topic of interest for some time. Horridge (1962), for example, showed that a headless cockroach could be conditioned to withdraw a leg. Kandel and Tauc (1965) also demonstrated classical conditioning with single nerves of sea slugs. An evolutionary perspective suggests itself in such studies; there is a beginning process of learning already evident in primitive animals.

There has also been interest at the biological level in what distinguishes the learning of humans from the learning of lower animals. A common belief has it that the learning ability of advanced species increases because basic nerve action becomes better, or faster. That doesn't appear to be the case—"all species . . . formed simple associations at about the same rate" (Diamond and Hall 1969, 252)—as is shown in research.

That means that rats and pigeons learn simple things at about the same rate as humans. So where does the great increase in human learning ability over other primates that operate according to the same learning principles come from? The answer given here is the learning of repertoires by humans that makes possible cumulative learning. That, of course, leads to another question for biology: Are there special biological structures that make cumulative learning possible? Or is it the huge number of learned connections in the human nervous system that provides cumulative-learning ability?

The point is that basic biological science areas need to be established, by name, as the basic end of *the learning sciences*.

The Brain and Learning

This call for new types of interactive research occurs in other areas as well. The development of brain-imagery instruments has made it possible to see brain structure and functioning. The brains of normal people and those with behavior disorders, for example, can be compared structurally and in terms of how they react to different conditions. Brain imagery has been used in the attempt to show that brain conditions are the cause of behavior.

That approach stems from the Great Scientific Error paradigm, which does not recognize how learning determines human behavior. Interestingly, however, brain imagery was also the means by which the concept of "brain plasticity" arose. Research has shown that learning experiences change the brain. That suggests the brain is not prewired and static, but rather is open to change by experience, in a way that agrees with and supports the present paradigm.

Furthermore, analyzed in this new paradigm, brain imagery provides a potential methodology for advancing a learning-biology interaction that should add to the joining of nature and nurture. This calls for a new type of brain-imagery research. The focus should be on the full picture, that is, on how learning produces changes in the brain and how those changes in the brain produce changes in behavior. A whole field should open to study how learning affects the brain. New questions of learning and brain anatomy and function can be expected. For example, studies show that the reading training that produces reading ability does so by changing the brain; adding a reading repertoire affects the brain. The brain development is not evident in those who do not read. We need a new focus on how learning causes important brain changes in addition to how brain changes produce human behaviors.

Undoubtedly the learning of language, a very basic repertoire, would leave a brain imprint as well. (That is probably one of the reasons why autistic children as a group have brain differences from normal children, because some autistic children have learned no language.) There are other, more specialized, repertoires that can be learned once a person has learned the language repertoire, for example, advanced mathematics, physics, history, and literature. Would learning each of the more specialized areas change the brain enough that changes would be detectable by brain imaging? It is unlikely that only the learning of a reading repertoire produces changes large enough to detect by brain imagery.

Are the brain-imagery technologies sensitive enough to distinguish the various types of basic repertoires in terms of how they affect the anatomy or functioning of the brain? For example, when a youngster learns to be a skilled athlete in tennis, or any other sport, will this result in distinguishing effects on her or his brain image? Does a professional dancer have a different type of brain anatomy or action than an Olympic weightlifter? How about a surgeon, a mathematician, a singer, and a violinist? Will having such different repertoires of sensory-motor skills be evident in brain differences? What is the sensitivity of the new technologies, and how can they be used to establish knowledge of the way learning affects the brain, and thus behavior?

Paleoanthropology

In this sampling we can also consider paleoanthropology, which gives us such a wonderful picture of human development. This science, too, has functioned mainly with the Great Scientific Error's lack of understanding of learning.

Rather, I suggest that human-evolution specialists need to consider the new theory of human selection, along with the conception that human behavior is the result of huge learning ability, a body capable of hugely variable behavior, and the huge human repertoire that has been learned by the prehuman hominids and the many generations of *H. sapiens*. Those experts must systematically consider whether learning was a part of human evolution. How does the human-selection learning theory compare with theories based solely on natural selection? Do the various facts of paleoanthropology support a human-selection/natural-selection theory versus just a natural-selection theory? Take, for example, theories concerning when language appeared in human evolution. Is that question answered better by considering language to have appeared about fifty thousand years ago through mutation? Or does the theory of human language learning provide a better conception for explaining the appearance of language in human evolution as a gradual learning-evolution process that must have begun a million or more years ago? Shouldn't paleoanthropological findings address the proposed effect of language accrual on human evolution itself?

Paleoanthropological evidence and interpretation have a strong influence on the popular conception of humanness, the Great Scientific Error conception. But the sciences that research human evolution do not study human behavior and its causes. They do not have or use a systematically constructed theory of human behavior and human nature. They do not recognize the importance of learning for humans and do not take that into account. As a consequence, theory and research occur that are wasted and mislead terribly in the popular understanding of humanness.

The fields of ethology and primatology, for example, aim to reveal the genetic-evolutionary causes of human behavior, even studying such things as ant soldiers displaying "altruistic" behavior or chimpanzees displaying "culture." Large amounts of time, effort, and money are spent trying to find out about human behavior and human nature by studying chimpanzees and gorillas. As engaging as the view is, it is largely a waste in terms of providing knowledge about humans. Human behavior and human nature are learned. They are not a product of traditional evolution. They did not derive from other animals. No other animal's behavior has a huge repertoire that is learned in good part through cumulative learning principles. Humans are unique in the enormity of their learned repertoires, and unique in their cumulative learning. Culture for humans involves enormously complex repertoires acquired over many generations of complex cumulative-learning experiences. Culture for chimpanzees involves

332   PART 6: A HUMAN PARADIGM

such things as a youngster observing its mother use a twig to fish ants out of an ant mound. Calling such chimpanzee behavior "culture" is misleading. The evidence that chimpanzee groups have learned some common behaviors does not qualify as a simpler form of human culture. Nor does it support the belief that human behavior has come from genetic mutations. Generally it should not be assumed that the behavior of any other animal can be studied as a means of knowledge of human behavior. Humans are different than all other animals in learning. Various biological sciences are needed in studying human behavior and human nature, but such study has to be projected using a new conception of human learning as well as of biology.

Basic Learning

The first generation of behaviorism—Ivan Pavlov's classical conditioning, Edward Thorndike's operant conditioning, and John Watson's behaviorism— was a great scientific contribution. The second generation of behaviorism, which included Clark Hull and B. F. Skinner, was also a great scientific advancement in refining the study of the basic principles in laboratory experiments conditioning simple human responses and in general theorizing about the principles applying to human behavior.

The start of our third generation consisted of our applications of the basic principles to real human behaviors, especially problems of human behavior. Our applied work mostly involved operant conditioning, and B. F. Skinner's theory was the most straightforward of that time. So it was taken as basic in the applied behavioral movement.

But his radical behaviorism was laid down many decades ago and does not provide the framework needed for the continued basic study of learning-behavior principles. His approach completely separates emotion and behavior, giving classical conditioning of emotion no importance. Radical behaviorism has neglected its study. That is completely wrong. The basic field must now focus on studying how classical conditioning and operant conditioning are inter-related. As outlined here, that theory change opens new vistas of research, actually the foundations for a new behavioral movement.

In addition, much more development is needed to make the learning principles a basic foundation for the broad study of human behavior. That means the basic field can no longer restrict itself to studying simple responses and simple environmental stimuli. Human behavior involves complex combinations of learned stimulus-response combinations, repertoires. The field must show what these combinations are in basic form and how they work. For example, sequences of responses are generally learned. How do humans learn word associations, as in learning to count? How do word associations work? Obviously the individual can't add, subtract, multiply, or divide without having learned the counting sequence. What are the behavioral constituents of such arithmetic operations and how do these repertoires work in learning algebraic ability? How do word associations function in grammatical rules? The *Journal of Child Language* has descriptions of many types of grammar that appear in first form in children. They should be analyzed in terms of what they consist of behaviorally and how they are learned. What is communicated behaviorally in communication and via what learning principles? What takes place behaviorally in advertisements? How do repertoires, once learned, affect the individual's ways of thinking (logical or not), problem solving, consciousness, and other cognitive processes. Obviously the individual cannot think logically without having learned sequences of logical verbiage. Knowing about cognitive abilities depends on knowing about the learned behavioral combinations that compose them, and that calls for experiments that characterize the learned mechanisms involved. Rather than inferring internal mental abilities, what is called for is the demonstration of the learned mechanisms of which those mental abilities are composed. Multiple responses are learned in various ways in the motor, emotion, and language acts of human behavior. Such combinations must be understood on a basic level to make analyses of complex human behavior. A large field of learning is needed to study how the complexities of behavior are learned, including the ways that repertoires cumulatively serve as foundations for learning other repertoires. Knowledge is needed about how learning produces repertoires and how one repertoire is basic in learning another repertoire. Learning-behavior analyses and study should address cognitive psychology's concerns with problem solving, perception, purpose, choice, and goals. The way to answer the behaviorism-cognitive schism is by showing that cognitive-mind processes actually consist of learned repertoires, an essential unification.

The basic field of learning has to be opened up widely to new studies of complex learning. Learning is central to humanness, and the field has to model basically the types of behavioral mechanisms involved in human learning.

OF THE LEARNING AND BEHAVIORAL SCIENCES

Child Development, Personality, and Abnormal Behavior

The field of child development exhibits so clearly that it is part of the Great Scientific Error.

The field follows the belief that biological maturation produces behavioral development, that human genetics produces inherent features of the mind that determine children's behavior development. This belief system projects research aimed at establishing the features of the mind in childhood. The example was given in the first chapter of Elizabeth Spelke's research with infants and young children for the purpose of establishing that babies are born with knowledge about such things as geometry already in their minds. Babies are said to have an innate capacity for numbers and mathematics because six-month-olds can respond differently to eight and sixteen objects and sixteen and thirty-two objects. This is described because many works in the field of child development fall within the same conceptual framework; they attempt to infer how the nature of the child's mind develops or advances from maturation by observing the behavior of children at different ages, believing that human nature is innate, a result of evolution. The traditional field of child development thus attempts to establish the chronologically determined growth of the mind-brain.

Can such research yield knowledge of how children learn, what they learn, or the role that what they learn plays in determining what else they will do and learn? Certainly not. Such research does not study learning, nor does it provide knowledge that is useful in raising children. The general belief is that a commonsense knowledge of learning is sufficient. After all, everyone sees all around how learning occurs. No need for use of a systematic scientific view. That is just as wrong as it would be to base knowledge of genetics on how behavioral similarities appear in families rather than on meticulous detailed study. Yet there is a huge lack of study of the ways that learning could cause child development. This is unimaginably in error. The learning of children must be studied meticulously and extensively, so much so that a science is made to address the subject matter.

It is in childhood, too, that the personalities of people begin to show themselves. Freud touched on that in his psychoanalytic theory, suggesting that different personality traits were determined by how parents dealt with their children's biological developmental needs. The large field of personality is devoted to considering theories of what personality is. There is also a large field devoted to constructing tests for measuring the different aspects of personality. Underlying this work is the belief that the determinants of personality are wired into the individual. Everyone has this innate potential personality structure that grows with biological maturation in interaction with life experiences. Take a textbook, *Personality: Contemporary Theory and Research* (Derlega, Winstead, and Jones 1999)—it has chapters on heredity and the biological bases of personality. A chapter does deal with personality development, however that only includes various theories, none of which systematically treats learning as a cause of personality development. Learning is not really treated; a science field that is devoted to doing so is needed. That should include analyzing the various personality traits that have been considered and measured in terms of the repertoires that compose them, how those repertoires are learned, and how those repertoires function in life and in further learning.

The field of abnormal behavior also operates primarily according to the Great Scientific Error. Most strongly it is believed that some biological problem underlies the various behavior disorders. There has been a great search for genetic explication of behavior disorders, for isolating causes by observing structure and function of the brain in those with those disorders, and for pharmaceutical products that treat the disorders. The field attempts to define the mental illnesses by the behaviors they produce, definitions that are used in diagnosing the behavior disorders. The drugs that yield improvement in abnormal behavior are taken as evidence that the drugs are treating the mental-illness causes.

There is a field for diagnosing behavior disorders, and it has important descriptions of the repertoires that compose them. But that is not understood. Nor is the need to study how those repertoires are learned understood. Nor the need to establish definitively how those repertoires lead to life actions that are disadvantageous and perpetuate the disorder. How do the actions of the autistic, ADHD, or dyslexic child have disadvantageous and perpetuating results? Even though behavior therapy and applied behavior analysis began within the field of learning, those two fields have not projected a program for the study of how learning produces the behavior disorders and how the abnormal repertoires function. Learning is largely ignored. A new abnormal psychology field is needed.

A Science of Learning Humanness

You can't find out about learning in child development by studying what children's behaviors or abilities generally are at different ages. You can't find out about learning personality by theories based on commonsense observation and constructing tests. You can't find out how the behavior disorders are learned by conducting genetic, brain-imagery, diagnostic, or pharmacological studies. Although much of what has been studied is essential, you can find out about how human behaviors are learned only by studying just how they are learned.

What is surprising is that no one has apparently realized that the only way to gain knowledge about the huge, wide-ranging learning of humans is by studying that learning. Deeply, systematically. There is no way to know how learning determines child development except by studying that learning. That goes for personality phenomena and the phenomena of abnormal behavior. Where does that learning begin, and where does much of that learning go on? In the home! That is where study must go. No other way.

Strangely, although everyone would agree that the child's learning experience is important, a most evident avenue of gaining knowledge about that learning is not being conducted at all. *From the perspectives given by the new paradigm, it can be seen that large-scale studies are needed in which trained observers are placed in homes to record what goes on that produces the child's learning. Beginning from birth.* We say that a child who unreasonably wants to get his or her way must have been "spoiled" at home. But the learning experiences of children have never been systematically studied. That goes for all important behaviors. Whether learning is the cause cannot thus be known. There are various differences in behavior that are exhibited in children during their development. Knowing why demands research in the homes where the nature of children's learning experiences, and the behaviors that result, can be systematically studied.

No doubt this position will elicit opposition. (Karl Minke has a very good solution to this problem, the use of cameras in the home versus an observing person.) But let me ask, does the child learn important things in the home? Everyone will answer yes. Do the things the child learns in the home play an important role in the child's behavioral development, in the child's development of personality, in the child's development of abnormal behaviors? A heavy opinion would answer yes. Have those numerous and important types of learning been studied in the home, systematically, deeply? The answer is no. Parent opinions have been taken, yes. Unfortunately, parents have only a commonsense understanding of learning and are not able to give a systematic understanding of their children's learning.

We need studies based on having trained observers (or cameras) placed in homes where they record the learning experiences the child has and also the behaviors the child thereby acquires. Then it would be possible to see the child's learning for all types of behavior, for emotional learning, for sensory-motor learning, and for language learning. The differences in the experience of the child because of the different behavior of parents would be observable, as would the effects of those differences on the development of the child's behaviors. This type of study should occur for the years of childhood.

Such study has to be done. How else can we begin to know what goes on with children and parents that produces the child's learning? Only by discovering the actual conditions of learning in childhood—how they affect the child's behavior development, how that behavior development then affects the parents' behavior, and how that interaction continues—will it be possible to discover what is involved in the cumulative learning of the basic repertoires.

This absolutely necessary type of research will be basic to fields other than child development. For example, many have said that personality develops in childhood in interaction with parents. But there is no study of the learning that goes on in the home that produces personality traits. The theorizing about personality development as affected by experience has no foundation without study of that experience and the way it affects the repertoires that compose personality.

The same is true of the field of abnormal behavior. Pretty much everyone recognizes that personal experiences can be involved in producing abnormal behavior. Again, it is madness to think that the learning causes of abnormal behavior can be ascertained without studying them. Could anyone question that a child who behaves like an autistic child, a mentally challenged child, an ADHD child, a conduct-disordered child, or a normal child has a different set of life experiences in the home? Of course not! But how do we find out just what those differences are? As soon as a child begins to exhibit behaviors different from others, the child is treated differently. And we know that different treatment will produce different behavioral development. After all, different children speak Mandarin, Castellano, Catalan, Greek, and Urdu because they receive different language experiences. Wouldn't that hold where children have language experiences that differ in grammaticality, logicality, in coinciding with reality, in richness, and in having various psychotic features? The different learning experiences that produce different languages is blatantly obvious, however. But those that produce different types of abnormal behaviors are not obvious. If we want to know the causes of abnormal behavior, we have to study the learning experiences involved, beginning in childhood but extending beyond.

The new paradigm's position is that learning a serious behavior disorder is a long-range process. It does not spring up overnight. First, repertoires are learned that, while not desirable, are not full-fledged disorders. For example, a child who will later be diagnosed as autistic may first learn a grunting-pointing-whining repertoire to get things he or she wants, rather than learning a language repertoire that has that function. Very importantly, that grunting-pointing-whining repertoire, when learned, will be an obstacle to learning language. Instead of saying the word for something, the child will simply point and whine. When the parent tries to get the child to say the word, the child simply becomes more vehement, so the parent relents and gives the child the object he or she wants. That of course strengthens the grunting-pointing-whining repertoire and increases its development. However, although undesirable, that does not constitute autism. But it may be on the road to the cumulative learning of autism. What about the cumulative learning that leads to ADHD, to conduct disorder, to schizophrenia, just to mention a few disorders. Are there cumulative learning paths to growing, elaborating abnormal repertoires? That question can only be answered by research that directly studies what has gone on.

How can we establish whether learning is involved as a cause of autism? How can we establish whether remedying the child's early language problems will help prevent the child becoming autistic? How can that be done for the various learning experiences and repertoire developments that may be involved in autism? What about the repertoires of ADHD children, of depressed individuals, of those with conduct, bipolar, and other disorders, not to mention normal repertoires? Don't we have to make contact with that learning where it occurs, in a meticulous, long-time way? And does that not demand that research go into homes? Research in the home for the study of child development—involving a large random sample of homes—would undoubtedly pick up homes in which some behavior disorder would develop. So the same research would provide central knowledge for the three areas: child development, personality, and abnormal behavior. Homes where children have already begun to show deficient and inappropriate basic repertoire development also could be selected for study.

Yes, of course, that makes sense. However, is such research possible? Will parents allow an observer into their home? It might be necessary to pay parents. Pay for participation in research is not unusual. Training observers would cost, as would cameras, as would the large amounts of time spent involved and the analysis of extensive data. The research would be costly. True, but building cyclotrons to research the particles in an atom cost billions and billions of dollars.

Research of all kinds is expensive, and great sums of money are spent on research that produces nothing. An article in *Science* titled "New Hints into the Biological Basis of Autism" (Stokstad 2001) referred to extensive and costly resources spent on brain-scan studies that differentiate in anatomy and function between autistic and normal children. The same large research expenditure is being made in pursuit of the belief that autism is caused by genetic defect. "Twin studies have shown that autism clearly has a genetic component. But pinning down the chromosomal regions that contain the genes that are involved—much less the genes themselves . . . has been a daunting task" (36). Actually, the twin studies are only suggestive, and the genetic studies as indicated have failed. Let me say it clearly. Great expenditures have been made on research inspired by the old paradigm's focus on biological causation that has not delivered. Shouldn't we try the different direction being proposed, with the promise it holds?

I will conclude this section with one other point, because it is characteristic of the new paradigm. The conception of human behavior and human nature that has been developed performs a unifying action. What now consists of three largely separated fields—child development, personality, and abnormal psychology—are unified in the proposed research. The three science areas would be supplied by the same study in the home. The field of child development would be concerned with how that development is learned. Personality would consider how personality traits are learned, and the same study would concern how abnormal repertoires are learned. The three fields should not be separated; they have the same basis.

The in-home study is only one of the types of research generated by the new paradigm, as shown in the preceding chapters. A new paradigm calls for new research and in this case new applications.

Other Human Science Areas

The field of psychology incorporates other human science areas, such as social psychology, cognitive psychology, educational psychology, clinical psychology, and psychological measurement. Presently those science areas operate as separate and independent. Mostly what is done in one area isn't communicated to the other areas. The new paradigm should be developed in each of these areas, and the result would be actual unification—interactive and communicating areas of study—generating new avenues of development in each.

Cognitive Psychology

Psychology as a science has basic divisions. Perhaps the most fundamental is that already referred to between those seeking explanation of human behavior within the individual versus those who look outside to the environment. This yields the nature-nurture split, part of which is the mind-learning split. Traditionally the mind had the center of attention, until Pavlov, Thorndike, and Watson made the experimental study of learning central. They brought a science to psychology versus the subjective introspections of the mentalists. However, in the mid-1950s, the mentalists became cognitivists, who had become experimental, too, producing many experiments about perceiving, thinking, remembering, problem solving, language, and other workings of the mind. In this approach, the assumed mind is central because it is thought to determine how these typically human behaviors occur.

This field of cognitive psychology has grown exponentially, energetically producing abundant areas of research. Following the belief that the characteristics of the mind are wired into the brain, the theoretical joining of the two has produced the field of cognitive neuroscience. The belief is that brain differences determine mind differences that in turn determine behavior differences. This belief reigns as basic principle through an important part of the science of psychology.

Not so fast. The new paradigm spreads a wet blanket, in a way. Many of the things studied in cognitive psychology and cognitive neuroscience are valuable. Studies of thinking and problem solving and grammar and such add knowledge of the way humans behave in important areas. The wet blanket is thrown over the biological route taken for explanation. The brain does not explain complex cognitive behaviors. Explanation does not lie in genetics. Nor does it lie in evolution—we can't study chimpanzee language or problem solving or culture and gain knowledge about human behavior and human nature.

The characteristics of people's thinking, language, problem solving, and other cognitive actions are determined by their learning of basic repertoires. A schizophrenic, depressed, or paranoid person, or a Yanomomi Indian, does not think like other people in our society, not because of a mental illness or racial brain difference, but because of having learned different basic repertoires. The cognitive phenomena studied are not definitions of the mind. The experiments actually are studying the behavior repertoires their subjects have learned. A problem that can be solved only by someone with a mathematics repertoire that includes algebra will not be solved by a Yanomomi Indian or someone who stopped his education at the sixth-grade level. A person will comprehend and remember read passages better if they are expert readers than if their reading repertoire is less developed.

Rather than looking for explanation in the brain, cognitive psychology and cognitive neuroscience should base themselves in the new paradigm. A great science area of cognitive behavioral study should be imbedded in the conception of the marvelous learning animal and the learning of basic repertoires. This type of development would help resolve the nature-nurture, mind-learning, division.

Educational Psychology

As another example of the level of advancement of the educational field with respect to human learning principles, take a recent article in *Developmental Psychology* (Justice et al. 2009) reporting important findings. What Justice, Turnbull, Bowles, and Skibbe did was find groups of children with poor language development at the ages of fifteen, twenty-four, thirty-six, and fifty-four months and groups of that age with normal language development. Then they measured how well these groups of children did in learning to read on entering school. They found that the group of children who still were language handicapped at the time they entered school did indeed have greater difficulty learning to read. These authors interpreted their results in terms of a critical stage of language development, inferring a mental (brain) process as the cause. This shows how important is the conception held by scientists. In this case, with the Great Scientific Error conception, these researchers did not understand about the children's differences in previously learned repertoires and how that can affect the performance in learning to read. But the explanation of their results lies in the human learning principles. Language is a basic repertoire. It provides the foundation for later learning of various kinds. A child with a rich and well-constructed language repertoire will do better in that later learning than a child with a lesser language repertoire. Learning to read depends upon having already learned language. The point is that a scientifically constructed theory of language enables a scientifically constructed understanding of why some children are better learners than others, for example in learning to read. Such explanation also instructs what should be done to prevent such school failures, which theorizing about the mind-brain cannot provide.

Take the field of educational psychology as an example. The nature of our schools, the way they operate, was derived from commonsense experience in trying to teach children. It did not utilize a scientific knowledge of learning and child development through learning. Basic changes in the nature of schools would result from constructing a new framework for education based on the new paradigm. For example, the knowledge of cumulative learning immediately indicates that children begin school with differently developed basic repertoires necessary for school learning. The child's language repertoire will determine how the child does in the group method of teaching. So will the repertoires that determine how the child will attend and work, as well as those that indicate the extent to which the child has the specific repertoires necessary for learning the new repertoires that will be taught. Measures could be constructed that would indicate which children would succeed in the group situation, so that the teaching situation could employ methods that would work for all. Children should be taught needed repertoires before being subjected to teaching situations that demand those repertoires. There are new ways to teach children using reinforcement when their attention flags, and those should be expanded in ways that would advantage all children and save money at the same time. William Butterfield and I (1965) showed how the most difficult of students can be successfully taught using reinforcement, and many studies have elaborated that discovery. But the great potential involved has not begun to be applied.

Actually, the new paradigm's conceptions of humans and the learning of humans provide a whole new framework for designing a new type of institution of education so that it would be effective for all children as well as less costly. Quickly, I will sketch such a design. Kindergarten would be built around rewarding activities, with movies, recesses, games, class periods, and such. Participation in those activities, however, would be used as rewards. Admission to them would be based on tokens the children had earned in learning activities in reading, writing, and numbers, taught by educational technicians. Children would enter school for a brief classroom period and then disperse into a brief learning activity where they would earn tokens. The tokens could then be used to gain admittance to one of the rewarding activities. The day would consist of alternating learning periods and rewarding periods. The teacher would supervise the whole operation and deal with problems and special cases. This type of design should be systematically constructed and tested in research.

Psychological Testing

Today there is no demand that the various personality tests should be related to one another conceptually. Tests can be constructed that overlap in their items, thus in the repertoires they measure, and yet be considered to measure different personality traits. As examples, the items on developmental tests overlap with the items on intelligence tests, and there is overlap among interest, values, needs, and attitudes tests. That occurs because the individuals who produce the tests use idiosyncratic concepts in their personality theories and are concerned with producing a different test rather than organizing the general field of study. The great many tests that result make a disorganized, bewildering field. When the various tests are analyzed and researched within the same theory, the knowledge they supply can be interwoven, mutually supportive, and can constitute a strong and simplified field. All tests should be analyzed in terms of the repertoires they measure, using the same principles and general theory language.

There are additional lines of needed personality research. One would tie together the findings that learning experience can change both brain action and personality-test performance. Showing such relationships will have the effect of bringing together the different approaches as well as deepening knowledge of personality. When such developments have taken place, the field of psychological testing will become much more effective in providing knowledge of human nature.

Social Sciences

Constructing frameworks in the various areas of psychology based on the new paradigm would not only offer productive foundations for those areas, but the result would be unification of knowledge that in itself is a great advancement. Darwin's theory and Mendel's genetics did that for biological sciences, which is one of the reasons for their great power. New sciences begin in great disunity and only gradually come to generate generally meaningful knowledge that brings together all the diverse, redundant, local efforts. The behavioral sciences in general are in that disunified state (Staats 2005).

Wouldn't it be advantageous were psychology, sociology, economics, political science, and history to have a common underlying conception of human behavior and human nature—especially if that conception explained human behavior better than the one commonly held? It was said more than thirty-five years ago that the principles of reinforcement were common throughout the behavioral sciences. For example, the principle of marginal utility (supply and demand) was analyzed according to the principles of satiation-deprivation. When an animal is satiated with food, the reward value of food goes down; when the animal has been deprived of food, food's reward value goes up. The economics principle mirrors that; when a commodity is scarce, its value climbs; and when that commodity is in generous supply, its value goes down (see Staats 1975). Today the science area called behavior economics revolves around demonstrating such reinforcement principles in a way that is relevant to economics.

That is an example, but the position of the new paradigm, even in its early forms, held that a conception of human behavior based on learning was important for all the behavioral sciences.

> Much of the work done in other behavioral sciences is based on a "psychology," a conception of man's behavior. For example, much of anthropology in recent years has utilized psychoanalytic terms and principles in an approach to many problems of analysis. As yet, however, an integrated set of learning principles has not been comprehensively applied in the various fields of social science. (Staats 1963, 54)

> What I do claim is that, no matter what we say our theories are, when we try seriously to explain social phenomena . . . we find ourselves in fact, and whether we admit it or not, using what I have called psychological explanations. (Homans 1964, 817)

Homans developed his sociology using the principles of reinforcement. Today the interest for using learning principles has largely been lost in favor of pursuing the Great Scientific Error paradigm. The noted political scientist Francis Fukuyama, for example, anchors his approach in innate, wired-in behavioral characteristics derived from evolution.

> *Human beings have an innate propensity for creating and following norms or rules. . . . Human beings have a natural propensity for violence. . . . Human beings by nature desire not just material resources but also recognition.* (Fukuyama 2011, 439–41)

Fukuyama also believes that cultural evolution causes change in social phenomena: "religious views evolve along with political and economic orders, moving from shamanism and magic to ancestor worship to poly- and monotheistic religions with highly developed doctrines" (443). However, he recognizes that cultural evolution is not just like Darwin's theory, since it is necessary "to understand the ways that political evolution differs from its biological counterpart" (443). His analysis includes and is based on a conception of human behavior and human nature, illustrating how the social sciences depend on such conceptions. His conception, however, is of the Great Scientific Error kind, also illustrating how valuable a paradigm based in learning—systematic, consistent, and unifying—would be in the social sciences. Human cumulative learning should replace cultural evolution in those sciences.

OF APPLIED SCIENCE AND APPLICATIONS

The new paradigm, moreover, also suggests applications, not just knowledge accrual. Studies aiming to advance knowledge about how the brain causes behavior, from child development to political science, don't have much in the way of applied value. If genetic changes have to occur before young men are no longer driven to fight wars, or new forms of government arise that solve human problems better, it will be a distant day before anything useful arises. The same is true with respect to dyslexia or autism. On the other hand, use of learning for treating problems of behavior has already proved of great value in the fields of behavior therapy, behavior analysis, and cognitive behavior therapy. Those fields, however, work with a delimited conception of humanness as learned. As a consequence they too have never studied the child's learning conditions at home that produce normal behavior as well as problems of behavior. How can prevention measures be advanced without knowledge of how learning experiences are the causes of normal and abnormal repertoires? That knowledge would produce a new science of abnormal behavior that would describe not only the behavior disorders but also how they arose, how to prevent them, and how to treat them. The new paradigm has avenues to be developed for application throughout its range of concerns.

Parents

Parenthood calls for knowledge of how learning determines child development. Thus we need a very large amount of study of how normal and abnormal basic repertoires are learned. When studies reveal the learning experiences that produce such repertoires, a monumental future work, this will call for additional applied study regarding how parents can be educated to prevent problems of learning that occur in children. Also called for will be applied study of how parents can be trained to treat their children who have already developed problems of behavior. That was indicated early on, beginning in the first work in the present program (see Staats 1963), and a number of behavioral studies have shown how parents can be trained in dealing with children's problem behaviors. That is only a beginning; the new paradigm calls for much additional application, including new applications geared toward prevention.

As knowledge of children's learning advances, there will be great changes in child raising and in the treatment of children with developmental problems. There will be a time, for example, when there will be only a fraction of the autistic children there are today. For it will be common knowledge how the child's learning gets onto the track that leads to autism and how, instead, that learning can get onto the normal track. Children's behavioral development will be more precisely watched by pediatricians. If the child has not said his or her first words by the age of fifteen months, the parents will receive training in how to train the child to language and other necessary repertoires. If the child does not display signs of love and affection for the parents at the age of six months, the parents and child clinical psychologists will be oriented toward studying the condition with the aim of ensuring the child has the experiences that are needed to produce that central aspect of the child's emotional-motivational development. If the child does not exhibit the skills of coloring in books, writing letters, and such at two or so years of age, the same type of diagnostic procedures will reveal the delayed development and procedures will be introduced to the parents by which to correct the problem. The child of three or so years who has not learned the skills of peer play will receive experiences that produce this kind of learning. The child who displays temper tantrums or who bothers other children in the group or who takes other children's toys or hits them will receive training by which these behaviors will lessen and disappear. The excuse that such behavior occurs because "he or she is just being a child," is incorrect, erroneous. Whatever behavior the child displays is learned. Undesirable behaviors shouldn't be learned and they shouldn't be ignored; they are not good for the child. If the problems are severe enough and the conditions are not changed, the child won't just grow out of them. As the principles of cumulative learning tell us, those undesirable behaviors may be the foundation repertoires on which more serious abnormal behavior is learned.

A DSM (diagnostic and statistical manual) that set forth the causative learning experiences for the various behavior disorders would be enormously more valuable than the contemporary DSM. That knowledge would include the behavioral repertoires that compose each disorder, the situations that elicit the abnormal behaviors involved, the learning conditions that produce the repertoires, the measures for preventing that learning, and the measures for producing the normal behaviors. Clinicians would then have the full knowledge they need for treating the child and advising parents how to do so.

The position is firm: cumulative learning that involves interactive experiences with others plays the central role in the acquisition of the abnormal repertoires that constitute a behavior disorder. There is no way that behavior disorders will be understood unless that learning is studied systematically, over time, and in detail. There is no way that treatment of behavior disorders will ever become efficient without that study and without that knowledge of learning, for treatment has to involve "reversal" of that abnormal learning. Treatment has to occur by learning, as is shown clearly by the advances in the treatments already made in our behavior-analysis, behavior-therapy, and cognitive-behavior-therapy fields. These efforts, however, are still based on a less-than-full understanding of human learning and the learning of normal and abnormal repertoires. A more complete knowledge will better inform parents, provide for prevention, enable the construction of diagnostic instruments, and enable the construction of therapies for remedying behavior disorders after they have been learned, even in cases where the normal behavior to be learned requires complex and long-term training (see Burns and Kondrick 1998).

There already are many behavioral studies attempting to work with children through their parents. Use of a widely expanded understanding of humanness in learning terms, the behavior repertoires, and cumulative learning will provide a more profound and wide-ranging applied science area dealing with parents and their children

Professional Practice: Psychotherapy

In a 1972 article I described how human behavior can be changed through language communication, as takes place in psychotherapy. The individual's emotion and motivation can be changed through language. The individual can be instructed regarding changing behaviors and learning new ones. The individual's own conceptions of self, of family and friends and people in general, and of life situations, of purpose and future, can all be changed through language. Prior to that time, the reigning behavioral position was that basic principles of conditioning should be applied, not verbal psychotherapy techniques. That was the Skinnerian position. Change did occur; verbal means of treating problems of behavior were introduced. Skinnerian-oriented cognitive-behavior therapies ultimately emerged. But the cognitive-behavior therapies have not been based on a full understanding of the learning of language and the functions that learning has. And that understanding of language was not joined with a general conception of human behavior and human nature. Moreover, the cognitive-behavior therapies have not had access to a theory of abnormal behavior that deals with human learning. The field, thus, while having valuable offerings, is incomplete, has error, and is only a beginning. A great deal of advancement lies ahead through use of the new paradigm.

Public Policy

Public policies are devised to aid the conduct of our lives. They are designed by people on the basis of an understanding of humanness. It is thus imperative that that understanding be true. The new paradigm should be considered in that light.

When conditions of learning have been deplorable for the parent, then the conditions of learning the parent provides for his child will also be deplorable. This is true for individuals, and it is true for groups. When groups of people have deplorable learning conditions, their behavior will be deplorable, and it will be perpetuated in their children.

If society is interested in improving the behavior of the children of any generation, then a central place to begin is in improving the learning conditions parents can provide for their children. For example, today there is a widespread acceptance of the fact that large portions of the [perenially poor] population have been subjected to deplorable learning conditions to the point where many parents are not able to provide either the appropriate physical conditions or the appropriate learning conditions for their children. If there is no intervention in the process whereby deficits in one generation are passed on to the next, this crucial social problem can be expected to continue.

The preceding would suggest that one way to break into this vicious cycle would be through training and support of the disadvantaged parent so that he could improve the physical conditions and the parental training his children will receive. It is suggested that following the type of principles and procedures that have been outlined herein, such parents could receive training that would enable them to markedly improve the manner in which they would themselves provide training for their children in the basic . . . repertoires. Rather than giving inadequate welfare payments, it is suggested that society provide a suitable living to the parent who undertakes training to make himself a more accomplished parent and in so doing also to make himself a more accomplished individual. Instruction in child-training principles and methods of the type described herein would be productive. However, in addition, the parent himself should receive intellectual training of various sorts. It is quite evident from various sources that the language-intellectual training of disadvantaged adults has left them with gaping deficits that are then passed on to their children. For example, the disadvantaged parent's language is [frequently] very poorly developed. It is ungrammatical, ungainly, sparse, and minimally functional. This is shown vividly in Hersey's *The Algiers Motel Incident* (1968) which includes abundant quotations from ghetto residents of Detroit. Only with great difficulty and inexactitude can these disadvantaged individuals describe events of everyday life. Individuals with such poorly developed language would have great difficulty with most academic learning. They would also be incapable, without some additional input, of training their children to better intellectual skills than they themselves have.

Such programs of parental training should not be conducted as a charity. They are no more charitable than free public education, or the education provided by state universities to middle-class students, which is largely publicly subsidized. The parent is the sole educational institution for the child in the very early years, and to some extent throughout childhood. It is sheer waste to provide school facilities for the child but not provide him with the basic behavioral skills with which to profit from the school. It is in society's interest to break into the vicious cycle of disadvantaged parents providing disadvantaged training for their children. It is in society's interest to help prepare the parent to assume his role of instructor of his child so that the child will be able to adjust to and contribute to the society.

This is proposed because the present attempt at the solution of . . . inadequacies in the child learning is to provide some sort of limited compensatory education for disadvantaged children. Programs such as Headstart are surely in the right direction. However, they are not based on an adequate understanding of early child learning. . . . Programs that begin when the child is three or four years of age and that only include an extra year or two of preschool for the child will not prove to be adequate. For one thing, such a program does not recognize that the child has much to learn before he is three. For the disadvantaged child there are great deficits in the learning of the basic repertoires during those first three years. . . .

To some extent this could be remedied by having compensatory education for the child begin at a much earlier age. Thus, it is suggested that nursery-school participation that included training experiences for aspects of the basic repertoire should commence with children one, two, and three years of age. Procedures should be utilized whereby the nursery-school situation could be used to produce basic behavioral repertoires in the child. . . . The most complete type of remediation of our poverty and cultural-deprivation problems, however, would include adults as well as children, the general environment for both, and would be of long duration. (Staats 1971a, 330–32)

This analysis was published over forty years ago in a book that should have raised various questions regarding public policies to solve the societal problem. The belief in the Great Scientific Error paradigm was too strong. Our society continued to dole out welfare to indigent families but do nothing to improve the ability of those families to better the training extended to the children involved. So the problem continues. Rather the parents in such families should be paid to learn themselves the basic repertoires necessary to impart to their children. It should be understood that parents in such a program would be working productively to improve their society. The children would instead be learning from their parents and from their early education the necessary basis for becoming productive members of society themselves instead of becoming indigent consumers of charity or of prison incarceration. Forty years later, conceptual support for such public-policy actions has been accruing, however.

> *To improve kids' grades, it's not enough to help low income kids to achieve academically—you also have to educate parents,* say researchers from the University of Michigan and the University of California, Los Angeles. The researchers looked at data from the Los Angeles Family and Neighborhood Survey, including reading and math assessments of 2,350 children and teens, their mothers' level of education, and neighborhood and family income. The greatest overall determinant to students' academic success was the low levels of mothers' education, the researchers found, though poverty or wealth became a larger influence as the kids got older. The finding highlights the importance of parental education to starting young children down the path to academic success. (American Psychological Association 2011, 14, italics added)

Clearly this framework opens the way for further applied research and ultimately for a change in public policy. One study should involve a large group of ghetto-dwelling families composed of a single mother with a young child. The mother should herself have done poorly in school, be living on charity or menial labor, and not be addicted to drugs or alcohol (other research should deal with such problems). The mothers should be enlisted in a program for learning reading and learning to read for pleasure and for knowledge of society; for work skills of a suitable sort; for knowledge of human behavior, of human nature, and of children; and for knowledge of raising a child to learn the necessary basic repertoires. The parents' own progress should be systematically assessed, as should their children's progress, for the purpose of rectifying lack of progress. Gaining knowledge of the problems that arise in successful conduct of such a program should be a goal, as well as the positive results of the program.

Clearly applied research on public policies that involve human problems are an important aspect of the new paradigm. Darwin's theory, with Mendel's genetics, constituted a new paradigm, a unifying theory that inspired research and theory across an astonishing number of science areas as well as the common beliefs regarding humans. Could a change of the contemporary conception of humanness also suggest and contribute to a wide range of sciences studying human behavior and human nature, as well as improve day-to-day beliefs and practices? *The Marvelous Learning Animal: What Makes Human Nature Unique* proposes that lies ahead.

CONCLUSION

Human actions are of consummate importance as natural phenomena. Human actions determine how children are raised and educated. Human actions determine whether there is feudalism, colonialism, communism, capitalism, democracy, imperialism, or despotism. Human actions determine whether economies flourish or stagnate, whether unemployment will be high or low, whether there will be people in the world who are poor, uneducated, unhealthy, and short-lived. Human actions determine whether people will be killed in war or live long, fulfilled lives of plenty. Human actions yield marvels of art, music, literature, science, and invention.

Humanness includes a vast number of phenomena. In their "gross," unanalyzed state, there are too many of those human phenomena to be encompassed within one book, or one person. Those phenomena are thus grouped grossly into fields. There is education that addresses itself to the phenomena of children in school and draws on the knowledge of various fields. Psychology has a broad range of fields such as developmental, social, personality, abnormal, experimental, educational, and clinical psychology. Sociology, political science, and economics also are social sciences that are concerned with aspects of human behavior. Ethics, philosophy, and history go into other aspects of human behavior. Cultural anthropology, paleoanthropology, and archeology also address human behavior phenomena, as do fields like paleontology and evolutionary psychology. Sciences like primatology conduct their studies of chimps, gorillas, and orangutans in part on the supposition that their behaviors provide knowledge of the causes of human behavior. Biological fields like genetics, evolutionary biology, and behavioral genetics look for causes of human behavior.

Actually this is a simplified picture of the various fields concerned with human behavior phenomena. Most of these fields pretty much act independently of the others, with their own concepts, principles, theories, methods, and findings. Little communication occurs—one is either a social psychologist or an experimental psychologist, a sociologist or a linguist, a paleoanthropologist or an economist—for their knowledge differences prevent communication. Clearly this unrelated plethora produces pretty much a mess of knowledge.

How can that be? Human behavior is a topic in each of these science areas. Each addresses the same human. The same basic principles should be functioning in each area. The principles added in certain areas should be in agreement with and derivable from those basic principles.

But that is the way sciences are at the beginning; their first goal is finding new phenomena. They study those phenomena as though they are different. No aim exists for relating the different phenomena by finding their common underlying principles. The action of a ball rolling down an inclined plane, the oceanic tides, and the movement of heavenly bodies all seemed to be unrelated phenomena until later, when Newton's gravitation showed their relationship in underlying principle. With explanation and broad application, what had been disordered features of the world were brought into relationship, ordered and simplified, which was much more useful.

The same wondrous thing can be expected for the many phenomena of human behavior. The task is much greater, of course, because human behavior covers so many phenomena that seemingly are fundamentally different. There are so many different conceptions, each directed to a few phenomena, perhaps only to one. Nevertheless, all of them, in their bewildering number and distinctness, are considered differently. They are yet to be ordered and simplified by the common principles that underlie them. Finding those underlying principles must be striven for. Establishing them, systematizing them, and disseminating them constitute most important goals. Understanding causes will become straightforward. Making things happen the way desired will become possible. Old errors in addressing the phenomena of human behavior will diminish.

That type of knowledge, however, will not be gained unless it is sought, unless huge time and effort are expended, and unless a new paradigm serves as the needed framework. I have described this book as a new paradigm. I aver that learning principles pave the way for a road to sciences of human behavior and human nature. But the road is long, endless—the knowledge to be gained by a science is never completed. There are many more major cities to be passed on that road before everyone begins to accept that it is a primary, broad avenue built to carry an enormous traffic of knowledge.

The marvelous learning animal and all the paradigm includes and calls for has overwhelming importance as the explanation of all human behavior and human nature phenomena. That is the central message of this book. And that has powerful meaning for a widespread number of things—thoughts and goals and actions. Here is a siren call for a general change in thinking about humans, human origin, human behavior, human nature, who we are, where we are going, and where we should be going. Yes, knowledge of evolution and genetics by Darwin and Mendel has great importance; all of the wide purview of biology does. But knowledge of learning does also. It has the same wide purview and demands the same study and application.

Yes, we humans are *the marvelous learning animal*. And knowledge of that changes everything: our spiritual beliefs; our beliefs about children and raising them; our beliefs about what makes us different from one another; our beliefs about abnormal behavior and how to prevent it and treat it; our beliefs about human origin; our beliefs about culture, public policies, sociology, economics, and political science; our beliefs about race and human equality and human progress; our beliefs about human nature. We have to think about ways to produce positive conditions for all, and by doing so produce positive conditions for us and those we love as well. And that means becoming concerned with things like world population, improving conditions for other people all over the world, and keeping Earth healthy and habitable and beautiful. And we have to think about conceptions and actions that stand in the way of working toward such goals. Yes, considering us to be marvelous learning animals sets us to study much that is new, to analyze in new ways, and to do the many things that the knowledge we produce indicates. Our unique ability demands that we continue the journey and improve the highway and countryside as we go.

ACKNOWLEDGMENTS

"**H**ow long did it take you to write your book?" a friend asked recently. My quick answer was "Over fifty years," thinking back to my formal study in psychology at UCLA, and later to faculty positions at Arizona State University, the University of London, the University of California, the University of Wisconsin–Madison, and the University of Hawaii. In addition to that support I had years of financial support from the Office of Naval Research, the National Institute of Mental Health, the National Science Foundation, and the Office of Education. My journey yielded over eighty journal articles, fifty chapters in others' books, three monographs, three edited books, and six previously authored books, each an advance into the new territory. That fifty-plus-year journey provided foundations for the present work.

But, actually, my journey began much earlier than fifty years ago. At the age of eight, for example, my mother, Jennie Yollis Staats, made an indelible impression on me in describing how she liked to observe people. General in our extended family was a deep interest in people and the conditions that affect people, as was indicated in such things as family get-together discussions of politics, my sister Jenny's subscriptions to a left-wing magazine and newspaper, my uncle Rupert Poston giving me at age fourteen H. G. Wells's *A Short History of the World*, and my brother Frank taking me at age eleven to the museum in Los Angeles that housed prehistoric fossils from the La Brea Tar Pits, with visual exhibits depicting prehistoric human and animal settings. My penchant for personal study actually began much earlier, importantly with a great grandfather who was a Talmudic scholar for a Jewish community in Tetiev, Russia, and a grandfather who became an atheist after his own study. How could their doings have affected me? Through my mother, if I was reading, for example, she kept me from any other bother, a central start in becoming a self-taught scholar too. My acknowledgments, thus, go also to those people responsible for my interest in natural explanations of human behavior and my questioning of belief based on tradition, religion, or other authority.

My own reading brought me to behaviorism and later to selecting a behaviorist major professor, Irving Maltzman, who introduced me to his view of the experimental field of problem solving, in which I did my dissertation. I was already informally applying behaviorism analyses to behaviors I encountered in life. By that time, the barn door was open to an explosion of ideas leading me to study of the learning of complex behavior with adults and children.

Carolyn Staats, my wife, became my first student, using my language-conditioning developments in her experiments for her doctoral dissertation. She contributed importantly as my first assistant on my grant-supported research and she later did valuable editing and literary searches for my first book before deciding to pursue her own distinguished career as a clinical psychologist. Her spousal gifts to me (and to our family) are too deep and broad and emotional to begin to describe, providing much that came from no other source, including such diversities as putting up with my home office and special hours, reaching mutuality in diet with someone raised as a vegetarian, traveling with me and supporting me widely in my professional demands, and sharing with a husband, in ways unusual in our era, the responsibility and love of bringing up two children.

Karl A. Minke was my longest collaborator, across various studies, beginning as an undergraduate and graduate student and then as a fellow faculty member. He also constructed his own research program, training many students and doing research in a continuing way. Montrose Wolf was a graduate-student assistant on some of my early studies with preschool children. His doctoral thesis extended my token reinforcement and time-out concepts with his extraordinary research program. He then went beyond what I called behavioral analysis with the founding of the *Journal of Applied Behavior Analysis*. Leonard Burns, as my graduate research assistant at the University of Hawaii, collaborated on and initiated central works. A meticulous and creative researcher at Washington State University, he has continued to make fundamental contributions to behavioral analysis.

Elaine Heiby, already a behavioral clinical psychologist when she came to UH, collaborated on several papers that introduced a theory of depression. Rocio Fernandez-Ballesteros, a noted professor at the University Autonoma de Madrid, came to UH and joined me in extending the behavioral-analysis program to the field of psychological measurement. Jesus Carrillo also spent a Fullbright year at UH and later, when he was a faculty member at the University Complutense of Madrid, he initiated a series of studies that extended the approach; and he translated my 1996 book to Spanish. Aimee Leduc of Laval University in Quebec also came to UH. She went on to do various studies in the growing framework and organize a society in Quebec for advancing my program. Georg

Eifert, after receiving his doctorate at the University of Frankfurt in Germany, came to UH, where he joined me in doing a theory paper on emotion that provides behavioral analysis in that important part of human behavior.

Peter Staats, while a professor at Johns Hopkins Medical School, formulated a theory of pain that expanded the framework's treatment of emotion, to which Hamid Hekmat of the University of Wisconsin contributed. He and Hekmat then did a series of studies that tested and extended the theory. Peter and I bounced ideas around, in a very productive way from the time he was in high school. Professor Hekmat also did fundamental studies for the framework on the conditioning of emotion, beginning in the 1970s, that contributed to the beginnings of cognitive behavior therapy. Jennifer Staats Kelley contributed so much to the approach as a child and later in the way she mothered her own children. With her knowledge as a child psychiatrist she helped edit several chapters of the present book, including the one on child development.

Many friends, too many to list here, have contributed in our discussions, usually disagreements. I will mention Robert Littman, a noted scholar, because in addition he was also my computer guru and was indispensable in teaching me how to operate in the digital copyediting process. Judy Lind helped me change my writing voice from the style of science to one more suited for a general audience."

One final note of gratitude: to Steven L. Mitchell for editing that improved my book's succinct meaning and help with the subtitle; to Jade Zora Ballard for her editing, surely adding to the expressive character of the book, while keeping my spirits up in the process; to Brian McMahon, Jennifer Tordy, and Lisa Michalski who also made valuable contributions, as has the entire Prometheus Books team; and to Michele Kowalsky, who employed her library science expertise to ensure the correctness of quotes and references, mahalo nui loa.

When did I begin writing this book? All told, I began it at birth, in learning my basic repertoires. Those repertoires had a life of their own and took me on my idiosyncratic behavioral journey. So my acknowledgments go to those who produced those original repertoires and to the many who have continued to make input to those repertoires. I believe that "you are what you learn," and that applies to me, and this book is a product.

REFERENCES

Allport, G. W. 1966. "Traits Revisited." *American Psychologist* 21: 1–10.

Allport, G. W., and H. S. Odbert. 1936. "Trait Names: A Psycholexical Study." *Psychological Monographs* 47 (no. 211).

Allport, G. W., P. E. Vernon, and G. Lindzey. 1951. *Study of Values*. Rev. ed. Boston: Houghton-Mifflin.

American Psychological Association. 2011. *Monitor on Psychology* 42 (1): 14.

Associated Press. 2007. "DNA Expert Says He Is Not a Racist." *Honolulu Star-Bulletin*, October 19, p. C10.

Azar, B. 2000. "Wanted: Behavioral Researchers with a Penchant for Genetics." *Monitor* 31: 36–39.

Azrin, N. H., and R. M. Foxx. 1974. *Toilet Training in Less Than a Day*. New York: Simon and Schuster.

Balter, M. 2001. "Scientists Spar over Claims of Earliest Human Ancestor." *Science* 291: 1460–61.

Balter, M. 2002. "'Speech Gene' Tied to Modern Humans." *Science* 297: 1105.

Barinaga. 2000. *Science* 288: 2116–19.

Barry, P. 2008. "Life from Scratch: Learning to Make Synthetic Cells." *Science News* 173, January 12, pp. 27–29.

Baughman Jr., F. A. 2001. "Regarding the Preschool ADHD Treatment." *Science* 291: 595.

Begley, S. 2007. "How the Brain Rewires Itself." *Time* 169, pp. 72–79.

Bijou, S. W. 1957. "Methodology for an Experimental Analysis of Child Behavior." *Psychological Reports* 3: 343–50.

Bower, B. 1997a. "Everybody's Talkin': Language's Great Innate Debate Continues to Make Noise." *Science News* 151, pp. 276–77.

———. 1997b. "The Ties That Bond." *Science News* 152, pp. 94–95.

———. 1997c. "Kids with Schizophrenia Yield Brain Clues." *Science News* 152, October 25, p. 261.

———. 2000. "Language Goes beyond Sight, Sound in Brain." *Science* 158: 373.

———. 2001. "Rumble in the Jungle: A Bitter Scientific Dispute Erupts around the Yanomami Indians." *Science News*, 159, pp. 58–60.

———. 2002a. "Autism Leaves Kids Lost in Face." *Science News* 161, June 29, p. 408.

———. 2002b. "Wild Chimps Rocked On." *Science News* 161: 195–96.

———. 2003a. "Smells like Emotion." *Science News* 163, p. 54.

———. 2003b. "Dyslexia's DNA Clue: Gene Takes Stage in Learning Disorder." *Science News* 164, March 15, p. 131.

———. 2004. "Words in the Brain: Reading Program Spurs Neural Rewrite in Kids." *Science News* 165, p. 291.

———. 2005. "Wild Gorillas Take Time for Tool Use." *Science News* 168, p. 253.

———. 2007. "Mean Streets: Kids Verbal Skills Drop in Bad Neighborhoods." *Science News* 172, pp. 388–89.

Boyd, R., and P. J. Richerson. 2005. "Solving the Puzzle of Human Cooperation." In *Evolution and Culture*, edited by S. Levinson. Cambridge, MA: MIT Press.

Brand, C. 1997. "Utopian Behaviourism—A Monument." *Personality and Individual Differences* 23 (6): 1094–95.

Brave, R. 2001. "Governing the Genome." *Nation*, December 10, pp. 18–24.

Breggin, P. R. 2001. "Questioning the Treatment of ADHD." *Science* 291: 595.

Bridget, M. 2000. "From Brain Scan to Lesson Plan." *Monitor* 31: 23–24.

Brody, J. 2009. "Personal Health: An Emotional Hair Trigger, Often Misread." *New York Times*, July 16, p. D7.

Brooks, D. 2011. "It's Not about You." *New York Times*, May 30.

———. 2012. "When the Good Do Bad." *New York Times*, March 20, p. A23.

Brown, R., and E. H. Lenneberg. 1954. "A Study in Language and Cognition." *Journal of Abnormal and Social Psychology* 49: 454–62.

Brunton, M. 2007. "The Brain: What Do Babies Know?" *Time* 169, pp. 94–95.

Burns, G. L., and P. A. Kondrick. 1998. "Psychological Behaviorism's Reading Therapy Program: Parents as Reading Therapists for Their Children's Reading Disability." *Journal of Learning Disabilities* 31: 278–85.

Buzan, D. S. 2004. "I Was Not a Lab Rat." *Guardian: G2*. March 12. Accessed March 5, 2012. http://www.guardian.co.uk/education/2004/Mar/12/highereducation.uk.

Caporael, L. 1997. "The Evolution of Truly Social Cognition: The Core Configurations Model." *Personality and Social Psychology Review* 1: 276–98.

Carey, B. 2010. "As Life Alters Genes, Illness May Emerge." *New York Times*, November 9, p. D7.

Carpenter, S. 2001. "Psychology Is Bound to Become More Darwinian, Says Eminent Primatologist." *Monitor* 32 (4): 90–91.

———. 2001b. "When at Last You Don't Succeed." *Monitor* 32 (1): 70–71.

Chase, A. 1977. *The Legacy of Malthus*. New York: MW Books.

Check, E. 2000. "Sex and the Single Fly." *Newsweek*, August 14, pp. 44–45.

Cloud, J. 2010. "Why Genes Aren't Destiny." *Time* 175 (2), pp. 49–53.

Collette-Harris, M. A., and K. A. Minke. 1978. "A Behavioral Experimental Analysis of Dyslexia." *Behavior Research and Therapy* 16: 291–95.

Colvin, G. 2009. *Talent Is Overrated. What Really Separates World-Class Performance from Everybody Else*. New York: Portfolio.

Coyle, D. 2009. *The Talent Code: Greatness Isn't Born. It's Grown. Here's How*. New York: Bantam.

Dawkins, R. 1976. *The Selfish Gene*. Oxford: Oxford University Press.

De Duve, C. 2002. *Life Evolving*. Oxford: Oxford University Press.

DeLoache, J. S., D. H. Uttal, and K. S. Rosengren. 2004. "Scale Errors Offer Evidence for a Perception-Action Dissociation Early in Life." *Science* 304: 1047–49.

Demb, Jonathan B., Geoffrey M. Boynton, and David J. Heeger. 1998. "Functional Magnetic Resonance Imaging of Early Visual Pathways in Dyslexia." *Journal of Neuroscience* 18 (17): 6939–51.

Dennis, A. 2009. "Rowan's Healing." *People*, May 25, pp. 86–89. (From *The Horse Boy: A Father's Quest to Heal His Son*, by Rupert Isaacson. Reprinted by permission of Little, Brown, New York.)

Derlega, V. J., B. A. Winstead, and W. H. Jones. 1999. *Personality: Contemporary Theory and Research*. Belmont, CA: Thompson/Wadsworth.

De Waal, F. 2001. *The Ape and the Sushi Master*. New York: Basic Books.

Diamond, J. 1992. *The Third Chimpanzee: The Evolution and Future of the Human Animal*. New York: Harper Perennial.

———. 1999. *Guns, Steel, and Germs*. New York: Norton.

Diamond, J. C., and W. C. Hall. 1969. "Evolution of Neocortex." *Science* 164: 251–62.

Dingfelder, S. F. 2007. "Move over Mice." *Monitor* 38 (3): 42.

DiVesta, F. J., and D. O. Stover. 1962. "The Semantic Mediation of Evaluative Meaning." *Journal of Experimental Psychology* 64: 467–75.

Dodes, L. M. 2008. "The Ongoing War against Addiction." *Newsweek* 151, March 17, p. 18.

Dostoyevski, F. 1993. *The Brothers Karamazov*. New York: Penguin Classics.

DSM-IV: Diagnostic and Statistical Manual of Mental Disorders. 4th ed. 1994. Washington, DC: American Psychiatric Association.

Dunbar, R. I. M. 2011. "Brains on Two Legs: Group Size and the Evolution of Intelligence." In *Tree of Origin: What Primate Behavior Can Tell Us about Human Social Evolution*, edited by F. B. M. de Waal. Boston: Harvard University Press.

Early, J. C. 1968. "Attitude Learning in Children." *Journal of Educational Psychology* 59: 176–80.

Edwards, A. 1953. *Edwards Personal Preference Schedule*. New York: Psychological Corporation.

Ehrlich, P. 2000. *Human Natures*. Washington, DC: Island Press/Shearwater Books.

Ervin-Tripp, S. 1971. "An Overview of Theories of Grammatical Development." In *The Ontogenesis of Grammar*, edited by D. I. Slobin. New York: Academic Press.

Eysenck, H. J., ed. 1960. *Behavior Therapy and the Neuroses*. London: Pergamon.

Ferster, C. B., and M. K. DeMyer. 1961. "The Development of Performances in Autistic Children in an Automatically Controlled Environment." *Journal of Chronic Disease* 13: 312–45.

Finley, J. R., and A. W. Staats. 1967. "Emotional Meaning Words as Reinforcing Stimuli." *Journal of Verbal Learning and Verbal Behavior* 6: 193–97.

Fordyce, E. E. 1990. "Learned Pain: Pain as Behavior." In *The Management of Pain*, 2nd ed., edited by J. J. Bonica, pp. 291–300. Philadelphia, PA: Lea Febiger.

Fox, P. T. 1996. "A PET Study of the Neural Systems of Stuttering." *Nature* 382, July 11, pp. 158–62.

Fox, R. 1989. *The Search for Society: Quest for a Bio-Social Science and Morality*. New Brunswick, NJ: Rutgers University Press.

Franks, C. M. 1960. "Alcohol, Alcoholism, and Conditioning." In *Behavior Therapy and the Neuroses*, edited by H. J. Eysenck. New York: Pergamon.

Frewer, A., and F. Hanefeld. 2000. "Not-so-Simple Minds." *Science* 289: 1878.

Fukuyama, F. 2011. *The Origins of Political Order*. New York: Farrar, Straus, and Giroux.

Galton, F. 1869. *Hereditory Genius: An Inquiry into Its Laws and Consequences*. London: Macmillan.

Gamez, P. B., P. M. Shimpi, H. R. Waterfall, and J. Huttenlocher. 2009. "Priming a Perspective in Spanish Monolingual Children: The Use of Syntactic Alternatives." *Journal of Child Language* 36: 260–90.

Ganon, T. S. 2008. "Finding Freedom through Complexity." *Science* 319: 1044–45.

Gelman, R. 1972. "Logical Capacity of Very Young Children: Number Invariance Rules." *Child Development* 43: 75–90.

Gibbons, A. 2010. "Breakthrough of the Year: *Ardipithecus ramidus*." Science 326: 1598–99.

Gimpel, G. A, and M. I. Holland. 2003. *Emotional and Behavior Problems of Young Children*. New York: Guilford Press.

Gladwell, M. 2005. *Blink*. New York: Back Bay Books; Little, Brown.

———. 2008. *Outliers*. New York: Little, Brown.

Goldberg, C. 2008. "DNA Anomalies Linked to Autism." *Honolulu Star-Bulletin*, January 24, p. A8.

Gould, S. J. 1977. *Ever since Darwin: Reflections in Natural History*. New York: W. W. Norton.

———. 1981. *The Mismeasure of Man*. New York: W. W. Norton.

———. 1996. *Dinosaur in a Haystack*. New York: Harmony Books.

Graeber, D. 2000. "What Did This Man Do to the Yonomami?" *In These Times* 24 (25), pp. 36–38.

Greenfield, J. 2002. "My Brother." *Time* 159 (18), p. 54.

Guisinger, S. 2008. "Competing Paradigms for Anorexia Nervosa." *American Psychologist* 63: 199–200.

Guthrie, E. R. 1935. *The Psychology of Learning*. New York: Harper.

Hall, G. S. 1923. *Life and Confessions of a Psychologist*. New York: Appleton.

Hamer, D., and P. Copeland. 1998. *Living with Our Genes: Why They Matter More Than You Think*. New York: Doubleday.

Harlow, H., and M. Harlow. 1966. "Learning to Love." *American Scientist* 54, pp. 244–72.

Harms, J. Y., and A. W. Staats. 1978. "Food Deprivation and the Reinforcing Value of

Food Words: Interaction of Pavlovian and Instrumental Conditioning." *Bulletin of the Psychonomic Society* 12 (4), pp. 294–96.

Hayes, S. C. 1998. "Resisting Biologism." *Behavior Therapist* 21: 95–97.

Hekmat, H., and D. Vanian. 1971. "Behavior Modification through Covert Semantic Desensitization." *Journal of Consulting and Clinical Psychology* 36: 248–51.

Herrnstein, C. F., and C. Murray. 1994. *The Bell Curve: The Reshaping of American Life by Differences in Intelligence*. New York: Free Press.

Hill, W. F. 1970. *Psychology: Principles and Problems*. New York: Lippincott.

Holden, C. 2008. "Meeting Briefs." *Science* 321: 486–87.

Homans, George C. 1964. "Bringing Men Back In." *American Sociological Review* 25 (5): 809–18.

Horridge, G. A. 1962. "Learning of Leg Position by Headless Insects." *Nature* 193: 697–99.

Hull, C. L. 1943. *Principles of Behavior*. New York: Appleton Century.

Hunt-Grubbe, Charlotte. 2007. "The Elementary DNA of Dr. Watson." *Sunday Times* (London), October 14.

Isaacson, R. 2009. *The Horse Boy*. New York: Little, Brown.

Jablonka, E., and M. J. Lamb. 2005. *Evolution in Four Dimensions*. Cambridge, MA: MIT Press.

Jane Goodall Institute. 2012. "Communication." Accessed February 27, 2012. http://www.janegoodall.org/chimp-central-communications.

Jaskowski, P., R. H. J. Van der Lubbe, E. Schlotterbeck, and R. Verleger. 2002. "Traces Left on Visual Selective Attention by Stimuli That Are Not Consciously Identified." *Psychological Science* 13: 48–54.

Jensen, A. R. 1969. "How Much Can We Boost IQ and Scholastic Achievement?" *Harvard Educational Review* 39 (1): 1–123.

———. 2006. *Clocking the Mind: Mental Chronometry and Individual Difference*. Oxford: Elsevier.

Johanson, D., and B. Edgar. 1996. *From Lucy to Language*. New York: Simon and Schuster Editions.

Judson, A. J., C. N. Cofer, and S. Gelfand. 1956. "Reasoning as an Associative Process. II. 'Direction' in Problem Solving as a Function of Prior Reinforcing of Relevant Responses." *Psychological Reports* 2: 501–507.

Justice, L. M., R. P. Bowles, K. L. P. Turnbull, and L. E. Skibbe. 2009. "School Readiness among Children with Varying Histories of Language Difficulties." *Developmental Psychology* 45 (2): 460–76.

Kamin, L. J. 1974. *The Science and Politics of IQ*. Potomac, MD: Erlbaum.

Kandel, E. R. 2007. "Maintain Mental Might." *Bottom Line Health* 21 (7): 8.

Kandel, E. R., and L. Tauc. 1965. "Heterosynaptic Facilitation in Neurons of the Abdominal Ganglion of *Aplysia Depilans*." *Journal of Physiology* 181: 1–17.

Kates, W., et al. 1998. "Neuroanatomical and Neurocognitive Differences in a Pair of

Monozygotic Twins Discordant for Strictly Defined Autism." *Annals of Neurology* 43: 782.

Katzir, T., and J. Pare-Blagoev. 2006. "Applying Cognitive Neuroscience Research to Education: The Case of Literacy." *Educational Psychology* 41: 53–74.

Kennedy, D., and C. Norman. 2005. "What Don't We Know?" *Science* 309: 75–102.

Kennedy, P. 1987. *The Rise and Fall of the Great Powers*. New York: Random House.

Klass, P. 2010. "When to Worry If a Child Has Too Few Words." *New York Times*, February 9, p. D5.

Kluger, J. 2007. "The Power of Birth Order." *Time* 170, pp. 42–48.

Kolb, B., R. Gibb, and T. E. Robinson. 2001. "Brain Plasticity and Behavior." In *Current Directions in Developmental Psychology*, by J. Lerner and A. E. Alberts, pp. 11–17. Upper Saddle River, NJ: Prentice-Hall.

Krasner, L., and L. P. Ullman. 1965. *Research in Behavior Modification*. New York: Holt, Rinehart, and Winston.

Kristof, N. D. 2009. "Would You Slap Your Father? If so You're a Liberal." *New York Times*, May 28, p. 23.

Kuhn, T. S. 1962. *The Structure of Scientific Revolutions*. 2nd ed. Chicago: University of Chicago Press.

Laffal, J., J. D. Lenkoski, and L. Ameen. 1956. "'Opposite Speech' in a Schizophrenic Patient." *Journal of Abnormal and Social Psychology* 52: 409–13.

Liberman, R. P. 2002. "Optimal Treatment of Schizophrenia: Drug-Behavioral Interactions or How Does the Rubber Hit the Road." *Behavior Therapist* 24: 225–26.

Lindsley, O. R. 1956. "Operant Conditioning Methods Applied to Research in Schizophrenia." *Psychiatric Research Reports* 5: 118–53.

Lovaas, O. I. 1977. *The Autistic Child*. New York: Irvington.

Lowy, J. 2004. "When a Child's Mind Is Abducted." *Honolulu Star-Bulletin*, January 24, p. C5.

Maxen, A. 2008. "Animal Origins." *Science News* 173 (7), pp. 99–100.

Mayr, E. 2001. *What Evolution Is*. New York: Basic Books.

McClearn, G. E., B. Johansson, S. Berg, N. L. Pedersen, F. Ahern, S. A. Petrill, and R. Plomin. 1997. "Substantial Genetic Influence on Cognitive Abilities in Twins 80 or More Years Old." *Science* 276: 1560–63.

McGrew, M. C. 1992. *Chimpanzee Material Culture*. Cambridge: Cambridge University Press.

McKinley Jr., J. R. 2009. "Hispanic Children Lose Ground in Early Years to Peers, Study Finds." *New York Times*, October 21, p. A19.

Meichenbaum, D. H., and J. Goodman. 1971. "Training Impulsive Children to Talk to Themselves: A Means of Developing Self-Control." *Journal of Abnormal Psychology* 77: 115–26.

Milius, S. 2000. "Why Is That Wasp Helping?" *Science News* 158, July 8, p. 31.

Miller Analogies Test. New York: Harcourt Assessment.

Mowak, M. A., N. L. Komarova, and P. Niyogi. 2001. "Evolution of Universal Grammar." *Science* 291: 114–18.

Murray, B. 2000. "From Brain Scan to Lesson Plan." *Monitor* 31: 23–24.

Murray, H. A. 1938. *Explanations in Personality*. New York: Oxford University Press.

Nagourney, A. 2008. *New York Times*, September 23, p. D6.

Newcombe, N. S. 2002. "The Nativist-Empiricist Controversy in the Context of Recent Research on Spatial and Quantitative Development." *Psychological Science* 13: 395–401.

New Shorter Oxford English Dictionary. 1993. Oxford: Clarendon Press.

New York Times. 2011. June 28, p. D3.

Nisbett, R. E. 2009. *Intelligence and How to Get It: Why Schools and Culture Count*. New York: Little, Brown.

Osgood, C. E., and G. J. Suci. 1955. "The Factor Analysis of Meaning." *Journal of Experimental Psychology* 50: 325–38.

Palkes, H., M. Stewart, and B. Kahana. 1968. "Porteus Maze Performance of Hyperactive Boys after Training in Self-Directed Verbal Commands." *Child Development* 39: 817–25.

Parker-Pope, T. 2007. "Science Unlocks Secrets of Love." *Honolulu Star-Bulletin*, February 14, p. A8.

Parsons, T., and K. B. Clark. 1966. *Societies Evolutionary and Comparative Perspectives*. Boston: Houghton-Mifflin.

Paul, P. 2007. "Tutors for Toddlers." *Time* 170, pp. 91–92.

Peterson, I. 1993. "Neural Networks for Learning Verbs." *Science News* 143 (90), February 27, p. 141.

Pinker, S. 1994. *The Language Instinct*. New York: Penguin.

———. 1999. *The Blank Slate*. New York: Norton.

———. 2011. *The Better Angels of Our Nature: Why Violence Has Declined*. New York: Viking.

Prichard, Z. 2006a. E-mail to the author. March 22.

———. 2006b. E-mail to the author. April 7.

Psychology Information Online. 2010. "Causes of Depression." Accessed March 25, 2011, http://www.psychologyinfo.com/depression/causes.html.

Public Radio International. 2009. "Mexico's Anti-poverty Program." *To the Point*. http://www.pri.org/stories/business/global-development/mexico-anti-povery-program.html.

Quill, W. G. 1999. "Subjective, Not Cognitive Psychology: The Revolutionary Theory of the Twenty-First Century." *Journal of Mental Imagery* 23, pp. 124–25.

Rawe, J. 2007. "ADHD Riddle Solved." *Time* 170, p. 49.

Reardon, D. 2000. "Diamond Success: It's in the Genes." *Honolulu Star-Bulletin*, June 12.

Reed Jr., A. L. 2001. "Horowitz's Provocation." *Progressive* 65 (3): 14–16.

Richerson, P. J., and R. Boyd. 2005. *Not by Genes Alone*. Chicago: University of Chicago Press.

Ronald, A., F. Happe, and R. Plomin. 2006. "Genetic Research into Autism." *Science* 311: 952.

Rondal, J. A. 1985. *Adult-Child Interaction and the Process of Language Acquisition.* New York: Praeger.

Rose, G. D., and A. W. Staats. 1988. "Depression and the Frequency and Strength of Pleasant Events: Exploration of the Staats-Heiby Theory." *Behavior Research and Therapy* 26: 489–94.

Rushton, J. P. 1999. *Race, Evolution & Behavior.* New Brunswick, NJ: Transaction Publishers.

Ryback, D., and A. W. Staats. 1970. "Parents as Behavior Therapy Technicians in Treating Reading Deficits (Dyslexia)." *Journal of Behavior Therapy and Experimental Psychiatry* 1: 109–19.

Sailor, W. 1971. "Reinforcement and Generalization of Plural Allomorphs in Two Related Children." *Journal of Applied Behavior Analysis* 4: 305–10.

Saltus, R. 1998. "Prescription for War." *Boston Globe*, September 22, p. C1.

Schivone, G. M. 2007. "Determining Justice in Our Current History: An Interview with Howard Zinn." *Z Magazine* 20, pp. 50–52.

Selden, N. R. W., B. J. Everitt, and L. E. Jarrard. 1991. "Complementary Roles for the Amygdala and Hippocampus in Aversive Conditioning to Explicit and Contextual Cues." *Neuroscience* 42: 335–50.

Seppa, N. 1998. "Exploring a Genetic Link to Smoking." *Science News* 153, March 7, p. 148.

Shaywitz, S., and B. Shaywitz. 2004. "Words in the Brain: Reading Program Spurs Neural Rewrite in Kids." *Science News* 165: 291.

Simon, C. 1997. "When Severe Mental Illness Is a Hereditary Risk." *Honolulu Advertiser*, March 16, p. B1.

Simos, P. G., J. M. Fletcher, E. Bergman, J. I. Breier, J. R. Foorman, E. M. Castillo, R. N. Davis, M. Fitzgerald, and A. C. Papanicolaou. 2002. "Dyslexia-Specific Brain Activation Profile Becomes Normal Following Successful Remedial Training." *Neurology* 58 (8): 1203–13.

Skalitza, A. 2011. Letter to the Editor. *New York Times*, August 16. Accessed February 27, 2012. http://www.nytimes.com/2011/08/17/opinion/children-with-autism.html?scp=1&sq=Skalitza%20Children%20with%20Autism&st=cse.

Skinner, B. F. 1938. *The Behavior of Organisms*. New York: Appleton.

———. 1957. *Verbal Behavior*. New York: Appleton-Century.

———. 1975. "The Steep and Thorny Way to a Science of Behavior." *American Psychologist* 30: 42–49.

Smith, H. W. 1952. *Man and His Gods*. New York: Universal Library.

Spier, R. E. 2002. "Toward a New Human Species?" *Science* 296 (5574): 1807–1809.

Staats, A.W. 1957. "Learning Theory and 'Opposite Speech.'" *Journal of Abnormal and Social Psychology* 55: 268–69.

———. 1963. *Complex Human Behavior*. New York: Holt, Rinehart, and Winston (with contributions by C. K. Staats).

———. 1968a. *Learning, Language and Cognition*. New York: Holt, Rinehart, and Winston.

———. 1968b. "Replication of the 'Motivated Learning' Cognitive Training Procedures with Culturally-Deprived Preschoolers." *Technical Report No. 22*. Madison, WI: Wisconsin Research and Development Center for Cognitive Learning.

———. 1971a. *Child Learning, Personality, and Intelligence*. New York: Harper and Rowe.

———. 1971b. "Linguistic-Mentalistic Theory versus an Explanatory S-R Learning Theory of Language Development." In *The Ontogenesis of Grammar*, edited by D. I. Slobin. New York: Academic Press.

———. 1972. "Language Behavior Therapy: A Derivative of Social Behaviorism." *Behavior Therapy* 3: 165–92.

———. 1975. *Social Behaviorism*. Homewood, IL: Dorsey Press.

———. 1983. *Psychology's Crisis of Disunity*. New York: Praeger.

———. 1996. *Behavior and Personality*. New York: Springer.

———. 2005. "A Road to, and Philosophy of, Unification." In *Unity in Psychology*, edited by R. L. Sternberg. Washington, DC: American Psychological Association.

Staats, A. W., B. A. Brewer, and M. C. Gross. 1970. "Learning and Cognitive Development: Representative Samples, Cumulative-Hierarchical Learning, and Experimental-Longitudinal Methods." *Monographs of the Society for Research in Child Development* 35 (8): 141.

Staats, A. W., and C. K. Staats. 1957. "Meaning Established by Classical Conditioning." *Journal of Experimental Psychology* 54: 74–80.

———. 1958. "Attitudes Established by Classical Conditioning." *Journal of Abnormal and Social Psychology* 57: 37–40.

Staats, A. W., C. K. Staats, and H. L. Crawford. 1962. "First-Order Conditioning of a GSR and the Parallel Conditioning of Meaning." *Journal of General Psychology* 67:159–67.

Staats, A. W., and D. R. Warren. 1974. "Motivation and Three-Function Learning: Food Deprivation and Approach-Avoidance to Food Words." *Journal of Experimental Psychology* 109: 1191–99.

Staats, A. W., and E. Heiby. 1985. "Paradigmatic Psychology's Theory of Depression." In *Theoretical Issues in Behavior Therapy*, edited by S. Reiss and R. Bootzin. New York: Academic Press.

Staats, A. W., and G. H. Eifert. 1990. "A Psychological Behaviorism Theory of Emotions: A Basis for Unification." *Clinical Psychology Review* 10: 1–40.

Staats, A. W., and G. L. Burns. 1981. "Personality Repertoire as a Cause of Behavior:

Specification of Personality and Social Interaction Principles." *Journal of Personality and Social Psychology* 43: 873–81.

———. 1982. "Intelligence and Child Development: What Intelligence Is and How It Is Learned and Functions." *Genetic Psychology Monographs* 104: 237–301.

Staats, A.W., J. R. Finley, K. A. Minke, and M. M. Wolf. 1964. "Reinforcement Variables in the Control of Reading Responses." *Journal of the Experimental Analysis of Behavior* 7: 139–49.

Staats, A. W., K. A. Minke, and P. Butts. 1970. "A Token-Reinforcement Remedial Reading Program Administered by Black Instructional Technicians to Black Children." *Behavior Therapy* 1: 331–53.

Staats, A. W., K. A. Minke, W. Goodwin, and J. Landeen. 1967. "Cognitive Behavior Modification: 'Motivated Learning' Reading Treatment with Sub-professional Therapy Technicians." *Behavior Research and Therapy* 5: 283, 299.

Staats, A. W., and O. W. Hammond. 1972. "Natural Words as Physiological Conditioned Stimuli: Food-Word-Elicited Salivation and Deprivation Effects." *Journal of Experimental Psychology* 96: 206–208.

Staats, A. W., and W. H. Butterfield. 1965. "Treatment of Non-reading in a Culturally-Deprived Juvenile Delinquent: An Application of Reinforcement Principles." *Child Development* 36: 925–42.

Staats, P., H. Hekmat, and A. W. Staats. 1996. "The Psychological Behaviorism Theory of Pain: A Basis for Unity." *Pain Forum* 5: 194–207.

Starkey, P., E. Spelke, and R. Gelman. 1996. "Numerical Abstraction by Human Infants." *Cognition* 36: 97–127.

Stavans, I. 2001. "What Spain Interrupted." *Nation*, October 15, pp. 37–42.

Sterelny, K. 2003. *Thought in a Hostile World*. Oxford: Blackwell.

Stokstad, E. 2001. "New Hints into the Biological Basis of Autism." *Science* 294 (5540), October 5: 34–37.

Straus, R. 2006. "20 People Who Changed Childhood." *Child*, October, pp. 107–14.

Strauss, M. S., and L. E. Curtis. 1981. "Infant Perception of Numerosity." *Child Development* 52: 1146–52.

Strong Jr., E. K., J. C. Hansen, and D. P. Campbell. 1985. *Strong Interest Inventory*. Palo Alto, CA: Stanford University Press.

Sulzer-Azeroff, B., S. Hunt, E. Ashby, C. Koniarsky, and M. Krams. 1971. "Increasing Rate and Percentage Correct in Reading and Spelling in a Class of Slow Learners by Means of a Token System." In *New Directions in Education: Behavior Analysis*, edited by E. A. Ramp and B. L. Hopkins, pp. 5–28. Lawrence: University of Kansas, Department of Human Development.

Tattersall, I., and J. H. Schwartz. 2000. *Extinct Humans*. New York: Westview Press.

Terman, I. M., and M. A. Merrill. 1937a. *Measuring Intelligence*. Boston: Houghton-Mifflin.

———. 1937b. *The Stanford-Binet Intelligence Scale*. Boston: Houghton-Mifflin.

Tierney, J. 2004. "The Study Finds Differences in Political Thinking." *Honolulu Star-Bulletin*, April 20, p. C6.

Tobias, P. V. 2003. "Encore Olduvai." *Science* 299: 1193–94.

Tomasello, M. 1999. *The Cultural Origins of Human Cognition*. Cambridge, MA: Harvard University Press.

Toulmin, S. 1972. "Human Understanding." Princeton, NJ: University of Princeton Press.

Toynbee, A. J. 1947. *A Study of History*. Oxford: Oxford University Press.

Tyre, P. 2006. "The Trouble with Boys." *Newsweek*, January 30, pp. 47–51.

Van Vugt, M., R. Hogan, and R. B. Kaiser. 2008. "Leadership, Followership, and Evolution." *American Psychologist* 63: 182–96.

Vedantam, S. 2010. "Shades of Prejudice." *New York Times*, January 18, p. D3.

Wargo, E. 2009. "Intelligence and How to Get It." *Observer* 22: 15–17.

Watson, J. B. 1930. *Behaviorism*. Rev. ed. Chicago: University of Chicago Press.

Watson, J. D. 2005. *Darwin: The Indelible Stamp*. Philadelphia: Running Press.

Wechsler, D. 1967. *Wechsler Preschool and Primary Scale of Intelligence*. New York: Psychological Corporation.

Weissberg, N. C., and D. R. Owen. 1999. "Behavior Therapy and Behavioral Genetics Are Not Enemies: A Response to Hayes." *Behavior Therapist* 22: 102–107.

Weng, J., J. McClelland, A. Pentland, O. Sporns, I. Stockman, M. Sur, and E. Theland. 2001. "Autonomous Mental Development by Robots and Animals." *Science* 291, January 26: 599–600.

Werker, J. F., and A. Vouloumanos. 2000. "Who's Got Rhythm." *Science* 288: 280–81.

White, A. D. 1955. *A History of the Warfare of Science with Theology in Christendom*. New York: Braziller.

Wickelgren, I. 2005. "Autistic Brains out of Synch?" *Science* 308: 1856–58.

Wiesenthal, J. 2008. "Madoff Golf Scores: Eerily Consistent." *Business Insider*, December 15. Accessed February 28, 2012. http://articles.businessinsider.com/2008-12-15/wall_street/30006783_1_ponzi-scheme-bernie-madoff-birdie.

Wikepedia. 2011. "Evolution of Human Intelligence: Social Brain Hypothesis." http://en.wikipedia.org/wiki/Evolution_of_human_intelligence.

Wilford, J. N. 2009. "Excavation Sites Show Distinct Living Areas early in Stone Age." *New York Times*, December 22, p. D3.

Wilson, E. O. 1975. *Sociobiology: The New Synthesis*. Cambridge, MA: Belknap Press of Harvard University Press.

———. 1998. *Consilience: The Unity of Knowledge*. New York: Alfred A. Knopf.

Wolf, M. M., T. Risely, and H. Mees. 1964. "Application of Operant Conditioning Procedures to the Behavior Problems of an Autistic Child." *Behavior Research and Therapy* 1: 305–12.

Wolpe, J. 1958. *Psychotherapy by Reciprocal Inhibition*. Stanford, CA: Stanford University Press.

Workman, L., and W. Reader. 2004. *Evolutionary Psychology: An Introduction*. 2nd ed. New York: Cambridge University Press.

Zeifman, D., and C. Hazan. 1997. "Attachment: The Bond in Pair-Bonds." In *Evolutionary Social Psychology*, edited by J. Simpson and D. Kenrick. Malwah, NJ: Erlbaum.

Zelaso, P. R., N. A. Zelaso, and S. Kolb. 1972. "'Walking' in the Newborn." *Science* 176: 314–15.

Zimmer, C. 2002. "Darwin's Avian Muses Continue to Evolve." *Science* 296 (5568): 633–34.

Zimmerman, D. W. 1957. "Durable Secondary Reinforcement: Method and Theory." *Psychological Review* 64: 373–83.

Zinn, H. 1980. *A People's History of the United States*. New York: Harper Perennial.

Zuk, M. 2002. *Sexual Selections: What We Can and Can't Learn about Sex from Animals*. Berkeley: University of California Press.

INDEX

aardvark, behavior of, 49

ability

    abilities setting humans apart from animals, 19, 35, 43, 47, 55, 57, 64, 65, 93, 166, 185, 255, 293–94, 304, 306, 316, 329

        evolution of, 44, 53, 57–58, 72, 94, 167, 258, 263, 290, 328

        progenitors of man, 263, 272, 273, 275–76, 278–80, 282, 288, 289–90, 291–92, 300, 316, 331

    attempts to identify racial differences, 28–29, 297, 299, 301–302, 303

    conceptual ability, 158, 263

    environmental experiences, impact on, 41, 72, 91, 94, 259

    and gender, 20–22, 308

    innate (natural) abilities, 29, 93, 161, 300, 304, 334, 344

    lack of ability due to inadequate learning experiences, 132–33, 155, 156, 219, 220, 227, 245, 246, 247–48, 298

    See also animals, and learning; athletic ability; human body, sensory organs; infants, learning ability of; learning; musical ability; perception and perceptural ability; reading, reading ability

abnormal behavior, 334–35

    abnormal cultural repertoires, 308–309

abnormal environment producing abnormal behavior, 211–12, 227

    abnormal learning environments, 215, 233, 242, 347

    as a learned response, 337–38

    often unrecognized, 213–14

and abnormal repertoires, 211, 213, 230, 247, 249, 250, 335, 338, 339, 345, 347

anti-learning as a basic abnormal repertoire, 247–49

attempts to identify, 210–11

behaviorist studies of, 148

biological factors in abnormal behavior, 250–51

causes of, 23

cumulative learning of, 247, 347

as deficit behavior, 214–15, 217, 221

emotional-motivational disorders, 234–42

inappropriate behavior, 118, 119, 213, 214–15, 227, 228, 234, 235, 242, 247, 251, 252

    inappropriate repertoires found in autistic children, 225, 229, 233

language-cognitive disorders, 215–34

    abnormal language behavior, 211–12

    language repertoires, 234, 247, 249–50, 296, 338

and motivation, 212, 235, 248

obsessions as learned behavior, 229

"opposite speech," 211–12

relationship to learning, 209–252
sensory-motor disorders, 242–44
use of verbal psychotherapy techniques, 348
using learning to treat problems of
behavior, 34–35, 345
abortion, 184
equated with murder, 187, 191, 308
Acheulian tools, 262–63
ADHD. *See* attention deficit hyperactivity
disorder (ADHD)
Africans, abilities of, 28–29, 297, 299,
301–302, 303
age
developmental psychology setting
normal ages for behaviors, 29,
108–109, 114, 121, 129, 132, 134,
140, 156, 178, 201
as a factor used in responding to a
child, 143, 198
not a factor in learning process, 30,
31, 114–15, 121, 128, 130, 155–
56, 179, 341
aggression, 18, 20, 21, 22–23, 118, 221
gender differences, 19, 62
and genetic variations, 60, 61, 270
See also violence
algebra. *See* numbers and counting
Algiers Motel Incident, The (Hersey), 349
Allport, Gordon, 175, 205
altruism, 15–16, 17, 19, 331
American Academy of Pediatrics, 233
American Psychiatric Association, 215
amoeba, behavior of, 71
amygdala. *See* brain
animals
abilities setting humans apart from
animals, 19, 35, 43, 47, 55, 57, 64,
65, 93, 166, 185, 255, 293–94,
304, 306, 316, 329

evolution of, 44, 53, 57–58, 72,
94, 167, 258, 263, 290, 328
progenitors of man, 263, 272,
273, 275–76, 278–80, 282,
288, 289–90, 291–92, 300,
316, 331
adaptiveness of, 77
avoidance of pain, 55–56
brain size of, 54
emotional responses
negative, 55, 78, 91, 318
positive, 55, 76, 87, 91, 105
and evolution, 289–92, 293–94
animal behaviors as products of
evolution, 18–19, 331
and food, 49, 53, 56, 71, 72, 84, 100,
267–68, 281, 313, 344
used in conditioning, 74–76, 77,
82, 84, 86–87, 91
lack of facial expressions in, 52
and learning, 73, 82–83, 311, 318–19
animal learning principles, 92–
93, 94
different in humans compared to
other animals, 47
emotional learning, 184–85
impact of time on, 311
learning ability of, 54, 65, 71,
255, 267–68, 271, 272, 300,
311, 316, 329
and motivation, 100
and personality, 60
physical features as environmental
causes, 60–61
sense organs of, 48, 49
and sex, 53, 56–57, 58, 77–78, 105,
281, 307, 311
use of by behaviorists as basis for
human behavior, 148
and vocal responses, 53

animism, 12, 210

anti-learning as a basic abnormal repertoire, 247–49

ants, behavior of, 15–16, 17, 313, 331

anxiety, 239–40, 244, 245

apes, behavior of, 36, 92, 262, 311

applied behaviorism. *See* behaviorism

applied science and understanding human behavior, 345

Ardipithecus ramidus, 257, 272, 273, 277, 280, 281, 285, 286, 288

Aristotle, 32, 41, 68

arithmetic. *See* numbers and counting

articulation organs. *See* human body, and ability to speak; speaking

artifacts of early human progenitors. *See* artistic and symbolic artifacts; tools, development of

artistic and symbolic artifacts, 264–65, 286, 288

Asians, abilities of, 28–29, 297, 299, 301–302

Asperger's disorder. *See* autism

association

    association learning, 72–73

    in conditioning, 74–75

association cortex. *See* cortex

athletic ability. *See* motor skills

attention as a reward/punishment for child, 104, 117–18, 121, 124, 129, 141, 221, 224, 225, 229, 233

attention deficit hyperactivity disorder (ADHD), 136, 142, 162, 212–13, 215, 244–46, 248, 252, 335, 336, 338

attention span, 24–25, 98, 132, 217, 342

    investigative attentiveness, 247

    more important than age in learning, 129–30

attitude testing, 190–91

auditory sense. *See* hearing, sense of

Australopithecus, 257–58, 261–62, 273, 274, 277, 280, 281, 285, 288

autism, 19–20, 23, 63, 112, 120, 123, 213, 222–34, 252, 296, 335, 337–38

    Asperger's disorder, 229

    diagnostic criteria for, 23, 223–25, 232–33

    learning principles and treatment, 35, 44, 127, 148, 225–28, 338–39, 345, 346

    prevention of, 233–34

    study of twins, 24–25

aversion therapy, 185, 236

Azrin, Nathan H., 116

Aztecs, 321–22

babbling as beginning of language, 29, 93, 121, 122–24, 155, 177, 233

baboons, behavior of, 59

balance as a sensory-motor skill, 48, 114, 197, 200

baseball, development of, 158–59, 160

basic ganglia. *See under* brain

basic repertoires (BRs). *See* repertoires, basic repertoires (BRs)

bed-wetting. *See* enuresis

bees, behavior of, 84

behavioral repertoires. *See* repertoires

behavioral sciences and understanding human behavior, 334–39, 343–44

behavioral traits, 18–19

    animal behaviors as products of evolution, 18–19

    and brain differences, 25

    and brain imaging, 23–26

    and children's development, 29–30

    Darwin seeing behavioral traits the same as physical traits, 14–16, 18, 29, 32, 36–37, 42, 172

    experimental analysis of behavior, 33

genetic causes of, 38
group differences in, 301
impact of sense of taste on, 48–49
learned or evolved, 19–20
walking as a behavioral skill of
    humans, 113–15
See also human behavior; personality
behavior-analysis movement
    author's use of behavior-analytical
        parenting, 97–100, 138
        for emotional-motivational
            development, 104–107
        for language-cognitive develop-
            ment, 123, 126–27, 129, 130–
            31, 155–56, 178–79
        for sensory-motor development,
            109–10, 113–17, 243
        teaching number concepts, 128
        use of time-outs, 118
    combining behavior analysis and
        behavior therapy, 34, 42, 44, 141,
        210, 228, 335, 345, 347
    See also marvelous learning animal
        paradigm
behavior disorders. *See* abnormal
    behavior
behavior economics, 344
behaviorism, 32, 35, 41, 73, 85–86, 147–
    48, 176
    applied behaviorism, 34–35, 332
    experimental analysis of behavior,
        33, 82
    missing link between behaviorism
        and study of human actions, 147–
        67
    overview of, 332–33
    radical behaviorism, 332
behavior modification, field of, 34, 42, 228
Bijou, Sidney W., 148
Binet, Alfred, 175

biology and learning, 70–73, 204, 297, 327
    biological factors in abnormal
        behavior, 250–51
    as foundation for understanding
        human behavior, 328–29
bipedal locomotion. *See* human body,
    locomotive structure
bipolar disorder, 240–41, 338
"blank slate," 32
Blank Slate, The (Pinker), 161
blue crabs, behavior of, 18
bodily humors and personality, 172
Boesch, Christophe, 270
bonding
    lack of in experiment with monkeys,
        100, 307
    learning to bond in infancy, 100–
        101, 104–108
    punishment detracting from positive
        bonding, 118
    time-outs not detracting from posi-
        tive bonding, 118–19
borderline personality disorder, 212
Bowlby, John, 101
Bowles, Ryan P., 341
brain, 26, 67–94
    amygdala, 24, 55
    basic ganglia, 55
    brain activity, 23, 26–27, 61, 93,
        128, 135, 298, 343
    brain differences, 134–35, 301
        in autistics and normals, 222–
            23, 224
        caused by defective biology, 220
        in Democratic and Republican
            subjects, 203
        in men and women, 62–63
        between readers and dyslexics,
            27, 134–35, 205–206, 216
    caudate nucleus, 24–25

cerebellum, 24–25
cingulate gyrus, 55
concepts about
  a genetically pre-wired brain,
    37, 54, 64
  human brain is plastic and
    changes throughout life, 37–
    38, 55, 64
connecting sensory and response
  organs, 47
cortex, 244
  association cortex, 55, 92
  cortical, subcortical, and pos-
    tural nature of personality
    (Allport's views), 205
  frontal cortex, 25
  visual cortex, 26
emotional responses in
  comparing elderly and people in
    their 20s, 30–31
  emotion centers, 100–101
energy consumed by, 43–44
hippocampus, 24
human nature in, 54–57
impact of learning on, 37, 93, 112,
  134–36, 343
  changes in the brain, 25, 26–27,
    134–36, 144
made for connecting complex input
  to complex response, 43
measuring classical conditioning
  responses, 78–79
midbrain, 55
mind/brain as cause of human
  behavior, 23–24, 161
  impact of freedom and will,
    294–96
  rejection of by behaviorists, 66
need for research on the brain and
  learning, 329–30

of a newborn, 57–58, 138, 139, 144–
  45
no longer evolving, 310, 315
personality prewired in the brain or
  learned, 203–204, 206–207
plasticity of human brain, 37–38
and the pleasure-pain center, 55–56,
  79
size of
  in early man, 258–59, 264, 268,
    273, 275, 282, 288, 292, 300
  growth of during childhood, 144
  non-upright primate brains cease
    expanding, 281
  size of human brain at birth, 57–
    58
thalamus, 55
See also neural networks and associations
brain imaging, 79
  and autism, 24–25, 222
  and behavioral traits, 23–26
  and dyslexia, 134–35, 216
    brain differences between
      readers and dyslexics, 27,
      205–206, 216
  learning produces brain changes,
    134–36
  need for research on the brain and
    learning, 329–30
  used to identify brain differences in
    behavior, 205–206
  use of to justify racism, genderism,
    poorism, 299
Brewer, Barbara, 227
Brinkman, Baba, 17
Brooks, David, 214, 309
Brown, Roger, 162
BRs. *See* repertoires, basic repertoires
  (BRs)

Burns, G. Leonard, 180–82, 194, 218
Butkos, Nick, 139
Butterfield, William, 217, 342
Buzan, Deborah Skinner, 97

Cartwright, Alexander Joy, 158
cats, behavior of, 84
    feral cat behavior, 61
    problem solving experiments of
        Thorndike, 81–82
cat scans. See CTs (cat scans)
Caucasians, abilities of, 28–29, 297, 299,
    301
caudate nucleus. See brain
cave paintings, 158, 264–65, 282, 286
Center for Reading and Language
    Research (Tufts University), 134
cerebellum. See under brain
chance and the future of humans, 320–23
Chase, Allan, 28, 297
cheetahs, behavior of, 18
Child (magazine), 142
child development
    attempts to apply behaviorism to,
        147–49
    belief that child's nature is innate,
        29–30
    capacity for learning, 137–38
        in newborns, 138–40, 141, 142,
        154–55, 224, 294
    emotional-motivational development,
        73–90, 100–108, 101–108
        children learning to be human,
            144–46
    experiences result in learning and
        neural network formation, 37
    gaining intelligence through learning,
        42
    and the "Great Scientific Error,"
        334–35

impact of a child being raised by a
    robot, 306–307
importance of providing reinforce-
    ment in child development, 85
"it takes a village" concept, 307
language-cognitive development,
    120–33
need for long childhood in humans, 58
needing to be a science of learning
    humanness, 335–39
need to change conception of chil-
    dren, 134–40
parents' conception of child
    effecting, 134, 141
parents need for training, 345–47,
    349
physical growth of the child, 143–44
sensory-motor development, 108–19
See also education; infants; parents
childhood autistic disorder. See autism
Child Learning, Intelligence, and Person-
    ality (Staats), 142
chimpanzees, behavior of, 49, 52, 53, 56,
    267–68, 311, 331–32
    brain size not expanding, 281
    study of to learn about humans, 270
    use of tools, 267, 273
choice, freedom of, 294–96
Chomsky, Noam, 120
cingulate gyrus. See under brain
Clark, Kenneth, 317
classical conditioning. See conditioning
clinical psychology, 147, 200, 339, 346,
    352
Clinton, Hillary, 307
cockroaches, behavior of, 84, 329
Cofer, Charles N., 161
cognitive psychology, 340–41
    cognitive-behavior therapies, 348
Collete-Harris, Martha, 218

commonsense knowledge, 164–66
competition, normal and abnormal repertoire, 249–50
Complex Human Behavior (Staats), 42
complex repertoires. *See* repertoires
conceptual ability, 158, 263
conditioning, 32, 296, 328–29, 348
classical conditioning, 73, 74–81,
85–88, 92, 157, 187, 332
conditioned and unconditioned stimulus, 75
deprivation-satiation and conditioning, 77–78, 100, 344
emotional responses relationship
with behavior, 85–87
emotional stimuli as incentives, 89–
90, 91
extinction of conditioning, 75–76, 82
language conditioning, 103, 187–88,
190–91
and lawfulness, 83–84
operant conditioning, 73, 81–88, 91,
92, 332
question of Skinner applying to
his children, 97
positive and negative conditioning,
76–77
use of to create positive emotional
responses, 162
use of to help infants learn, 139–40
use of to treat psychotic patients, 34
See also learning
conduct-disordered child, 234, 242, 252,
337, 338
consistency in time-outs, 119
cortex. *See under* brain
creativity, 31, 301, 307, 310
in animals, 268
human body allowing, 57
and learning repertoires, 161–64, 202

CTs (cat scans), 23
culture, 351
abnormal cultural repertoires, 308–
309. *See also* abnormal behavior
chimpanzee behavior not culture,
331–32
cultural relativity, 309
and cumulative learning, 270, 286–
88, 294
evolution of, 39, 261–65
belief that culture evolves, 160–
61
explosion of human culture,
282–85
Fukuyama's views on, 344–45
and human-selection theory,
285–86
if evolution of mind is complete,
can there be progress, 315–22
genetics not explaining human culture, 310
human-selection theory of cultural
development, 285–86, 331
importance of to humanness, 304–309
See also artistic and symbolic artifacts; humanness, progress toward;
tools, development of
cumulative learning. *See* learning

Darwin, Charles, 13–20, 35, 42, 165, 172,
205, 255, 316, 327, 345
confusing physical and behavioral
traits, 14–16, 18, 29, 32, 36–37,
42, 255
See also evolution, theory of; natural
selection
Darwin, Erasmus, 13
Dawkins, Richard, 16–17
Dawson, Geraldine, 224
deafness, 50

deconditioning, 76, 82
deer, behavior of, 60
deficit behavior. *See* abnormal behavior; human behavior
deliberate behavior, 162
Demb, Jonathan B., 26
depression, 237–41, 338
deprivation-satiation and conditioning. *See* conditioning
developmental psychology, 25, 31–32
    example of "Great Scientific Error" in, 29–30
Developmental Psychology (journal), 341–42
de Waal, Franz, 36
Diagnostic and Statistical Manual of Mental Disorders, The (American Psychiatric Association), 215, 346–47
Diamond, Jared, 38, 274–75
dimorphism, 62–63
    See also gender differences; sex
"disgust test," 184
DiVesta, Frances J., 162
divine creation, 289–92
    and foreordination of humans, 320
"DNA Anomalies Linked to Autism" (article), 222
dogs, behavior of, 37, 48, 200
    emotional responses in, 56
    lack of facial expressions in, 52
    salivation experiments of Pavlov, 74–75, 76, 77, 80, 87, 187
    vocal sounds of, 53
dolphins, behavior of, 49, 268
    vocal sounds of, 53
dopamine, 30
Doubleday, Abner, 158
Down's syndrome, 220
dyslexia, 26–27, 35, 135, 136, 215–19, 234, 283

dyslexic child having negative attitude toward school, 215
and negative emotional responses, 248
and reading, 149–51
    brain differences between readers and dyslexics, 27, 134–35, 205–206, 216
    use of tokens as reinforcement, 129

early man, evolution of. *See* evolution, theory of, evidence of human evolution; hominids; hominins; *Homo sapiens*
education, 26–27
    school performance
        dyslexic child having negative attitude toward school, 215
        impact of poor language development skills, 341
        and intelligence, 182–83
        need to educate parents to increase children's performance, 350
        ways schools could improve learning opportunities, 342
    See also child development; learning
educational psychology, 26, 341–42
Edwards, Allen L., 191
Ehrlich, Paul, 36–37
Einstein, Albert, 31, 202
elderly. *See* gerontology
elephants, behavior of, 60
emotional-motivational development, 247
    emotional-motivational disorders, 234–42
        lack of empathy and affection. *See* autism
        personality tests, 189–95
        repertoires, 153–54, 157, 294
            affecting cognitive behavior, 162

affecting interests, 192–93
attitudes, 190–91
emotion motivation, 188–95
as facets of personality, 178,
184–95
interests, 191–93
needs, 194–95
testing of, 195
values, 193–94
See also punishment; rewards,
impact of
emotional responses, 76
and classical conditioning, 74, 79–80
comparing elderly and people in
their 20s, 30–31
emotional characteristics of language, 108, 177, 186–87
food words eliciting response,
78, 86–90
emotional stimuli as incentives, 89–
90, 91. *See also* rewards, impact of
emotion motivation, 187
emotional-motivational development, 100–108
and personality, 188–95
to food, 55, 77–78, 79, 82, 84, 86–
90, 101, 177, 185, 186, 187–88
humans showing through facial
expressions, 52
impact of lack of on infants, 100–
101, 105–106. *See also* autism
and learning, 73–90, 101–108, 293
emotional learning of the animal
variety, 184–85
emotional learning through language, 186–88
learning to respond emotionally,
55–56, 185
use of repertoires to learn, 157
measuring by interest tests, 192–93

to music, 78, 89, 102, 108, 138, 149,
153, 157
relationship with behavior, 85–87.
See also human behavior
and sensory organs, 55
to sex, 79, 82, 87, 185, 186, 189
See also attitude testing; negative
emotional responses; positive
emotional responses
empathy, 223
encopresis, 242–44
See also toilet training
enuresis, 117, 242–44
See also toilet training
environment
abnormal environment producing
abnormal behavior, 211–12, 221,
227, 237, 250–51
finding right environment, 231–
32
often unrecognized, 213–14
and conditioning in animals, 76
environmental experiences impact on
abilities, 41, 72, 91, 94, 259
hidden environment, 221
and attention deficit hyperactivity disorder, 245
and autism, 232–33
and mental retardation, 219–21
impact of on human behavior, 38, 40,
41, 161
physical features as environmental
causes, 60–61
of progenitors of man, 272
epigenetics, 209
equality, 299–300
evolutionary concept of, 300–304
estrous cycle, 58–59
ethology, 331
eugenics, 173, 279

euphoria, 240

European Journal of Developmental Psychology, 29

evaluative meaning, 186

evolution, theory of, 255–92, 345

  in animals, 293–94

    viewing humans as animals, 255, 293

  assuming emotions are genetic, 100

  biology and learning, 70–73

    evolution of basic learning principles, 91–93, 94

  confusing physical and behavioral traits, 14–16, 18, 29, 32, 36–37, 42, 255

  development of human legs and hands, 50–51

  evidence of human evolution, 293–94. *See also* hominids; hominins; names of individual species, i.e., *Ardipithecus ramidus, Australopithecus, Homo sapiens, etc.,*

    beginning of human evolution and learning, 288

    belief that intelligence is an evolved trait, 255

    brain no longer evolving, 310, 315

    changes in ways scientists study, 269–70

    culture as evidence, 261–65

    development of language, 267–68, 273–75

    Homo sapiens as endpoint of human evolution, 321

    human body, 47–65

    human progress not occurring through evolution, 309

    human selection theory, 271–77

    infinite learning and continuing evolution, 309–315

    missing-link learning, 275–77

    potential for additional human progress, 315–22

    progress toward humanness, 265–68

    reasons why it is different from animal evolution, 289–92

    stages in development of *Homo sapiens*, 256–60

    view that violence is part of human evolution, 309

  evolutionary concept of equality, 300–304

  infinite learning and continuing evolution

    compacting and archiving knowledge, 314–15

    infinite learning across space, 310–11

    infinite learning across time, 311–13

    role of specialization in, 313–14

  and the learning-biology relationship, 328–29

  non-upright primates brains cease expanding, 281

  not allowing for divine creation, 289

  not applying to learning, 67–70

  opponents to evolution believing in "intelligent design," 289–90

  and quick cultural development, 305–306

  and racism, 296–300

  See also natural selection; neo-Darwinism; social Darwinism

evolutionary psychology, 16–17

Evolution of Human Behavior (journal), 16

experimental analysis of behavior, 33, 82

experimental psychology, 41

external nature of human behavior. *See* human behavior

extinction of conditioning, 75–76, 82

eye-hand coordination, 110–11, 154, 197, 198, 273

Eysenck, Hans, 34

face, human, 51–52
    changes to the face in progenitors of man, 258, 260
    infants reacting to faces, 104–105, 109, 139, 140, 224

Fassel, Ian, 138–39

feeling. *See* touch, sense of

Finley, Judson, 217

Fisher, Helen, 79

five-factor theory of personality, 175

Fletcher, Jack M., 134–35

food
    and animals, 49, 53, 56, 71, 72, 84, 100, 267–68, 281, 313, 344
        used in conditioning, 74–76, 77, 80, 82, 84, 86–87, 91, 187
    in archaeological and fossil research, 261, 265, 272–73, 276, 280
    and brain size, 281
    causing emotion-eliciting and reinforcing, 55, 77–78, 79, 82, 84, 87, 101, 177, 185, 186, 187–88
        food words eliciting response, 78, 86–90
    use of in bonding, 104, 106
    and sensory loss, 50
    use of in learning process, 123, 179, 182

foreordination of humans, 320–23

fossils and the evolution of humans, 256, 257–60, 262, 265, 288, 291

Foxx, Richard M., 116

Frankenstein, Dr. (fictional character), 70

freedom and choice, 294–96

Freud, Sigmund, 68, 101, 116, 173

frontal cortex. *See* cortex

fruit fly, behavior of, 18

Fryer, Roland, 219

Fukuyama, Francis, 344–45

functional behavioral repertoires. *See* repertoires

Galton, Francis, 15, 28, 205

Gelfand, Sidney, 161

gender differences, 19–20
    in ability, 20–22, 308
    genderism, 299–300
    physical differences, 62–64
    psychological differences, 61–62

general anxiety disorder, 239

generalists, humans as, 47, 56–57, 64–65

geometry
    babies having innate knowledge of, 29, 334. *See also* numbers and counting
    geometric design test (Wechsler Preschool and Primary Scale of Intelligence), 181, 200

gerontology
    example of "Great Scientific Error" in, 30–31
    and the plasticity of the human brain, 38

giraffes, behavior of, 19, 48, 49

Gladwell, Malcolm, 40

Goddard, H. H., 28, 297

Goodman, Joseph, 162

gorillas
    behavior of, 266–67, 331, 352
    physical features, 50, 257, 277, 281

Gould, Stephen Jay, 28, 297, 316, 320

grammar, 120, 127–28, 274, 284, 333
    grammar. *See also* writing

See also language-cognitive
    development
gravitation, theory of, 68, 69, 75, 83–84,
    165, 315, 352
"Great Scientific Error" paradigm, 12–13,
    20–32, 79
    alternative to, 36–44
    belief that the brain is wired for
        "handedness," 113
    biological determination of human
        behavior, 137, 234, 329
    Darwin seeing behavioral traits the
        same as physical traits, 14–16, 18,
        29, 32, 36–37, 42, 255
    denial of parental influence on devel-
        opment of child, 142
    found in sciences
        in child development, 133, 334–35
        in developmental psychology,
            29–30
        in field of abnormal behavior,
            209, 335
        in gerontology, 30–31
        in the philosophy of science,
            31–32
    lack of understanding of learning,
        329, 331
        belief that infant development
            driven by biology not
            learning, 110, 120, 134
        impact of on children learning
            to speak, 123
        impact of on toilet training, 117
        leaving learning out of consider-
            ation, 32–34, 344–45
    need to change conception of chil-
        dren, 134–40
    prewiring of the brain
        belief that the brain is wired for
            "handedness," 111

seeing personality as prewired
    in the brain, 203–204
and racism, 27–29, 296–300
view of autism, 222, 225, 229
view of dyslexia as a mental illness,
    219, 261
weaknesses of, 36–37
Gross, Michael, 227
group-behavioral repertoires. *See* reper-
    toires
group cumulative learning. *See* learning
Guthrie, Edwin R., 85

"handedness," 111–13, 136, 144
hands, human. *See* human body
Harlow, Harry and Margaret, 100, 105,
    307
Harvard Laboratory, 30
Head Start preschool programs, 130–31
hearing, sense of, 50, 55, 57
    See also human body, sensory
        organs; sensory-motor develop-
        ment
Heiby, Elaine, 237
*heidelbergensis. See Homo heidelber-
    gensis*
Hekmat, Hamid, 241
heredity and personality, 205–206
Hersey, John, 349
hidden environment. *See* environment
hippocampus. *See under* brain
Hippocrates, 172
*History of the Warfare of Science with
    Theology in Christendom, A* (White),
    287
Homans, George, 344
hominids, 20, 275, 290, 300, 309, 331
    beginning of hominid line, 257
        apelike qualities disappearing in,
            262–63

culminating in *Homo sapiens*,
20, 258, 285, 291
development of hominid species
overlapping in time, 258–59,
261
cultural development of, 264
tool making, 262–64, 272
size of hominid heads and brains, 58,
264, 282
walking and upright stance of, 271–
72, 274
hominins, 44, 259, 265, 268, 275, 277,
278–79, 286, 300, 306–308, 317, 321
Homo erectus, 258–59, 262–63, 285
Homo ergaster, 262–63
Homo habilis, 258–59, 261–62, 281, 288
Homo heidelbergensis, 259
Homo neanderthalensis, 259, 287
Homo sapiens, 44, 65, 211, 277, 282–83,
316, 319
capacity for learning, 277, 279, 286,
291, 310
first appearance of, 200, 258, 268,
283, 284, 292, 305
Homo sapiens as endpoint of
human evolution, 321
speed of evolution of, 67–68,
259, 282, 285
as a generalist, 57
growth in sensory-motor repertoires,
200–201
impact of being raised by robots,
306–307
specialness of, 57, 200–201, 266
stages in human development, 256–
60
See also evolution, theory of, evidence of human evolution
homosexuality, 20, 24, 112, 189, 190, 236
homunculus, 172

Horridge, G. A., 329
horses, behavior of, 19
Hull, Clark, 32–33, 82, 85, 255, 332
human behavior, 12, 36–37, 164
altruism as a human act, 15–16, 17
behaviors come in groups, 173–74
biologically oriented conception of,
137, 297
complex activities that are not
behaviors, 149, 153
Darwin seeing behavioral traits the
same as physical traits, 14–16, 18,
29, 32, 36–37, 42, 63–64, 172, 255
deficit behavior, 214–15, 217, 221
deliberate vs. impulsive, 162
emotional-motivational development,
100–108
emotional responses relationship
with behavior, 85–87. *See also*
emotional responses
environmental impact on, 32, 38, 40,
41, 161
and evolution, 19–20
attempts to link differences to
race, 28
not studied by evolutionary psychology, 17
experimental analysis of behavior,
33, 82
explained through the marvelous
learning animal paradigm, 353.
See also marvelous learning
animal paradigm
external nature of human behavior,
161, 164
facial expressions impact on, 52–53
foundations for understanding
applied science, 345
behavior sciences, 334–39
psychology fields, 339–43, 347–48

sciences, 328–34
social sciences, 343–45
and freedom of choice, 294–96
genetic causes of, 309–10
human body as essential part of
human nature, 57
impulsive behavior, 21, 28, 162, 212,
248, 297
inappropriate behavior, 118, 119,
213, 214–15, 227, 228, 234, 235,
242, 247, 251, 252
inappropriate repertoires found
in autistic children, 225, 229,
233
internal nature of human behavior,
161, 164
early Greek theories about, 172
and learning
children learning to be human,
144–46
learned behaviors, 31, 114, 128,
146, 229, 280, 333
learning as cause of, 34–35, 37,
93–94, 331, 344
and motor behavior, 73
using learning to treat problems
of behavior, 34
mind/brain as cause of, 23–24, 161
impact of freedom and will,
294–96
rejection of by behaviorists, 33
obsessions as learned behavior, 229.
See also abnormal behavior
parental treatment impact on, 68
and personality. See also personality
as a cause of behavior, 33, 189,
195, 201, 206–207
changes in personality as a
result of learning, 201–203,
207

consistency of behavioral characteristics, 174
and the selfish gene theory, 16–17
sensory-motor repertoires, impact on,
197–98
social behavior, 15, 53, 142, 144,
191, 201, 219, 270, 300, 317
changes in social practice as
progress, 319–20, 321–22
lack of social skills. See autism
study of as a "young" science, 161
use of animals by behaviorists as
basis for study of, 148
variability of, 171–208
wide variety of sciences and social
sciences concerned with, 351–52
See also behavioral traits; humanness, progress toward
human body, 47–65, 93–94, 257
and ability to speak, 51, 53–54. See
also speaking
allowing flexibility in location, 49
body differences and motor performance, 198–200
effect on gender differences, 62–64
as an essential part of human nature,
57
evolution of, 260. See also evolution,
theory of, evidence of human evolution
face and articulation organs. See also
speaking
hands
and eye-hand coordination, 110–
11, 154, 197, 273
"handedness," 111–13, 136, 144
thumb, 51, 257, 292
locomotive structure, 50–51
bipedal locomotion, 51, 258, 260,
265, 271, 272, 280, 290–91

delays in walking as indication
of few learning opportunities,
114, 115
upright locomotion as a human
trait, 113–15
motor performance and body differences, 199–200
muscular characteristics of humans,
50–54
and kinaesthetic stimuli, 55
neural networks and associations, 26,
52, 72, 80, 91, 134–35, 139, 175,
199, 328
and human evolution, 56, 268,
270
pain receptors, 48
physical features, 52–53
Darwin seeing behavioral traits
the same as physical traits,
14–16, 18, 29, 32, 36–37, 42,
63–64, 172
as environmental causes, 60–61
face, 51–53
and psychological differences in
humans, 61–62
physical growth of the child, 143–44
physical response to depression, 237
response organs, 47, 56, 67, 71, 72,
80, 166, 211, 328
sensory organs, 47–50, 55, 57, 67,
71–72
and emotional responses, 55
and eye-hand coordination, 110–
11
humans showing their generality, 57
sensory abilities of humans, 50,
55, 181, 199–200
sexual biology and bonding, 58–59.
See also sex

skeletal and muscular characteristics,
50–54
See also brain; sensory-motor development
human creation, 289–92
human evolution. *See* evolution, theory
of, evidence of human evolution;
hominids; hominins; *Homo sapiens*
human nature. *See* human behavior
humanness, progress toward, 265–67, 331
abilities setting humans apart from
animals, 19, 35, 43, 47, 55, 57, 64,
65, 93, 166, 185, 255, 293–94,
304, 306, 316, 329
Christian theory about, 255
cultural developments, 261–65. *See
also* culture
artistic and symbolic artifacts,
264–65
explosion of human culture,
282–85
human-selection theory, 285–86
importance of culture, 304–309
tools, 261–64, 274, 277–78,
280, 285, 306, 309
and evolution
evolution of language in early
man, 267–68, 273–77
if evolution of mind is complete,
can there be progress, 315–22
physical development, 256–60,
271–72
evolutionary concept of equality,
300–304
impact of being raised by robots,
306–307
learning as the new paradigm of
"humanness," 327–53
need to use various scientific and
social sciences to improve, 351–52

science of learning humanness, 335–39

stages in human development, 316–17

human progress not occurring through evolution, 309

human-selection theory of cultural development, 285–86, 331

hummingbirds, behavior of, 18

hypomania, 240

hypothalamus, 55

id, 173

impulsive behavior. *See* human behavior

inappropriate behavior. *See* abnormal behavior; human behavior

incentives. *See* rewards, impact of

inequality, 299–300

infants

   bonding with, 100–101, 104–108

   brain of a newborn, 57–58, 138, 139, 144–45

   early phases of learning to walk, 114

   learning ability of, 138–40, 141, 142, 154–55, 224, 294

   sensory-motor development in, 108–10

   See also babbling as beginning of language; child development

infinite learning and continuing evolution

   compacting and archiving knowledge, 314–15

   across space, 310–11

   across time, 311–13

   role of specialization in, 313–14

innate (natural) abilities. *See* ability

intelligence, 312

   and behavioral repertoires, 179

   as a complex activity, 149

   and eugenics, 173

   in identical twins, 25, 28

   impact of learning on, 42, 68, 205

   measuring of, 33, 175, 301–302

      black-white differences, 303

      misdiagnosing mental retardation, 220

      as predictor of school performance, 182–83

      Revised Stanford-Binet Intelligence Scale, 178, 180

      training in basic repertoires increasing scores, 177, 178, 179, 180–83, 206

      Wechsler Preschool and Primary Scale of Intelligence (WPPSI), 181, 200

   and personality, 42, 178–82, 204–205

   and scientific racism, 28–29

   theories about, 183–84

      Galton's theory of, 15

      Nisbett's theory of, 182

   as a trait that evolved, 255

"intelligent design," 289, 291

interests, 191–93

internal nature of human behavior. *See* human behavior

investigative attentiveness. *See* attention span

Iraq War, 22

Isaacson, Rupert, 231–32

"Is Cash the Answer?" (*Time*), 219

"I Was Not a Lab Rat" (Buzan), 97

Jablonka, Eva, 36

jaguar, behavior of, 49

jelly fish, behavior of, 328

Johns Hopkins University, 156

Journal of Applied Behavior Analysis, 228

Journal of Child Language, 333

Judson, Abe J., 161
Judson, Olivia, 17
Justice, Laura M., 341

Kahana, Boaz, 162
Kandel, Eric R., 72, 137, 329
Kenyanthropus, 257
kinaesthetic stimuli, 55
Kondrick, Patricia Ann, 218
Kramer, Arthur, 38
Krasner, Leonard, 34
Kristof, Nicholas, 184
Kuhn, Thomas, 11, 36, 43

Lamb, Marion, 36
language-cognitive development, 120–33
    complexity of language, 127–28
    and emotions
        emotional characteristics of lan-
            guage, 78, 86–90, 108, 177,
            186–87
        language and love, 105–106
        language conditioning and emo-
            tional responses, 103, 187–88
        learning emotion through lan-
            guage, 186–88
    and grammar, 120, 127–28, 274,
        284, 333
    importance of bonding to language,
        104
    importance of in development of lan-
        guage, 275–78
    lack of language. *See* autism
language-cognitive disorders, 215–34
    abnormal language behavior,
        211–12
    language repertoires, 234, 247,
        249–50, 296, 338
    language-conditioning, 103, 187, 191
        study, 187–88, 190–91

language not a behavior, 149, 153
language repertoires, 202, 205, 294,
    296, 330
    aiding in counting and number
        repertoires, 156, 330. *See also*
        numbers and counting
    author teaching own children,
        123, 126–27, 129, 130–31,
        155–56, 178–79
    in children, 124, 128, 153–54,
        155, 158, 177, 181–82, 183,
        201, 227, 250, 341, 342. *See
        also* babbling as beginning of
        language
    development of in progenitors
        of man, 262, 267–68, 273–76,
        277, 306, 313
    as facets of personality, 178
    importance of, 154, 156–57
    and intelligence tests, 177, 179,
        180, 183, 206
    word-imitation repertoire, 154–
        55
    word-reading repertoire, 150
language skills, 68
learning and language
    how language is learned, 120
    impact of poor language devel-
        opment skills on school per-
        formance, 341
    impact of sensory organs on, 50
    impact on negative emotional
        responses on, 88, 102–103,
        107, 108, 145, 187, 190–91,
        217, 238, 248
    importance of learning in the
        development of language,
        275–78, 279
    language learning able to begin
        at birth, 139

learned or evolved, 19–20

linguistic method for learning language, 120

use of positive emotional responses to encourage, 108, 122–23, 125–26, 131, 140, 145, 162, 187, 224, 231, 244, 246, 247, 248

verbal imitation, 124

verbal-labeling responses, 125, 179–80, 227

verbal-motor learning, 125–27

measured in intelligence testing, 179

verbal-motor repertoire, 149–51

See also reading; speaking; writing

Lascaux cave (France). See cave paintings

lawfulness, 83–84, 123

Leakey, Mary, 280, 288

learning, 30–31

and abnormal behavior, 209–252. See also abnormal behavior

abnormal learning environments, 211, 212, 213, 215, 233, 242, 347

as a learned response, 337–38

often unrecognized, 213–14

accelerated pace of, 322

and animals, 71, 73, 82–83, 92–93, 311, 318–19

animal learning principles, 92–93, 94

different in humans compared to other animals, 47

emotional learning, 184–85

impact of time on, 311

learning ability of, 54, 65, 71, 255, 267–68, 271, 272, 300, 311, 316, 329

association learning, 72–73

and attention span. See also attention deficit hyperactivity disorder (ADHD); attention span

more important than age in learning, 129–30

basic principles of, 73–90, 147–67

See also conditioning; cumulative learning; repertoires

and behavior. See also behavioral traits; human behavior

causing human behavior, 34–35, 37, 93–94, 331, 344

timing of behavior development, 98

behaviorism, views on, 32–33

and biology, 70–73

evolution of basic learning principles, 91–93

and the brain, 37, 67–94, 112, 134–36, 343. See also brain

changes in the brain, 25, 26–27, 134–36, 144

and children's development, 29–30. See also child development; infants

capacity for learning, 93, 138–42, 154–55, 224, 294

impact of a long childhood on learning, 58

parents needing training, 345–47, 349

and conditioning, 32. See also conditioning

connecting complex input to complex response, 43

culture as learned, 305–306. See also culture

cumulative learning, 154–58, 164–67

of abnormal behaviors, 247

accelerated pace of, 284–85
in artistic and symbolic artifacts,
    286, 288
as a basis for creativity, 163
of complex repertoires, 163,
    293, 300, 331–32
as a cultural process, 270
group cumulative learning, 158–
    61
religious values as, 286–87
role of in abnormal repertoires,
    347
sensory-motor repertoires result
    from, 197
speed of, 318
and unlearning autism, 229–32
and emotions, 55–56, 73–90, 101–
108. *See also* emotional responses
group cumulative learning, 158–61
growth of learning
    increased by ability to commu-
        nicate across space, 310–11
    over time and generations, 310–
        13
    through specialization and divi-
        sion of labor, 313–14
and humanness
    need for a science of learning
        humanness, 335–39. *See also*
        humanness, progress toward
    as the new paradigm of "human-
        ness," 327–53
impact of sensory organs on, 50. *See
    also* sensory organs
infinite learning and continuing evo-
    lution, 309–15
    compacting and archiving
        knowledge, 314–15
    infinite learning across space,
        310–11

infinite learning across time,
    311–13
role of specialization in, 313–14
and intelligence, 68, 205. *See also*
    intelligence
lack of ability due to inadequate
    learning experiences, 132–33, 155,
    156, 219, 220, 227, 245, 246, 247–
    48, 298
learned behaviors, 31, 114, 128, 146,
    229, 280, 333
and learned natures, 60
missing-link theory (learning), 176–
    201, 206–208, 294. *See also*
    repertoires
    and human evolution, 275–77
and motor behavior, 73
negative emotional responses, impact
    on learning, 88, 102–103, 107, 108,
    145, 187, 190–91, 217, 238, 248
paired-associate learning, 150
and personality, 201–203, 207. *See
    also* personality
    as the basis of the missing-link
        theory of personality, 176–201
positive emotional responses used to
    encourage learning, 108, 122–23,
    125–26, 131, 140, 145, 162, 187,
    224, 231, 244, 246, 247, 248
and progress, 315–23
    as progress through generations,
        158–61
research questions that need to be
    answered, 333
school performance
    dyslexic child having negative
        attitude toward school, 215
    impact of poor language devel-
        opment skills, 341
    and intelligence, 182–83

need to educate parents to increase children's performance, 350

ways schools could improve learning opportunities, 342

and sexual biology of humans, 59

specialization as aid to learning, 313–14

talent vs. learning, 199–200

trial-and-error learning, 318

verbal-motor learning, 125–27

measured in intelligence testing, 179

verbal-motor repertoire, 149–51

versatility and learning ability, 65

word-learning ability, 155

See also emotional-motivational development; language-cognitive development; punishment; repertoires; rewards, impact of; sensory-motor development

left-handedness, 111–13

Legacy of Malthus, The (Chase), 28, 297

Lenneberg, Eric, 162

limbic system, 55, 79

See also brain; pleasure-pain center

Lindsley, Ogden R., 148

linguistic method for learning language, 120

lions, behavior of, 18–19, 48–49, 60, 61, 199–200

locomotive structure of humans. See human body

Los Angeles Family and Neighborhood Survey, 350

Lovaas, O. Ivar, 228

love in humans, 59

as a learned response, 105–106. See also bonding

need for in training children, 99

Lucy (early hominid), 273, 280, 306

macaque monkeys, behavior in, 100

Madoff, Bernard, 173–74

magnetic resonance imaging. See MRIs (magnetic resonance imaging)

Man and His Gods (Smith), 287

mania, 234, 240

Manning, Peyton and Eli, 115

marginal utility, principal of, 344

marvelous learning animal paradigm, 92, 93, 94, 137–38, 140, 144–45, 166–67, 183–84, 185, 196–97, 308, 351, 353

as an emotional being, 186–88

interpretation of abnormal behavior, 211–52

See also behavior-analysis movement

Marx, Karl, 316

masturbation, 235–37

mathematics. See numbers and counting

McGrew, William, 270

measuring. See testing (psychological measurement)

Meichenbaum, Donald, 162

memory, 24, 151, 162, 197, 270, 295

Mendel, Gregor, 13, 42, 165, 327

Mendeleev, Dmitri Ivanovich, 163

mental illness. See abnormal behavior

mental retardation, 219–21

Merzenich, Mike, 38

microcephaly, 220

midbrain. See brain

Miller Analogies Test, 205

mind/brain as cause of human behavior, 23–24, 161

impact of freedom and will, 294–96

rejection of by behaviorists, 33

Minke, Karl, 132, 180, 217, 218, 225, 335

Mismeasure of Man, The (Gould), 28, 297

missing-link theory (learning), 176–201, 206–208, 294

and human evolution, 275–77
  See also learning, cumulative
    learning; repertoires
monkeys, behavior of, 100, 307
motivation
  and abnormal behavior, 212, 235,
    248
  in animals, 100
  and attitude, 191
  and incentives, 102
  love and hate as motivation for
    behavior, 86
  and the selfish gene theory, 16, 17
  See also emotional-motivational
    development; rewards, impact of
motor skills, 293
  athletic ability, 115–16, 156–57,
    197–98, 203
    changes in high-jumping skills
      and techniques, 312
    as a complex activity, 149
    "natural athlete," 196
    superior performances by black
      Americans in, 303
  based on successive, cumulative
    learning, 196–97
  body differences and motor perform-
    ance, 198–200
  impact of learning on, 68
  individual differences in, 196
  motor behavior and learning, 73
  tests of, 200–201
  See also sensory-motor development
Mousterian tools, 263
"mouthing" in infants as a reward, 110–
  11
Movellan, Javier, 138–39
MRIs (magnetic resonance imaging), 23,
  24–25
Murray, Henry A., 191

muscular characteristics of humans. *See*
  human body
musical ability, 157

National Football League, 115
National Institutes of Health (NIH), 209
National Science Foundation, 38
Native Americans, 299, 311
"natural athlete." *See* motor skills, athletic ability
natural selection, 15, 31, 265, 278
  human selection theory, 271–77
  impact on brain size, 281
  language as a selection device, 275–
    77
  and learning, 91–93, 300
  not distinguishing human and animal
    evolution, 289–90
  and race, 297
  for survival and reproduction advan-
    tage, 316
  use of to explain behavioral charac-
    teristics, 299
  vs. human-selection theory, 331
  See also evolution, theory of
nature-nurture schism, 67, 92, 340, 341
Nazi Germany, 296, 309
Neanderthals. *See Homo neanderthalensis*
needs, 58, 100, 116, 164, 173, 188, 189,
  194–95, 230, 233, 234, 319, 334, 343
negative emotional responses, 293
  in animals, 55, 78, 91, 318
  choosing positive stimuli over nega-
    tive, 89–90, 189
  deprivation increasing strength of
    emotional response, 78
  in humans, 77, 80, 90, 184, 185, 188,
    190, 238–39
    and anti-learning repertoire, 248
    anxiety, 239–40

bipolar disorder, 240
and bonding, 105, 107
depression, 237–38
impact on learning and lan-
    guage, 88, 102–103, 107, 108,
    145, 187, 190–91, 217, 238,
    248
to pain, 101, 241–42
physical reactions to, 77, 78, 80
and politics, 90, 203
as punishment, 90, 102–103,
    107, 127, 189, 319
reactions to abortion, 187
unlearning wired-in responses,
    55
used to treat paraphilia, 236
See also emotional responses; posi-
    tive emotional responses
negative reinforcement, 322
impact of on children learning to
    speak, 125
not to be used in toilet training, 117
See also punishment
neo-Darwinism, 39
nerve net in jellyfish, 328
neural networks and associations, 26, 37–
    38, 52, 72, 80, 91, 134–35, 139, 175,
    199, 328
and human evolution, 56, 268, 270
in simple organisms, 91, 328
neurobiology, 27
New Guineans use of language, 274–75
"New Hints into the Biological Basis of
    Autism" (Stokstad), 339
New Shorter Oxford English Dictionary,
    151
Newton, Isaac, 69
New Yorker (magazine), 11
New York Times (newspaper), 193, 214,
    229

NIH. See National Institutes of Health
    (NIH)
Nisbett, Richard E., 40, 182
Noah (an autistic child), 230
nonfunctional repertoires. See repertoires
numbers and counting, 128
    age children can learn to count
        according to Piaget, 156
    and algebra, 154, 156, 184, 333,
        340–41
    babies having innate knowledge of
        geometry, 29, 334
    not a behavior, 153
    number repertoire as part of measure
        of intelligence, 178, 181, 183
nurture. See nature-nurture schism

obsessions as learned behavior, 229
obsessive-compulsive disorder, 239
occupational behavior skills, 200
Odbert, H. S., 175
Olduwan tools, 262, 263
O'Neal, Shaquille, 115
operant conditioning. See conditioning
"opposite speech," 211–12
Orronin tugenesis, 257
Osgood, Charles, 186
Outliers (Gladwell), 40

pain, 241–42
    feeling positive emotion from others'
        pain, 88, 236
    pain receptors, 48
    pleasure-pain center, 55–56, 79
paired-associate learning, 150
paleoanthropology, 256, 288
    as foundation for understanding
        human behavior, 330–32
Paleolithic tools. See tools, development
    of

paleontology, 256
Palkes, Helen, 162
pandas, behavior of, 48–49
panic disorders, 239
paradigms, 36–44
    learning as the new paradigm of
        humanness, 327–53. *See also*
        humanness, progress toward;
        learning
    as a term, 11–13
    See also "Great Scientific Error" par-
        adigm; marvelous learning animal
        paradigm
paranoid schizophrenia, 234
paraphilias, 235–37
parents
    author's use of behavior-analytical
        parenting, 97–100, 138
        for emotional-motivational
            development, 104–107
        for language-cognitive develop-
            ment, 123, 126–27, 155–56,
            178–79
        for sensory-motor development,
            109–10, 113–17, 243
        teaching number concepts, 128
        teaching reading, 129, 130–32
        use of time-outs, 118
    conception of children and child
        development, 134, 141
    getting attention from parent as a
        reward for poor behavior, 117–18,
        119, 213–14
    impact of parental treatment on
        human behavior, 68
    impact on behavior traits their chil-
        dren display, 143
    role in helping increase child's intel-
        ligence, 182
    schools need to educate parents to

        increase children's performance,
        350
    training needed in child develop-
        ment, 345–47, 349
        materials to help parents learn
            about learning, 142
Parsons, Talcott, 317
Pavlov, Ivan, 32, 69, 74–75, 76, 80, 83,
    86, 185, 187, 255, 332, 340
Pearson, Karl, 28, 297
pedophilia, 235
perception and perceptural ability, 132–
    33, 162
    dyslexics showing deficit in percep-
        tual ability, 218–19
periodic table of elements, 163
personality, 171–208, 305, 334–35
    abnormal personalities and learning,
        209–52. *See also* abnormal
        behavior
    and behavior, 33, 189
        as a cause of overt behavior,
            195
        consistency of behavioral char-
            acteristics, 174
        as an internal cause, 201, 206–
            207
    biological causes of, 204
    bodily humors and personality, 172
    changes in personality as a result of
        learning, 201–203, 207
    cortical, subcortical, and postural
        nature of personality (Allport's
        views), 205
    differences in animals, 60
    emotion motivation, 188–95
    familial similarities in behavior, 174
    impact of heredity on, 205–206
    measuring of, 175–76, 177–78, 189–
        95

personality traits that determine social behavior, 191
personality traits, 178–82, 339
    emotional-motivational repertoires as a facet of, 178, 184–95. *See also* emotional-motivational development
    and intelligence, 42, 178–82, 204–205. *See also* intelligence
    language-cognitive repertoires as a facet of, 178. *See also* language-cognitive development
    seen as causes of behavior, 189
    sensory-motor repertoires as a facet of, 178. *See also* sensory-motor development
    prewired in the brain or learned, 203–204, 206–207
    research questions that need to be answered, 343
    subcortical nature of personality, 205
    theories about, 207–208
        early Greek theories about, 172
        five-factor theory, 175
        Freud's theory of, 173
        missing-link theory (learning), 176–201
        Watson's theory, 175
Personality: Contemporary Theory and Research (Derlega, Winstead, and Jones), 335
PET scans (positron emission tomography), 23
phenolketonuria, 220
philosophy of science, example of "Great Scientific Error" in, 31–32
phobias, 239
physical features. *See* human body
Piaget, Jean, 156

pigeons, behavior of, 150, 329
Pinker, Steven, 161, 317
plasticity of human brain, 37–38
play as a bonding mechanism, 106
pleasure-pain center, 55–56, 79
Plomin, Robert, 38
poorism, 299–300
positive emotional responses, 76–77, 100–101, 237–38, 244, 245, 293
    in animals, 55, 76, 87, 91, 105
    choosing positive stimuli over negative, 89–90, 189
    classical conditioning, 76, 87, 91, 157
    deprivation increasing strength of emotional response, 78
    feeling positive emotion from others' pain, 88, 236
    in humans, 55, 78, 88, 102, 103, 188, 189, 272–73, 306
        abnormal responses, 88, 237–39, 240, 241, 245
        to beauty and attractiveness, 251
        to food, 79, 87, 90, 101, 104, 185, 186
        to love, 79, 86, 105
        to music, 89, 157
        to politics, 203
        to sexual activity, 59, 185, 186, 235
        to sports, 108, 197
        used to encourage learning and language, 108, 122–23, 125–26, 131, 140, 145, 162, 187, 224, 231, 244, 246, 247, 248
        use of to bond with child, 102, 104–106, 107, 124
    physical reactions to, 80
    See also emotional responses; negative emotional responses

positron emission tomography. *See* PET scans (positron emission tomography)
postural nature of personality, 205
Prichard, Z., 39
primates, behavior of, 19, 50
    See also apes, behavior of; baboons, behavior of; chimpanzees, behavior of; gorillas; hominids; hominins; human behavior; monkeys, behavior of
primatology, 331
problem solving (reasoning), 25, 161, 164, 175, 183, 234, 284, 333, 340
    problem solving experiments of Thorndike, 81–82
progress toward humanness. *See* humanness, progress toward
propaganda, 187
psychology
    definition, 12
    as foundation for understanding human behavior, 339–43
    See also behaviorism; child development; clinical psychology; cognitive psychology; developmental psychology; educational psychology; evolutionary psychology; experimental psychology; human behavior; psychotherapy; social psychology; testing (psychological measurement)
psychopathology, 23
psychotherapy, 347–48
public policy, needed for improving learning for children, 348–51
punishment, 88, 89, 90, 102–103, 104, 107, 118–19, 125, 127, 189, 214, 234, 319
    attention as a reward/punishment for child, 104, 117–18, 121, 124, 129, 141, 221, 224, 225, 229, 233
        banning of corporal punishment in Sweden, 317
        detracting from positive bonding, 118
        emotional stimuli as incentives, 89–90
        and negative reinforcement, 322
            impact of on children learning to speak, 125
            not to be used in toilet training, 117
        See also motivation; rewards, impact of

rabbits, behavior of, 60
racism, 27–29, 132, 296–300
    and perceptions about black Americans in sports, 303
    scientific racism, 28, 297–98
radical behaviorism, 332
rats, behavior of, 37, 87, 329
reading, 128, 129–33
    and brain activity, 26
    brain differences between readers and dyslexics, 27, 134–35, 205–206, 216
    not a behavior, 149, 153
    reading ability, 26, 27, 129, 136, 150, 218, 245, 298, 330
    reading repertoires. *See also* repertoires
        as a complex repertoire, 129, 149, 216, 302
        and dyslexia, 149–51, 218
        impact of on brain, 330
        related to perceptual ability, 218–19
        word-imitation repertoire, 154–55
        word-reading repertoire, 150
    See also language-cognitive development

reasoning. *See* problem solving
(reasoning)
reinforcement. *See* rewards, impact of,
148, 149, 235
religious values, 304
affecting behavior, 194
children learning about, 107–108
concerning divine creation, animal
evolution, and human creation,
289–92
as cumulative learning, 286–87
and the repertoires relating to, 157
repertoires
abnormal repertoires, 211, 213, 230,
247, 249, 250, 335, 338, 339, 345,
347
deficits in repertoires as cause
of abnormal behavior, 214–
15. *See also* abnormal
behavior
anti-learning repertoire, 248
basic repertoires (BRs), 153–54, 161,
166–67, 250–51, 332–33
changes in personality as a
result of learning, 201–203,
207
culture as learned basic repertoires of a group, 305–306
determining which are needed
for learning tasks, 183
examples of, 161–64
finding right environment to be
able to teach, 231–32
slow learning leading to diagnosis of mental retardation,
219–20
training in basic repertoires
increasing intelligence scores,
177, 178, 180–83, 204–205
as a basis for creativity, 163

behavioral repertoires, 151, 152, 155,
164, 179, 278–79, 280, 350. *See
also* behavioral traits; human
behavior
and behavioral disorders, 215,
227, 308, 347. *See also*
abnormal behavior
functional behavioral repertoires, 152–53
complex repertoires, 129, 130, 157,
159, 216, 224, 293, 294, 302
causing individual differences in
behavior, 207
cumulative learning of, 163,
293, 300, 331–32
reading as a complex repertoire,
129, 149, 216, 302
and religious values, 157
and cumulative learning, 158–61,
164–66
emotional-motivational repertoires,
153–54, 157, 294. *See also*
emotional-motivational
development
affecting cognitive behavior,
162
affecting interests, 192–93
attitudes, 190–91
emotion motivation, 188–95
as facets of personality, 178,
184–95
interests, 191–93
needs, 194–95
testing of, 195
values, 193–94
functional behavioral repertoires,
152–53
group-behavioral repertoires, 279,
281, 291, 294, 298, 318
and culture, 305–306

enhanced by group members
having spare time, 322
impact of freedom and will, 294–96
impulsive or deliberate actions
dependent on repertoires, 162
inappropriate repertoires found in
autistic children, 225
language repertoires, 202, 205, 294,
296, 330. *See also* language-
cognitive development
and abnormal behavior, 234,
247, 249–50, 296, 338
aiding in counting and number
repertoires, 156, 330. *See also*
numbers and counting
author teaching own children,
123, 126–27, 129, 130–31,
155–56, 178–79
in children, 124, 128, 153–54,
155, 158, 177, 181–82, 183,
201, 227, 250, 341, 342. *See
also* babbling as beginning of
language
development of in early man,
262, 267–68, 273–76, 277,
306, 313
as facets of personality, 178
importance of, 154, 156–57
and intelligence tests, 177, 179,
180, 183, 206
word-imitation repertoire, 154–
55
word-reading repertoire, 150
learning one accelerates learning of
others, 155
nonfunctional repertoires, 151–52
reading repertoires. *See also* reading
as a complex repertoire, 129,
149, 216, 302
and dyslexia, 149–51, 218

impact of on brain, 330
related to perceptual ability,
218–19
word-imitation repertoire, 154–55
word-reading repertoire, 150
repertoire competition, 249–50
repertoires related to religion, 157
sensory-motor repertoires, 153–54,
156–57, 294. *See also* sensory-
motor development
as facets of personality, 178,
195–201
resulting from cumumlative
learning, 197
in sports, 158–59
tests of motor skills, 200–201
slow beginnings of learning repertoires, 160
use of long-term learning of a repertoire with dyslexics, 149–51
verbal-motor repertoire, 149–51
research questions that need to be
answered, 333–39, 346–47, 350–51
in the area of public policy, 348–51
response organs, 47, 56, 67, 71, 72, 80,
166, 211, 328
rewards, impact of, 30–31, 149–50
attention as a reward/punishment for
child, 104, 117–18, 121, 124, 129,
141, 221, 224, 225, 229, 233
getting attention from parent as
a reward for poor behavior,
117–18, 119, 213–14
on children learning to speak, 121–
22, 125
different impacts for children who
learn different emotions, 102–103
emotional reinforcement, 86–87
emotional stimuli as incentives, 89–
90, 91

food words eliciting response in hungry subjects, 78, 86–90
sex eliciting positive emotional responses, 79, 82, 87, 185, 186, 189
importance of providing reinforcement in child development, 85
motivation and incentives, 102
"opposite speech," 211–12
problem solving experiments of Thorndike, 81–82
reinforcement, 148, 149, 235
use of "mouthing" in infants as a reward, 110–11
use of tokens as reinforcement, 34, 141, 342. *See also* reading
    with autistic children, 225–26, 227–28
    with dyslexics, 217–18
    to teach reading, 129–32
See also motivation; positive emotional responses; punishment
right-handedness, 111–13
Ritalin®, 136
robots raising a child, 306–307
Rowan (an autistic child), 230–32
Rushton, J. Philippe, 28, 297–98
Ryback, David, 218

Sailor, Wayne, 128
salivation experiments of Pavlov, 74–75, 76, 77, 80, 87, 187
satiation. *See* conditioning, deprivation-satiation conditioning
savanna in Africa
    human progenitors move to, 51, 62–63, 257, 259, 271–72, 273, 277–78, 290
    lions on the savanna, 18, 61
schizophrenia, 24, 44, 112, 234

school performance. *See* education
Schutz, Richard, 217
Schwartz, Jeffrey, 268, 290
science, 164–66
    as foundation for understanding human behavior, 328–34
        applied science, 345
        behavior sciences, 334–39
        psychology fields, 339–43, 347–48
        social sciences, 343–45
    philosophy of science, 31–32
    scientific racism, 28, 297–98
    scientific theories use of repertoires and cumulative learning, 164–66
Science (journal), 339
Science of Learning Centers (National Science Foundation), 38
scientific racism, 28, 297–98
sea slugs, behavior of, 329
self-esteem, 197–98, 240, 246
selfish gene, 16–17
selfishness, impact of learning on, 319
self-preservation, 17
sensory-motor development, 108–19, 198, 247
    balance, 48, 114, 197, 200
    eye-hand coordination, 110–11, 154, 197, 198, 273
    kinaesthetic stimuli, 55
    repertoires, 153–54, 156–57, 294
        anti-learning as a basic abnormal repertoire, 248
        as facets of personality, 178, 195–201
        resulting from cumulative learning, 197
        in sports, 158–59
        tests of motor skills, 200–201
    sensory-motor disorders, 242–44

toilet training as example of, 116–17.
See also toilet training
verbal-motor learning, 125–27
measured in intelligence testing,
179
verbal-motor repertoire, 149–51
See also motor skills
sensory organs. *See* human body, sensory
organs
sex, 311
and animals, 53, 56–57, 77–78, 105,
281, 307, 311
and humans, 171, 239, 308, 319
behavioral disorders, 212, 235–
37
estrous cycle, 58–59
homosexuality, 20, 24, 112, 189,
190, 236
psychosexual development,
Freud's views, 116
sex eliciting positive emotional
responses, 79, 82, 87, 185,
186, 189
sexual reproduction, 279, 289,
291, 320
sexual satiation, 78
and tactile stimulation, 185
sexual identity, 20
See also gender differences
sharks, behavior of, 48
Shaywitz, Sally and Bennett, 26, 135
sight. *See* visual sense
Simon, Theodore, 175
Sirois, Sylvain, 29
skeletal characteristics of humans. *See*
human body
Skibbe, Lori E., 341
Skinner, B. F., 32, 33, 82, 85–86, 90, 97,
148, 149, 255, 332, 348
smell, sense of, 50, 55

See also human body, sensory
organs; sensory-motor development
Smith, Homer W., 287
social behavior, 15, 53, 142, 144, 191,
201, 219, 270, 300, 317
changes in social practice as
progress, 319–20, 321–22
lack of social skills. *See* autism
monkeys raised without mothers
unable to interact, 105, 307
social Darwinism, 15–16, 17, 279
social psychology, 147, 191, 251, 339,
352
social sciences and understanding human
behavior, 343–45
social sexual contact, 53
sociobiology, 15–16, 17
Sociobiology (journal), 16
speaking
children learning to speak, 92–93,
121–24
babbling as beginning of language, 122–24, 155, 177, 233
need to learn names of objects,
124–26
creating verbal-motor connections,
125–27
physical features allowing for
speech, 51, 53–54
and verbal imitation, 124
verbal-labeling responses, 125, 179–
80, 227
verbal-motor learning, 125–27. *See
also* sensory-motor development
measured in intelligence testing,
179
verbal-motor repertoire, 149–51
See also language-cognitive
development

specialization as aid to learning, 313–14
specificity-generality, 56–57
Spelke, Elizabeth, 30, 334
sports
    changes in high-jumping skills and
        techniques, 312
    superior performances by black
        Americans in, 303
    See also motor skills; sensory-motor
        development
Staats, Carolyn, 217
Staats, Peter, 241
Stanford-Binet Test of Intelligence, 178,
    180
Stewart, Mark, 162
Stover, Donald O., 162
Strong Interest Inventory (Strong,
    Hansen, and Campbell), 192
"Study of Values, The" test, 193
subcortical nature of personality, 205
Suci, George, 186
supply and demand, 344
Sylvan Learning Systems, Inc., 218
symbolic artifacts of early human progen-
    itors, 264–65, 286, 288

tactual sensibility. See touch, sense of
Talent Code: Greatness Isn't Born. It's
    Grown. Here's How, The (Coyle), 40
talent vs. learning, 199–200
taste, sense of, 48–49, 55, 57
    See also human body, sensory
        organs; sensory-motor develop-
        ment
Tattersall, Ian, 268, 290
Tauc, Ladislav, 72, 329
Tay-Sachs disease, 220
Temple, Elise, 216
terrible twos developmental stage, 20,
    137, 144, 206

testing (psychological measurement),
    190–91
    attitude testing, 190–91
    of the emotion-motivation repertoire,
        195
    of intelligence, 33, 175, 301–302
        black-white differences, 303
        misdiagnosing mental retarda-
            tion, 220
        as predictor of school perform-
            ance, 182–83
        Revised Stanford-Binet Intelli-
            gence Scale, 178, 180
        training in basic repertoires
            increasing scores, 177, 178,
            180–83
        Wechsler Preschool and Primary
            Scale of Intelligence
            (WPPSI), 181, 200
    interest tests, 192
    Miller Analogies Test, 205
    of motor skills, 200–201
    of personality, 175–76, 177–78, 189–95
        of personality trait that deter-
            mines social behavior, 191
        "Study of Values, The" test, 193
    ways to improve, 342
thalamus. See under brain
theory-construction methodology, 25,
    314–15
Third Chimpanzee, The (Diamond), 38
Thorndike, Edward, 32, 69, 73, 81–82,
    85, 86, 255, 332, 340
thumb, human. See human body, hands
Tiller, George, 191
Time (magazine), 219
"time-outs," 34, 68, 141, 214, 228
    as benign discipline training, 117–19
    in a group setting, 225
    recognized by Child (magazine), 142

Titanic (ship), 16

toilet training, 136, 137, 226, 227, 230, 231, 232, 243–44

    as an example of sensory-motor development, 116–17

    toilet repertoire, 152

tokens as rewards. *See* rewards, impact of

Tomasello, Michael, 269–70

tools, development of, 261–64, 272–73, 274, 277–78, 280, 285, 306, 309

    chimpanzees crude use of tools, 267, 273

touch, sense of, 49, 55, 57

    See also human body, sensory organs; sensory-motor development

Toulmin, Stephen, 31

Toynbee, Arnold, 316–17

training sessions, length of, 98–99

    See also learning

trial-and-error learning, 318

Tufts University Center for Reading and Language Research, 134

Turnbull, Khara L. Pence, 341

"20 People Who Changed Childhood" (*Child* Magazine), 142

twin studies

    and autism, 24–25

    and intelligence, 28, 205

    why they may not be useful, 251

Ullmann, Leonard, 34

unconditioned stimulus, 75

University of California at San Diego, 138–39

University of Hawaii, 226

upright locomotion as a human trait, 259, 290–91, 306

    See also human body, locomotive structure

utopia, 320–23

values, 193–94

variability of humans, 50, 52, 53, 56, 171, 199, 293, 294

Verbal Behavior (Skinner), 149

verbal imitation, 124

verbal-labeling responses, 125, 179–80, 227

verbal-motor learning, 125–27

    measured in intelligence testing, 179

    verbal-motor repertoire, 149–51

    See also language-cognitive development

versatility of humans, 43, 64, 65, 291

violence

    efforts to link to race, 28

    gender differences, 20

    and progress, 317

    and racism, 298

    view that it is part of evolution of humans, 309

    See also aggression

visual cortex. *See* cortex

visual-discrimination skill, 198

visual sense, 55, 57

    See also human body, sensory organs; sensory-motor development

visual stimuli, 37

vocabulary. *See* verbal-labeling responses

vocal responses in humans. *See* speaking

Vouloumanos, Athena, 268

walking

    delays in walking as indication of few learning opportunities, 114, 115

    upright locomotion as a human trait, 113–15, 259, 306

Watson, James, 29

Watson, John B., 32, 33, 35, 85, 175, 255, 332, 340

Wechsler Preschool and Primary Scale of Intelligence (WPPSI), 181, 200
Werker, Janet, 268
"When a Child's Mind is Abducted" (Lowy), 222
White, A. D., 287
White Boys Can't Jump (movie), 303
will and freedom, 294–96
Williams disease, 220
Wilson, Edward O., 15–16, 17
Wolf, Maryanne, 134, 150
Wolf, Montrose, 228
Wolpe, Joseph, 34
Woods, Tiger, 173, 297
words
    hurtfulness of words, 191
    word-imitation repertoire, 154–55

word-motor units, 179
word-reading repertoire, 150
    See also language-cognitive development
WPPSI. *See* Wechsler Preschool and Primary Scale of Intelligence (WPPSI)
writing, 128, 132
    grammar, 120, 127–28, 274, 284, 333
    use of basic repertoires to improve, 181
    See also language-cognitive development

zebras, behavior of, 48
Zinn, Howard, 22